Climate Politics and the Climate Movement in Australia

Climate Politics and the Climate Movement in Australia

Verity Burgmann and Hans A Baer

MELBOURNE
UNIVERSITY
PRESS

MELBOURNE UNIVERSITY PRESS
An imprint of Melbourne University Publishing Limited
187 Grattan Street, Carlton, Victoria 3053, Australia
mup-info@unimelb.edu.au
www.mup.com.au

First published 2012

Text design by Phil Campbell
Cover design by Phil Campbell
Typeset by J&M Typesetting
Printed and bound by CPI Group (UK) Ltd, Croydon, CR0 4YY
Cover photo by Verity Burgmann, 'Switch off Hazelwood, Switch on Renewable Energy' protest, Hazelwood Power Station, Latrobe Valley, Victoria, Australia, 10 October 2010.

National Library of Australia Cataloguing-in-Publication entry

Burgmann, Verity, Baer, Hans A.

Climate politics and the climate movement in Australia /
Verity Burgmann and Hans Baer.

9780522861334 (pbk)
9780522861341 (epub)
9780522861358 (pdf)

Climatic changes—Government policy—Australia.
Climatic changes—Economic aspects—Australia.
Climatic changes—Social aspects—Australia.

363.738740994

Contents

Abbreviations

ABARE	Australian Bureau of Agricultural and Resource Economics
ABC	Australian Broadcasting Corporation
ABS	Australian Bureau of Statistics
ACA	Australian Coal Association
ACCI	Australian Chamber of Commerce and Industry
ACF	Australian Conservation Foundation
ACOSS	Australian Council of Social Service
ACTU	Australian Council of Trade Unions
AES	Australian Election Study
AGO	Australian Greenhouse Office
AIG	Australian Industry Group
AIGN	Australian Industry Greenhouse Network
ALP	Australian Labor Party
AMWU	Australian Manufacturing Workers' Union
ANU	Australian National University
AP6	Asia Pacific Partnership on Clean Development and Climate
ARG	Australian Research Group
ASAP	as soon as possible
AYCC	Australian Youth Climate Coalition
BAU	Business As Usual
BBC	British Broadcasting Corporation
BCA	Business Council of Australia
BOM	Bureau of Meteorology
BZE	Beyond Zero Emissions
CAC	Climate Action Canberra
CAG	Climate Action Group
CARG	Climate Action Research Group
CASPI	Climate Adaptation Science and Policy Initiative
CCLP	Centre for Climate Law and Policy
CCS	carbon capture and storage
CCSD	Centre for Coal in Sustainable Development)
CEO	Chief Executive Officer
CFCs	chlorofluorocarbons
CFMEU ME	Construction Forestry Mining and Energy Union, Mining and Energy Division
CMO	climate movement organisation
CO_2e	carbon dioxide equivalent concentration
CPRS	Carbon Pollution Reduction Scheme
CRA	Conzinc Riotinto of Australia Ltd
CSG	coal seam gas
CSIRO	Commonwealth Scientific and Industrial Research Organisation
CST	concentrating solar thermal

DEWR	Department of the Environment and Water Resources
EITE	emissions-intensive trade-exposed
ENGO	environmental non-government organisation
EPA	Environment Protection Authority
ETS	Emissions Trading Scheme
EU	European Union
EV	Environment Victoria
FOE	Friends of the Earth
GDP	Gross Domestic Product
GFC	Global Financial Crisis
GHG	greenhouse gas
GLW	*Green Left Weekly*
GNP	Gross National Product
GST	Goods and Services Tax
IPCC	Intergovernmental Panel on Climate Change
MC	Master/Mistress of Ceremonies
MCA	Minerals Council of Australia
MJ	megajoule
MSSI	Melbourne Sustainable Society Institute
NGO	non-government organisation
NOAA	National Oceanic and Atmospheric Administration
NTEU	National Tertiary Education Union
PNG	Papua New Guinea
ppm	parts per million
PR	public relations
PV	Photovoltaic
QPSU	Queensland Public Sector Union
RECs	Renewable Energy Certificates
RET	Renewable Energy Target
RSPT	Resource Super Profits Tax
SCA	Safe Climate Australia
SCCC	Southern Cross Climate Coalition
SLF	Sustainable Living Foundation
SMH	*Sydney Morning Herald*
STE	solar thermal electricity
T10	Transition Decade
UCAN	Union Climate Action Network
UNEP	United Nations Environment Programme
UNFCCC	United Nations Framework Convention on Climate Change
UNSW	University of New South Wales
VTHC	Victorian Trades Hall Council
WMC	Western Mining Corporation
WWF	World Wide Fund for Nature/World Wildlife Fund
ZCA2020 Plan	*Zero Carbon Australia Stationary Energy 2020 Plan*

Acknowledgements

The authors wish to thank the climate movement activists, researchers and scholars who helped us in some way, especially: Ian Angus, Fiona Armstrong, Jody Betzien, Sue Bolton, Simon Butler, Peter Christoff, Ben Courtice, Chris Breen, Jenny Curtis, Robyn Eckersley, Jim Falk, John Bellamy Foster, Boris Frankel, James Goodman, Jennifer Kent, Dave Kerin, Judy McVey, Andrew Milner, David Milner, Jane Morton, Jack Mundey, Mike Nicolaides, Dick Nichols, Anne O'Brien, Rebecca Pearse, Bronwyn Plarre, Susan Price, Thomas Reuter, John Rice, Jack Roberts, Stuart Rosewarne, Ariel Salleh, Merrill Singer, David Spratt, Philip Sutton, Ted Trainer, Luke Van Der Meulen, Cam Walker and Erik Olin Wright.

For advice and assistance with publication, the authors are grateful to Jeremy Moss, Andrew Schuller and Peter Spearritt.

At the University of Melbourne: the School of Social and Political Sciences provided infrastructural support; the Faculty of Arts granted a sabbatical leave to each of the authors; and the cost of indexing this publication was supported by the Faculty of Arts Publication Subsidy Scheme.

The team at MUP have been most efficient and thorough: Elisa Berg, Olivia Blake, Madeleine Davis, Diane Leyman, Frith Luton, Collette Vella and Penelope White.

1

The Politics of Survival

Climate change is the hottest topic of the twenty-first century. In the 130 years that global temperatures have been recorded, the three warmest years have been 1998, 2005 and 2010.[1] With the first decade of the new millennium completed, meteorological statistics reveal 2001–2010 was the warmest decade on record, the global temperature 0.46° C above the 1961–1990 average.[2] On 1 October 2011, Britain experienced the hottest October day ever, with 30° C recorded in Yorkshire.[3] In Australia, this first decade of the twenty-first century was 'easily' the hottest on record. In 2012 a major review by the CSIRO and Bureau of Meteorology reported that Australia's annual daily average temperatures have increased 0.9° C since 1910, each decade since the 1950s has been warmer than the previous decade, and temperatures will increase by 1 to 5 degrees across Australia by 2070 compared with the last two decades of the twentieth century.[4]

From the 1970s onwards, scientists acknowledged the looming problem of the 'greenhouse effect': the warming of the earth's surface and lower atmosphere due to the emission of greenhouse gases such as carbon dioxide (CO_2), methane, nitrous oxide, ozone, and various chlorofluorocarbons (CFCs). These absorb some of the thermal radiation leaving the earth's surface, acting as a partial blanket for this radiation and increasing the earth's surface temperature. Some greenhouse gases are emitted naturally and were present before humans, but many are emitted by anthropogenic (human-related) activities, especially the burning of fossil fuels and deforestation.[5]

The World Meteorological Organization organised the First World Climate Conference in Geneva in 1979.[6] In 1988 the first major international meeting of scientists and national policymakers on climate change was held in Toronto, leading to the establishment of the Intergovernmental Panel on Climate Change (IPCC). In 1992 at Rio de Janeiro 154 countries plus the EU signed the United Nations Framework Convention on Climate Change (UNFCCC), which aimed to stabilise greenhouse gas atmospheric concentration 'at a level that would prevent dangerous anthropogenic interference with the climate system'.[7] The Third Conference of Parties to the UNFCCC adopted the Kyoto Protocol in 1997, which set legally binding targets for emissions reductions by 2012 on the part of developed countries, responsible for most of the historical and current emissions. The only two listed countries that rejected the Protocol were the USA and Australia.[8]

Soon after the new century began, on 13 July 2001, more than 1500 scientists from over 100 countries issued the Amsterdam Declaration, which warned governments that: 'Anthropogenic changes to Earth's land surface, oceans, coasts and atmosphere and to biological diversity, the water cycle and biogeochemical cycles are clearly identifiable beyond natural variability ... Global change is real and is happening now.' These scientists collectively urged governments, public and private institutions, and people of the world to agree that: 'An ethical framework for global stewardship and strategies for Earth System management are urgently needed. The accelerating human transformation of the Earth's environment is not sustainable. Therefore, the business-as-usual way of dealing with the Earth System is not an option.'[9]

The Climate Justice Movement

The need for such an 'ethical framework' and the issue of 'climate justice' began to be pressed urgently by representatives of people most likely to be adversely affected by climate change and least complicit in causing the problem—and least able to protect themselves from the consequences. Climate change is not a simple (unpleasant) experience in which all humans equally have contributed to the problem and together suffer the consequences. Far from it. South African writer Hein Maras calls on us to 'Shelve the abiding fiction that disasters do not discriminate—that they flatten everything in their path with "democratic" disregard.'[10]

Generally speaking, those people who contribute least to global warming are both facing the most severe consequences and have the least capacity to cope. Former UN Secretary-General Kofi Annan, now President of the Global Humanitarian Forum, stated in 2009:

> The first hit and worst affected by climate change are the world's poorest groups. Ninety-nine per cent of all casualties occur in developing countries. A stark contrast to the one percent of global emissions attributable to some fifty of the least developed nations. If all countries were to pollute so little, there would be no climate change.
>
> The effects of pollution driven by economic growth in some parts of the world are now driving millions of people into poverty elsewhere …
>
> Where does a fisherman go when warmer sea temperatures deplete coral reefs and fish stocks? How can a small farmer keep animals or sow crops when the water dries up? Or families be provided for when fertile soils and freshwater are contaminated with salt from rising seas?[11]

As Robyn Eckersley notes, this inverse relationship between vulnerability, on the one hand, and responsibility and capacity, on the other, constitutes 'the primary injustice' of climate change.[12] The climate justice movement campaigns to highlight this fundamental unfairness that those most affected by accelerating climate-related disasters around the world are generally the least responsible for causing disruptions in the climate.[13] This is true in both the geographical and sociological senses.

In terms of geography, countries with extremely low emissions will be amongst the worst affected by climate change. For example, Bangladesh and low-lying Pacific islands such as Kiribati and Tuvalu are at the most serious risk of inundation by rising sea levels; agricultural lands in South America will be subject to desertification; and between 75–250 million Africans will suffer from increased 'water stress'.[14] Andrew Revkin evokes the notion of a climate divide: 'In almost every instance, the people most at risk from climate change live in countries that have contributed the least to the atmospheric buildup of carbon dioxide and other greenhouse gases linked to the recent warming of the planet. Those most vulnerable countries also tend to be the poorest.'[15]

That these countries are poor guarantees they have a reduced ability to cope with their disproportionate experience of the adverse effects of climate change. As Brian Tokar points out: 'the effects of chaotic global warming are most felt by those people who are least able to adapt or compensate for these changes, especially the roughly half of the world's population that live on less than two dollars a day.'[16] Jeremy Moss articulates the climate justice perspective that 'an intuitively plausible view about who pays for climate change is that those whose countries caused the problem in the first place should pay for the costs of adaptation or mitigation'. He notes too that, locating the blame for climate change with currently developed countries is also often entangled with other issues of historical injustice that have been generated by previous interactions between countries.[17]

However, geography is not the only basis to the climate divide. Not only *between* but also *within* countries, there is, in general, an inverse correlation between complicity and suffering. Individuals in any country, whether developing or developed, have differing environmental impacts and differing access to resources that will determine the extent to which they will suffer from climate change. It is a simple rule of science and sociology that wealthier individuals create more—and poorer people create fewer—emissions. Thus the individuals most complicit in causing global warming will suffer least because they have the greatest capacity to 'weather' the consequences.

This is already evident in the playing out of emergency situations created by freak weather events. It was no accident that, in September 2005, the wealthier residents of New Orleans were more likely to escape the drowning city, often in gas-guzzling SUVs, while the poorer residents were more likely stranded without transport to contribute disproportionately to the death toll. Naomi Klein describes the divide: 'The economically secure drove out of town, checked into hotels and called their insurance companies. The 120 000 people in New Orleans without cars, who depended on the state to organize their evacuation, waited for help that did not arrive.'[18]

Disasters and emergencies aside, the everyday experience of a warmer planet will not be endured equally either. Those with the most financial resources will buy their way out of unsafe or simply unpleasant environments: with more heat-proof housing; greater ability to pay increased prices for water, food and other essentials; greater access to medical care for heat-related illnesses; or relocation to cooler or higher areas. Poorer people, wherever they live, will have fewer options, an obviously unjust outcome since they have contributed least to global warming.[19]

A report in July 2011 provides an extreme example of a climate divide. Italy's rich are buying their own islands 'as rising seas and environmental damage spoil some of the trendiest bathing spots in the country'. In 2010, 150 islands, from Croatia's coast to the Caribbean and the South Seas, were bought by Italians; and demand for islands in 2011 would outstrip that of 2010. It is a good time to buy, because others are doing it tough. 'The choice of islands has never been better, as governments seek to refill coffers emptied by recession.'[20] Unfortunately for the health of the planet, the world is, to a large extent, for sale—to the very rich.

A persistent theme in climate justice movement pronouncements is the problem of capitalism and commodification. Capitalism has created the inequality between humans that ensures the worst effects of climate change will not be experienced by the people most privileged by capitalism, wherever they reside. More than that, capitalism (and state capitalism in countries such as China) is also the major cause of climate change—its focus on profit-making encourages use of the cheapest production methods possible, regardless of ecological consequences and effects on workforces in terms of unemployment, lower wages and harsher working conditions. In Tokar's words: 'Corporations almost invariably prefer to lay off workers, outsource production, or move factories overseas than to invest in environmentally meaningful improvements.'[21]

This tendency has been especially marked during the past few decades, the period of corporate globalisation. With neo-liberal ideology hegemonic, governments that might have restrained and regulated corporations have been discouraged, to say the least. Capitalism not only dictates the production methods that enhance the greenhouse effect but also encourages the contributing factor of overconsumption, also in the interests of profit-making. Shorthand, this destructive cycle could be termed, as Hans Baer does, 'the capitalist treadmill of production and consumption'.[22]

Indigenous peoples are frequently the front line resisting the spread of capitalism into every nook and cranny of the globe. Tokar claims the demand for climate justice was first expressed by Indigenous Environmental Network founder and director Tom Goldtooth in the mid-1990s; and the concept of 'climate justice' gained international attention following a meeting in Durban in 2004 that included representatives of social movements and indigenous peoples' organisations in various countries.[23] The Second International Indigenous Forum on Climate Change in 2000 announced:

> Earth is our Mother ... not a commodity ... our cultures, and the territories under our stewardship, are now the last ecological mechanisms remaining in the struggle against climate devastation ... Climate change is a reality and is affecting hundreds of millions of our peoples and territories, resulting in famine, extreme poverty, disease, loss of basic resources in our traditional habitats and provoking involuntary displacements of our peoples as environmental refugees. The causes of climate change are the production and consumption patterns in industrialised countries and are therefore, the primary responsibilities of these countries.[24]

Yet indigenous peoples are frequently discounted in international discussions about the climate crisis.[25] The climate justice movement aims to challenge that discrepancy and prioritise the voices of the most affected communities.[26]

This climate justice movement also has affinities with the global justice movement—also called the anti-corporate or anti-capitalist movement—that erupted on the world stage late last century and was announced at the 'Battle of Seattle' in November 1999.[27] Climate

Justice Action, a global network of climate justice activists and NGOs, explicitly regards climate action as the new focus for the global justice movement; and the tactics of the global justice movement a template for climate justice activities.[28] Like the global justice movement, the climate justice movement struggles against corporate control of the global economy: it identifies the capitalist treadmill of production and consumption as the major force contributing to emissions and emphasises that neo-liberal, market-oriented corporate globalisation has heightened capitalism's tendency to exploit both humans and nature. European Climate Justice Action explains: 'Climate Justice means linking all struggles together that reject neo-liberal markets ... we cannot prevent further global warming without addressing the way our societies are organized—the fight for climate justice and the fight for social justice are one and the same.'[29] Both movements connect with progressive segments of labour movements and far-left groups; and also with the most vulnerable populations of the planet, not just indigenous peoples but also peasants and other workers displaced and disoriented by the globalisation of capitalism.

The climate justice movement argues that capitalist market mechanisms that have caused global warming cannot solve the problem. The India Climate Justice Forum declared at its climate justice summit in New Delhi on 26–27 October 2002:

> We, representatives of the poor and the marginalized of the world, representing fishworkers, farmers, Indigenous Peoples, Dalits, the poor and the youth, resolve to actively build a movement from the communities that will address the issue of climate change from a human rights, social justice and labour perspective. We affirm that climate change is a human rights issue ... We reject the market principles that guide the current negotiations to solve the climate crisis: Our World is Not for Sale![30]

Capitalist 'solutions' to climate change would aggravate its unjust impacts.

Market-based mitigation strategies, notably carbon trading and carbon offsetting, cause 'secondary injustices', according to Eckersley, which will exacerbate the 'primary injustice' of climate change. Why should those with the capacity to pay have the right to pollute more than those who lack the capacity to pay? 'Carbon trading thus becomes a form of neo-liberal environmental governance that leaves the allocation of scarce environmental resources to the market, rather than to democratic negotiation or equitable rationing.' There is also the potential for the abuse of market power.[31] For such schemes to be palatable from a justice perspective, she argues they must offer real benefits, or at least no net disadvantage, to developing countries as well as marginalised and low-income social groups.[32] This does not appear to be happening. As Tokar points out, 'Corporate "solutions" to global warming are expanding commodification and privatization, whether of land, waterways, or the atmosphere itself, largely at the expense of those same affected communities.'[33] Two examples illustrate the problem: biofuels and carbon offsetting. Both please the affluent in the global North and fill the coffers of corporations, but further impoverish the global South.

The production of biofuels or agrofuels is good for global agribusiness, but running our cars on ethanol fermented from corn and diesel fuel made from soybeans and other food crops has contributed to the worldwide food shortages that have brought starvation and food riots to over thirty countries in the past few years.[34] The amount of corn needed to produce ethanol for one SUV tank contains enough calories to feed a hungry person for a year.[35] Researchers have also documented an expanding legacy of disturbing environmental and human rights impacts from development of agrofuels around the world.[36] And it does not work. A series of studies beginning in 2008 suggest the climate consequences of converting pasture and forest land to production of fuel crops are severe enough to make most agrofuels net contributors to global warming.[37]

Carbon offsets are increasingly popular in the global North. Nearly every time you buy tickets for an airplane flight, someone is out to sell you offsets to alleviate your conscience: a feel-good exercise for jet-setters. Like Robert Goodin's comparison between the sale of pollution permits and the medieval Roman Catholic Church's sale of indulgences,[38] Tokar is scathing about the 'new global mythology' of carbon offsets:

> Carbon offsets have become the postmodern version of the indulgences the Catholic church used to sell in the Middle Ages to buy your way out of sin. But on a global scale, with corporations instead of individuals as the main players, they have become a scam of gigantic proportions. Rather than promoting innovative measures to reduce energy use and sequester carbon in poor countries, as they are usually advertised, carbon offsets are instead subsidizing the already routine destruction of byproducts from China's rising production of ozone-destroying hydrofluorocarbons, minor retooling of highly polluting pig iron smelters in India, and methane capture from a notoriously toxic landfill in South Africa, to cite only a few examples.[39]

In Durban in 2004 climate justice activists drafted the Durban Declaration on Carbon Trading, which stated that the North's desire for carbon offsets resulted in adverse impacts such as the widespread conversion of forests to monoculture plantations.[40] Northern enthusiasm for the carbon sink capacities of rainforests also contributes to the poverty and displacement of rainforest communities through ill-conceived conservation strategies, which insist on removing all humans. In particular, as Ana Isla reveals, women from the peasant communities forced out of the forests move to urban areas and become cheap labour or cheap sex.[41]

The climate justice movement represents those most affected by climate change and responses to climate change. It aims to counter false solutions to global warming with specific alternatives, such as taxation on industrial-scale emissions, altered utility and transport policies, public funds for solar and wind energy, and aggressive reductions in energy consumption throughout the industrialised world.[42] In mid-2009, climate activists gathered as part of an emerging international Climate Justice Action network and agreed on an ambitious alternative agenda to deal-making at UN level. Their declaration read: 'We cannot trust the market with our future, nor put our faith in unsafe, unproven and unsustainable technologies. Contrary to those who put their faith in "green capitalism", we know that it is impossible to have infinite growth on a finite planet.' The statement called for leaving fossil fuels in the ground, popular and community control over production, reducing the North's overconsumption, respecting indigenous and forest people's rights, and reparations for the ecological and climate debts owed by the richest countries to those who are most affected by resource extraction and climate-related disasters.[43]

Climate Politics on the Rise

Less radical forms of climate movement activism have also erupted around the globe. Notwithstanding the strength of the movement that denies the scientific consensus about global warming, people of all political persuasions have responded to the climate change issue by demanding mitigation policies on the part of their respective governments. This broader climate movement generally avoids explicit discussion of the nature of production and consumption—and the ethical issues raised by climate change. Yet it often inadvertently draws considerable attention to the problematic nature of capitalism by insisting that 'business as usual' is unsustainable.

The year 2006 was significant for increased international awareness of the global warming problem, for two main reasons. Firstly, there was former US Vice-President Al Gore's blockbuster documentary film *An Inconvenient Truth*, based on his book of the same title. Secondly, there was the Stern Review in the United Kingdom, which was not the first economic report to discuss the effect of global warming on the world economy, but was the largest and most widely discussed report of its kind.

An Inconvenient Truth, which contends that global warming is the most serious threat human civilisation has ever faced, did much to popularise the scientific consensus. It was watched by many millions of people globally and became the third-highest grossing documentary of all time, taking more than $US50 million and selling over 1.5 million DVDs.[44] It won two Academy Awards in 2007, including for Best Documentary.[45] An online Nielsen survey in April 2007 polled 26 486 internet users across forty-seven countries in North America, Europe, Asia Pacific and the Middle East on attitudes to climate change. The survey found that *An Inconvenient Truth* significantly influenced viewers' awareness of climate change and altered their behaviours.[46] Max Boykoff, James Martin Fellow at the Environmental Change Institute at the University of Oxford, commented on how the film helped 'to shift the focus for many people from whether there is a problem to what to do about it'.[47] Gore visited Australia several times in the immediate wake of his film's release, lecturing about the seriousness of climate change. He told a Sydney audience in October 2007: 'This is like a freight train coming at our kids.'[48]

The advice of the world's scientists was formally accepted in the Stern Review—but with no special effects. British Chancellor of the Exchequer Gordon Brown had asked Sir Nicholas Stern, a former Chief Economist of the World Bank and chief economic adviser to the UK Government, to lead a major review of the economics of climate change. It was prepared by a team of economists at the Treasury; independent academics were involved as consultants, and the scientific content was reviewed by experts. *Stern Review: the Economics of Climate Change* (Stern Review) became a 700-page report released on 30 October 2006. It began by acknowledging the scientific consensus that pointed to increasing risks of serious, irreversible impacts from climate change associated with business-as-usual (BAU) paths for emissions.[49] Climate change threatens the basic elements of life for people around the world—access to water, food production, health, and use of land and the environment.[50] The impacts are not evenly distributed: 'The poorest developing countries will be hit earliest and hardest by climate change, even though they have contributed little to causing the problem. Their low incomes make it difficult to finance adaptation. The international community has an obligation to support them in adapting to climate change.'[51] If and when the damages appear, it will be too late to reverse the process, so we are forced to look a long way ahead.[52]

An important reason for its impact is that the Stern Review approached the problem not from a green agenda but from a profit-oriented cost-benefit analysis. The central message was that 'The benefits of strong, early action on climate change outweigh the costs.'[53] Mitigation was 'a highly productive investment'.[54] It implied strongly that any logical businessperson should favour immediate far-reaching regulation and restraints.

> The evidence shows that ignoring climate change will eventually damage economic growth ...Tackling climate

> change is the pro-growth strategy for the longer term, and it can be done in a way that does not cap the aspirations for growth of rich or poor countries. The earlier effective action is taken, the less costly it will be.[55]

The Stern Review called for global emissions to be cut by a quarter by 2050.[56] Stern used standard scientific terminology in calculating costs and benefits: 'carbon dioxide equivalent concentration' (CO_2e), because not all greenhouse gases are carbon dioxide; and 'parts per million' (ppm).[57] The annual cost of achieving stabilisation between 500 and 550 ppm CO_2e is around 1% of global GDP, which was 'fully consistent with continued growth and development, in contrast with unabated climate change, which will eventually pose significant threats to growth'.[58] The transition to a low-carbon economy would bring new opportunities for growth.[59] Establishing a carbon price, through tax, trading or regulation, is an essential foundation for climate-change policy.[60] The problem required:

> co-operation between countries, through international frameworks that support the achievement of shared goals ... a partnership between the public and private sector, working with civil society and with individuals. It is still possible to avoid the worst impacts of climate change; but it requires strong and urgent collective action. Delay would be costly and dangerous.[61]

The Stern Review's growth-oriented approach, while guaranteeing its polite and popular reception, also ensured it contributed little to the climate justice debate. The Review ignored problems such as the constraint of finite global natural resources and the role of poverty as a catalyst of population growth, resulting in greater energy and resource demands. Climate justice activists were sceptical too about Stern's recommended mitigation strategies, reliant on market mechanisms, such as emissions trading schemes. Movement activists pointed out that the EU's emissions trading scheme (ETS), in operation since January 2005, has produced huge new subsidies for highly polluting corporations, without demonstrable reductions in pollution.[62]

Early in 2007 the IPCC released its Fourth Report, which predicted serious consequences for the planet from rising temperatures.[63] Global emissions due to human activities have increased 70% between 1970 and 2004.[64] 'Human influences' have:

> very likely contributed to sea level rise during the latter half of the 20th century; likely contributed to changes in wind patterns, affecting extra-tropical storm tracks and temperature patterns; likely increased temperatures of extreme hot nights, cold nights and cold days; more likely than not increased risk of heat waves, area affected by drought since the 1970s and frequency of heavy precipitation events.[65]

By this stage, there was increasing acceptance internationally that climate change was serious and anthropogenic; and that a period of climate instability would become a period of climate change catastrophe if drastic measures were not taken to reduce emissions.

'Climate change is the great moral challenge of our generation.' These were the much publicised Opening Remarks to the Australian National Climate Change Summit held in Canberra on 31 March 2007 by Kevin Rudd, soon to become Prime Minister. This summit heard from leading climate change scientists. Professor Tony Haymet, former head of CSIRO atmospheric research and now head of the Scripps Institute of Oceanography at the University of California, said the evidence of anthropogenic global warming was irrefutable. 'We must reduce our emissions. It is not an option.' He also observed: 'Australia will be able to lead the world on a couple of technologies … we have the best solar researchers in the world.' Tom Burke, environmental adviser to Rio Tinto and the British government, said climate change was a bad problem getting worse and not acting now would be a moral failure unmatched in human history. 'This is an issue on which we cannot afford policy failure. There is no rewind button on climate change. We can't get it wrong and go back and say "oh dear that was a mistake". Let's do it differently next time around.'[66]

Climate scientists are attempting to define a safe temperature increase limit and how to secure it. James Hansen, Director of NASA's Goddard Institute of Space Studies, argued in 2007 that increases above 1.7° C would be 'highly disruptive', therefore humanity needs to reduce atmospheric CO_2 below 390–350 ppm to avoid irrevocable damage to human societies and the planet.[67] Others recommended limiting temperature rise to 2° C by stabilising CO_2e at 445 ppm, which would require a drop of 89% in emissions from a 1990 baseline.[68] The Copenhagen Diagnosis asserts that, if global warming is to be limited to 2° C above pre-industrial values, global emissions need to peak by 2020 then decline rapidly.[69]

In 2008 Stern admitted the latest research indicated emissions needed to be reduced more sharply than recommended in the Stern Review: to halve by 2050 to keep CO_2e below 500 ppm, which would cost around 2% of GDP.[70] In 2009, he reiterated his argument that this would be money worth spending in his *A Blueprint for a Safer Planet*, but argues that aiming to stabilise emissions at 400 ppm CO_2e is unrealistic, because humanity has already reached a level of 430 ppm CO_2e and will be at 450 ppm CO_2e within a decade.[71]

With annual emissions growth since 2000 increasing at rates at the upper end of IPCC scenarios, over 140 delegates from the science community, government, NGOs and business gathered at a '4 degrees and beyond' conference at the University of Oxford 28–30 September 2009. Fearing that the IPCC's Fourth Report had understated the seriousness of the situation, given more recent findings, the conference aimed to: assess the consequences of a change in global temperature above 4° C for a range of systems and sectors; and explore the options for avoiding climate changes of this magnitude.[72]

Shortly after this conference, the denialist cause heightened its offensive, especially in the United States, Canada, the UK and Australia, aided by the December 2009 Copenhagen debacle and two other incidents around the same time. First there was 'Climategate': the stealing of emails from the University of East Anglia's Climate Research Unit that allegedly demonstrated senior scientists were manipulating land surface temperatures to show global warming. This was followed by 'Glaciergate': the 2007 IPCC report quoted a WWF report that stated, inaccurately, that Himalayan glaciers could disappear by 2035. The East Anglia scientists were cleared by an independent inquiry and there were no other howlers uncovered in the 3000-page IPCC report, but, as Clive Hamilton observed, these incidents were 'a big coup for climate denialists around the world and were beaten up for all they were worth by some media outlets'.[73]

In mid-2010, the *State of the Climate Report*, produced jointly by the British Met Office and the US National Oceanic and Atmospheric Administration (NOAA), reported 'clear and unmistakeable signs of a warming world', with each decade since the 1970s hotter than the preceding one, and greenhouse gases were the glaringly obvious explanation.[74]

On 12–14 July 2011, the University of Melbourne held a conference on 'Four Degrees or More? Australia in a Hot World', which highlighted an appearance by Hans Joachim Schellnhuber, Director of the world-renowned Potsdam Institute for Climate Impact Research. His public lecture on 12 July, 'Climate change: the critical decade', was disrupted by members of the far-right Citizens' Electoral Council waving a noose in front of him and saying 'Welcome to Australia'.[75] His lecture painted a bleak picture of Australia in 2100 if drastic global mitigation measures are not taken. He revealed a growing number of climate scientists believe the world is likely to be four degrees warmer by 2100 if emissions increase at the rate they have in recent decades. Humanity, he argued, therefore needed to develop a new social contract that would entail new and substantial international and national climate strategies over the next decade to shift the world away from the possibility of climatic catastrophe.[76]

The Failure of Democratic Politics and the Significance of Climate Activism

Politicians and governments around the world have failed to develop such a contract. The fiasco of the UNFCCC Summit held in Copenhagen in December 2009 highlights this problem. This was supposed to design an international treaty for multilateral climate change mitigation. Climate scientists had optimistic expectations. Alden Meyer, director of strategy and policy for the Union of Concerned Scientists, said optimistically: 'Never have you had this civil society pressure. Never have you had world leaders coming in for a decision-making meeting like this.' Climate activists flocked to Copenhagen, hopes high. Archbishop Desmond Tutu told their rally: 'They marched in Berlin, and the wall fell. They marched in Cape Town, and the wall fell. They marched in Copenhagen—and we are going to get a deal.' The *Calgary Herald* noted that the Copenhagen protest was the centrepiece to demonstrations that took place around the world at the same time. 'Australia, the developed world's worst per-capita polluter, saw as many 50 000 people taking to the streets nationwide, according to organizers.'[77]

Greenpeace observed of the summit: 'This represents the best chance we have of reversing current emissions trends in time to prevent the climate chaos that we are hurtling towards.'[78] This opportunity was not grasped. In a secret climate agreement dubbed the 'Danish text' leaked to the *Guardian* on 8 December, various developed countries suggested the Kyoto Protocol be scrapped in favour of a new treaty that would permit developed countries to double their emissions. This ensured the summit was conducted in a climate of suspicion on the part of developing countries towards the rest of the world, so time to broker a meaningful and mandatory multilateral agreement ran out.

At the close of the summit, US President Obama proposed the Copenhagen Accord. Signed by 190 countries, it agreed global warming must be kept below a 2° C rise to avoid dangerous climate change. Developed countries also agreed to commit US$100 billion to help developing and poor countries finance the massive changes required to tackle the causes and effects of climate change. Supporters of the Accord emphasised this was the first time so many nations had agreed on the need to reduce emissions.[79] However, the Accord is voluntary and sets no reduction targets; and Venezuela, Cuba, Nicaragua, Bolivia and Sudan opposed the Accord. The demonstrations that had earlier expressed hope, now articulated anger at the impotence of world leaders to arrive at definite agreement. Venezuela President Hugo Chavez echoed two slogans of climate activists: 'Change the system, not the climate' and 'If the climate were a bank, they would have saved it by now.'[80] With 100 000 people on the streets of Copenhagen engaging in direct action, including an attempted occupation of the summit location to challenge false solutions and confront rising corporate influence over UN proceedings, it was a 'Seattle moment' for the climate justice movement.[81]

Bolivian President Evo Morales invited those assembled to participate in a different kind of climate summit in Cochabamba. Morales, Bolivia's first fully indigenous head of state in the 470 years since the Spanish invasion, stated in his opening speech that capitalism was to blame for global warming.[82] Held 19–22 April 2010, with 35 000 people representing over 140 countries, this World People's Conference on Climate Change and the Rights of Mother Earth issued a 'World People's Agreement', the combined product from seventeen working groups, and now a formal document within UN negotiations. It states: 'Today, our Mother Earth is wounded and the future of humanity is in danger.' The agreement includes a comprehensive set of principles, rooted in indigenous views of harmony, complementarity and anti-colonialism; proposes a Universal Declaration on the Rights of Mother Earth; condemns carbon markets, as well as the commodification of forests for carbon offsets; calls to protect the rights of climate migrants; and demands the creation of an International Climate and Environmental Justice Tribunal to judge and penalise activities that promote climate change. It proposes that emissions reduction targets aim to stabilise emissions below 350 ppm to limit temperature increase to no more than 1° C.[83]

The failure of the Copenhagen Summit confirmed the worst fears of climate movement activists of all sorts and persuasions that the world's political leaders are simply not up to the great moral challenge of our time. Even when politicians acknowledge the reality of climate change, rhetoric is rarely matched by really effective action. Other world leaders are expressly hostile, influenced by climate change deniers, often funded by fossil fuel and traditional power generation corporations. Apart from lobbying politicians, they also confuse the public into believing the scientific consensus is in dispute. Climate scientists have started to expose the fact that scientists who provide reports for climate change deniers are paid by major oil companies, and that the papers they publish do not meet normal standards of academic scrutiny. For example, in January 2007 the US Union of Concerned Scientists reported that huge multinational oil company Exxon had paid US$16 million between 1998 and 2005 to a network of forty-three public relations groups to pacify public concern over scientific findings on climate change.[84] Deniers call themselves 'climate change sceptics'; climate movement activists call them 'climate change deniers'.

Frustration at the lack of action on the part of world leaders has caused the eruption around the world of the climate movement and its spectacular growth since the beginning of this century, in both developed societies, such as Europe, the USA and Australia, and developing societies, such as India and South Africa.[85] The climate justice movement is an important component of this international climate movement—and is of particular interest for its analysis of the causes and consequences of global warming, and the necessary and necessarily far-reaching solutions. The broader climate movement of which it is a part is diverse, encompassing a wide array of political positions and proposed solutions, but is united in its contempt for climate change deniers and its impatience with the timid approaches of corporations and governments. It urges the Precautionary Principle: that the cost of responding to the advice of climate scientists, if wrong, is considerably less than the cost of ignoring them, if they are right.

This climate movement is the most significant social movement of our times; and is likely to remain so for the foreseeable future. While the climate justice wing of the movement also owes much to the global justice movement, the climate movement as a whole has roots in the environmental movement from the 1960s onwards, which warned about silent springs and the limits to growth. Because of the urgency around global warming, environmentalism has become increasingly oriented towards climate activism: the issue has become not simply preventing undesirable environmental degradation but preventing unbearable environmental devastation; the climate change issue confirms the true significance of environmentalism. There are four ways in which climate activism is the most important movement in world history.

Firstly, the climate movement is universalist. Unlike other movements, it is not acting on behalf of a circumscribed constituency, such as women, indigenous peoples or workers; it contests current structures and power relations on behalf of all species and ecosystems. 'There is no planet B' said the signs waved by the 100 000 protesters from across the world in Copenhagen on 12 December 2009 when world leaders confirmed the talks would not lead to a binding legal agreement.[86] While the climate justice movement emphasises the particular interest of those who will be most adversely affected by climate change, this does not detract from the central message of the climate movement as a whole, that all will suffer to varying degrees from global warming and therefore all have a stake in climate change mitigation.

Secondly, the climate movement is a truly global movement: the problem it confronts is one affecting the entire planet; and solutions must be imposed globally to be effective. Members of the international 'NGO community' drafted a Copenhagen Climate Treaty: 'a just, effective, science-based agreement, in time to keep global average temperature rise far below the danger threshold of 2° C'. It stated optimistically: 'Reaching this understanding about climate change between 192 countries will mean that the world has started to learn how to manage its planet. Failure to agree on a strong, effective deal in Copenhagen will accelerate the demise into competing smaller entities, resource wars, disruption, refugees, and natural catastrophes.'[87]

Thirdly, this movement's opposition—at least that of the climate justice wing—is fundamental, because it challenges both the material foundation of our society and our basic attitudes, oriented as they are towards possessions and acceptance of very unequal distribution of resources. The NGO community's proposal at Copenhagen stated:

> Climate change is not just a human tragedy but changes the very basis of survival on this planet. We know that our window of opportunity for limiting climate change is closing and therefore unprecedented international cooperation and commitment is required. We need to, and we can, progress much faster, catalysing the world onto a low-carbon development pathway that is ambitious, effective and fair and ensures that the right to survival for the most vulnerable is not sacrificed.[88]

The climate justice movement is telling the world that the capitalist treadmill of production and consumption cannot continue and the more affluent people's lifestyles are unsustainable. The broader climate movement too is adamant that business-as-usual capitalism must be challenged.

Fourthly, the climate movement, like the environmental movement before it, raises a principle alien to democratic politics—inter-generational justice—which insists it is part of the evolutionary responsibility of the current generation not to leave an impoverished environment for the next generation. It suggests a new moral imperative: 'Do unto the next generation as you would have the past generation had done unto you.'[89] This principle of inter-generational justice is one to which liberal democratic politics is ill-suited to respond, given that the people who will most feel the effects of global warming are future generations and those currently too young to vote. Policies to protect the environment or mitigate climate change are not the normal stuff of parliamentary politics, for most politicians are simply not interested in the welfare of future voters. At the Copenhagen Summit, Greenpeace International mounted an exhibition of huge photographs of the world leaders present, digitally enhanced to represent their appearance in 2020, apologising for their inaction. A gigantic, greyer and more wrinkled Kevin Rudd was saying: 'I'M SORRY. WE COULD HAVE STOPPED CATASTROPHIC CLIMATE CHANGE ... WE DIDN'T.'[90]

It is possible that liberal democratic politics—because of its unresponsiveness to voters of the future—is simply incapable of enacting policies adequate and timely enough to prevent global warming. How can citizens of the future influence the politicians of the present to resist the power of those with a vested interest in ignoring climate science? Australia's richest person, mining magnate Gina Rinehart takes her cues from her immensely wealthy father, Lang Hancock, who boasted that the power of government 'could be broken by obtaining control of the media and then educating the public'.[91] Climate instability and global warming is already with us and many present-day voters listen to the climate science and reject the education sponsored by the likes of Rinehart; but the most drastic consequences of climate change have not yet been experienced. Therefore, in pressing the claims of future generations, the climate movement faces an uphill battle to persuade politicians to exercise political will and show climate leadership against the power and propaganda of those who pursue profit at the expense of the planet. This generational aspect—which will be played out over time—could be seen as the 'fourth dimension' of climate justice: the inherent unfairness of the fact that future generations, who have not contributed at all to the global warming problem, will inherit a planet less habitable than now.

Climate Politics and the Climate Movement in Australia

That climate politics is a phenomenon that has come to pervade Australian society is manifested in parliamentary deliberations, mass media reports and daily conversations. A distinction needs to be made, however, between climate politics and the climate movement. The climate movement is part of a larger whole that is climate politics, which occurs wherever climate change is at issue; but there is much climate politics where the movement is effectively excluded. Indeed, the huge gap between people's expectations for effective climate change mitigation policies and the meagre offerings of governments reveals the extent to which the climate movement barely penetrates the centres of power in Australia and elsewhere.

Climate politics operates at three different levels within Australian society, which we classify as 'from above', 'in the middle' and 'from below'. Power in climate politics is exercised most obviously from above; the climate movement operates most energetically from below. Our analysis of the layers or levels of climate politics is reflected in the structure of this book, with Part I concerned with climate politics from above, Part II with climate politics in the middle and Part III with climate politics from below.

Acting from above are the most powerful forces within Australian society: corporations and their peak bodies; and federal and state governments, Labor and Coalition. This is the level at which the climate movement is largely absent, though it haunts politicians and corporate leaders, who note its growing influence on attitudes and behaviours.

The actors in the middle include: the Australian Greens; the major environmental non-government organisations (ENGOs); academic research centres; think-tanks; public intellectuals; and the trade union movement. Actors in the middle frequently reach out to the burgeoning movement from below and participate in ground-level climate movement activities; and occasionally have formal and informal input into the very top level of climate politics, for example in official consultations or off-the-record conversations with relevant Ministers or public servants. Since the 2010 election, it is arguable that the Greens also operate from above as a party whose support is crucial to the Gillard government in the House of Representatives and which also holds the balance of power in the Senate, a situation likely to continue in the future.

Acting from below, largely below the radar of mainstream media except when huge demonstrations or spectacular protest tactics invite attention, are the activists of the grassroots climate movement: the many and varied local, regional and nationwide climate movement groups and networks, and far-left groups to whom global warming confirms their worst suspicions about global capitalism. This lowest level is the primary location of the climate action movement; and grassroots organisation and mobilisation the source of its vitality.

Mark Diesendorf provides a detailed classification of climate movement organisations.[92] His scheme is comprehensive and his listings informative. However, the loose layering in our scheme enables us to distinguish between actors in the climate movement that are established organisations with varying degrees of status and influence, and those that operate at the grassroots level, because they are newly emergent groups arising in spontaneous response to perceived climate crisis or groups traditionally marginalised within the political system. Our broader categorisation also permits investigation of the myriad links between actors within each level, and to note the particular significance of connections that cross these boundaries, for instance the grassroots activism of the more radical ENGOs, such as Friends of the Earth (FOE).

Our scheme acknowledges, too, that social movements are invariably diffuse and disparate, heterogeneous in their objectives and their strategies. The Australian climate movement, like the international climate movement of which it is a small part, is no exception to this rule. This layering enables us to unpick the myriad strands of the movement to reveal the significant differences within the movement about what is required to prevent serious climate change and how best to achieve these ends.

The academic literature on social movements sometimes conveys the impression of homogeneity and single-mindedness on the part of social movements. Manuel Castells defines social movements as being 'purposive collective actions whose outcome, in victory as in defeat, transform the values and institutions of society'.[93] For Dieter Rucht, a social movement is 'an action system comprised of mobilised networks of individuals, groups and organisation which, based on a shared collective identity, attempt to promote social change, predominantly by means of collective protest'.[94] There is a sense in which any social movement is singular and can be treated as an entity—and these definitions encapsulate this important attribute—but it is also important in social movement analysis to recognise divisions within each singular movement. Alberto Melucci, in his influential discussion of social movements, expressly distances himself from approaches that treat social movements as 'homogenous entities'. Sociological analysis, he claims, must reject 'the assumption of collective action as unified datum' and 'discover the plurality of perspectives, meanings and relationships which crystallize in any given collective action'.[95]

Part of the reason for movement heterogeneity is that social movements tend, over time, to develop institutionalised forms that become less confrontational in their demands and strategies. There is an evolutionary process that social movements often go through if they achieve some level of recognition and success. A social movement often starts as informal networking, then collective action such as protests occurs, then organisations are established, then eventually there is a decline in protest activity and organisations form long-term agenda.[96] Some climate movement groups tend to be more bureaucratic and procedural and others more spontaneous, egalitarian and participatory. Our analysis identifies a major tension within the climate movement between moderate ENGOs, the institutional legacy of the earlier environmental movement, and the freshly minted and militant new grassroots climate movement organisations. Indeed, while the grassroots climate movement 'from below' is motivated largely by frustration with the outcomes of climate politics 'from above', it is also inspired by a desire to strengthen the resolve of climate movement actors 'in the middle', such as the trade union movement, ENGOs and the Greens.

That said, institutionalisation is a tendency, not a hard-and-fast rule for social movements; a continuing demonstration of this is the experience of FOE, a longstanding organisation, now a leading force within the grassroots climate movement and pressing a climate justice agenda. Moreover, newness does not guarantee radicalism: some of the most recent climate movement organisations, such as

Safe Climate Australia, are the least confrontational amongst climate actors 'in the middle'; and even grassroots groups might exhibit tendencies towards bureaucratisation and compromise.

In presenting an analysis of the modus operandi of the climate movement, we consider questions relevant to the study of all social movements. Does the heterogeneity of the movement help or hinder it? Is progress in the corridors of power dependent on the strength of the movement at the bottom? Do extremist activities such as militant protests at coal-fired power stations alienate public opinion; or do they effectively publicise the issue and move the spectrum of debate so that 'moderate' climate activists are advantaged by appearing to be the voice of reasonable compromise? And so on. To answer such questions—and provide useful information and analysis generally—we examine Australian climate politics from a social scientific perspective, including analysis of primary sources and observational research methods, and with a commitment to promoting meaningful climate change mitigation strategies based upon climate justice principles.

Notes

1 Magdoff and Foster, p. 2; World Meteorological Association; Garnaut, *The Garnaut Review 2011*, p. 5; Diesendorf, *Greenhouse Solutions*, back cover.
2 Commonwealth of Australia Department of Climate Change and Energy Efficiency.
3 *Observer*, 2 October 2011, p. 1.
4 Garnaut, *The Garnaut Review 2011*, p. 19.
5 Strictly speaking, scientists distinguish between the 'natural greenhouse effect' and the 'enhanced greenhouse effect'. The natural greenhouse effect is caused by water vapour, carbon dioxide and other minor gases, all present before humans. The 'enhanced greenhouse effect' is the added effect caused by the gases present in the atmosphere due to human activities such as deforestation and the burning of fossil fuels (Houghton, pp. 20–21). However, climate movement spokespeople tend to use simply 'greenhouse effect', while acknowledging that some of this effect is natural, some anthropogenic. This simpler terminology will be used throughout this book.
6 Sample.
7 Article 2, quoted in Barnett, p. 1365.
8 Barnett, p. 1366.
9 UNEP.
10 Hein Maras, 'A plague of inequality', *Mail & Guardian* (Johannesburg), 19 May 2006, quoted in Klein, p. 406.

11 'Introduction', Global Humanitarian Forum, p. ii.
12 Eckersley, 'Just Carbon Trading?', p. 100.
13 Tokar, p. 45.
14 IPCC, Working Group II Report, 'Impacts, adaptation and vulnerability', http://www.ipcc.ch, cited in Tokar, p. 23.
15 Andrew C Revkin, 'The climate divide: reports from four fronts in the war on warming', *New York Times*, 3 April 2007, quoted in Tokar, p. 19.
16 Tokar, p. 17.
17 Moss, p. 11.
18 Klein, p. 408.
19 IPCC, Working Group II Report, 'Impacts, adaptation and vulnerability', http://www.ipcc.ch, p. 393, cited in Tokar, p. 24.
20 Day.
21 Tokar, pp. 312.
22 Baer, 'Global warming'.
23 Tokar, pp. 45–6.
24 Second Indigenous International Forum on Climate Change.
25 International Indian Treaty Council.
26 Tokar, p. 45.
27 There is a huge literature on the global justice movement, but for a spectacular account of the 'Battle of Seattle', see St Clair.
28 Lawton.
29 'What does climate justice mean in Europe? A discussion paper', posted to Climate Justice Action listserv, 26 March 2010, quoted in Tokar, p. 52.
30 India Climate Forum.
31 Eckersley, 'Just carbon trading?', p. 103.
32 ibid., p. 104.
33 Tokar, p. 45.
34 ibid., p. 29.
35 LR Brown, 'Supermarkets and service stations now competing for grain', *Earth Policy Institute Update*, 13 July 2006; CF Runge and B Senauer, 'How biofuels could starve the poor', *Foreign Affairs*, vol. 86, no. 3, May/June 2007, pp. 41–53, quoted in Tokar, p. 29.
36 Tokar, p. 29.
37 J Fargione et al, 'Land clearing and the biofuel carbon debt', *Science*, vol. 319, no. 5867, 29 February 2008, pp. 1235–8; T Searchinger et al, 'Use of U. S. croplands for biofuels increases greenhouse gases through emissions from land use change', *Science*, vol. 319, no. 5867, 29 February 2008, pp. 1238–40, both available from www.sciencepress.org and cited in Tokar, p. 29.
38 Goodin.
39 L Lohmann, 'Carbon trading: a critical conversation on climate change, privatization and power, development dialogue', 48, Dag Hammerskjold Center, September 2006, cited in Tokar, p. 41.
40 Tokar, p. 46.
41 Isla, p. 201.
42 Tokar, pp. 49–50.

43 ibid., pp. 44–5.
44 Wilkinson, p. 21.
45 Topsfield, p. 9.
46 http://nz.nielsen.com/news/GlobalWarming_Jul07.shtml (viewed August 2010).
47 ibid.
48 Wilkinson, p. 21.
49 Stern, *Stern Review Executive Summary*, p. iii.
50 ibid., p.vi.
51 ibid., xxvi.
52 ibid., p. vii.
53 ibid., p. v.
54 ibid., p. xi.
55 ibid., p. v.
56 ibid., p. xi.
57 This is the ratio of the number of moles of a constituent in a given volume to the total number of moles of all constituents in that volume (Houghton, p. 422).
58 Stern, *Stern Review Executive Summary*, p. xii–xiii.
59 ibid., p. xvi.
60 ibid., p. xviii.
61 ibid., p. xxvii.
62 Tokar, p. 85.
63 IPCC, *Fourth Assessment Report*, 2.2.
64 ibid., 2.1.
65 ibid., 2.4.
66 *Age*, 31 March 2007, p. 1.
67 Hansen, p. xi.
68 Li, p. 60.
69 Alison et al, p. 7.
70 Jowit and Wintour.
71 Stern, *A Blueprint for a Safer Planet*, pp. 54, 150.
72 International Climate Conference.
73 Morton, 'Warming's icy debate'; *Weekend Australian*, 5–6 June 2010, Inquirer p. 5.
74 http://www.metoffice/gov.uk/corporate/pressoffice/2010/pr20100728.html; http://www.ncdc.noaa.gov/bams-state-of-the-climate/2009.php; *Daily Telegraph*, 28 July 2010. http://www.telegraph.co.uk/earth/environment/climatechange/7914611/Met-Office-report-global-warming-evidence-is-unmistakable.html (all viewed 29 July 2010).
75 Turnbull, p. 5; ABC, 'The carbon war'.
76 Chandler.
77 Cryderman.
78 Greenpeace International.
79 ACTU, 'Campaign: Clean Energy Jobs'.
80 *GLW*, 20 January 2010, p. 3.
81 Tokar, p. 64.

82 http://www.ens-newswire.com/ens/apr2010/2010-04-20-02.html (viewed 22 April 2010).
83 http://pwcc.wordpress.com/2010/04/26/peoples-agreement (viewed 25 August 2010); Hogan, p. 6.
84 http://www.ucsusa.org; Brownhill and Turner, p. 246.
85 McKibben, *Fight Global Warming Now.*
86 Cryderman.
87 David Suzuki Foundation et al, p. 3
88 ibid., p. 3.
89 *Conservation News*, June 1990, p. 1.
90 http://www.greenpeace.org/international/Global/international/planet-2/report/2009 (viewed 2 September 2010).
91 Lang Hancock, *Wake Up Australia*, 1979, quoted in Benedek, p. 3.
92 Diesendorf, *Climate Action*, esp. pp. 7, 140, 160–161.
93 Castells, p. 3.
94 Rucht, p. 207.
95 Melucci, p. 25.
96 Vromen et al, p. 267.

2

Climate Change in Australia

Australia stands in a peculiar relation to the climate justice issue. In contrast to the usual inverse relationship between complicity and suffering, Australia is both highly complicit in global warming, because of its high per capita contribution to emissions, and also extremely vulnerable to its adverse effects—and well before rising sea levels become a serious issue. This chapter considers Australia's disproportionate contribution to climate change and how seriously climate change will affect Australia.

'A Sunburnt Country'

Australia is almost as large in area as the forty-eight contiguous US states, but overall it is a much drier and harsher country. Generations of Australian school children have learnt this fact reciting Dorothea Mackellar's poem 'My Country' with its well-known line 'I love a sunburnt country'. Although Australia is bordered by three oceans—the Indian, Pacific and Southern—it is often described with good reason as the driest inhabited continent on the planet. Only about 10% of Australia's land mass is arable and much of that is marginal when compared with other landmasses.[1] Australia is the developed country most vulnerable to the direct impacts of climate change and also to the indirect impact from neighbour countries that are stressed by climate change.[2]

Australia's susceptibility to the worst effects of global warming is exacerbated because it is affected more than any other continent by the Southern Oscillation: the swings between El Niño and La Niña, which shape both temperature and precipitation variability. El Niño reduces Australian rainfall for up to 7 years, creating droughts, and the worst bushfires have tended to occur during El Niño years. By contrast, during La Niña, there is more rain and even heavy storms and flooding. Global warming appears to have increased the intensity of the Southern Oscillation. Also, since the late twentieth century, the Indian Ocean Dipole results in a warming of the west and reduced spring rains in south-western Australia.[3]

The 2007 IPCC Report delineated some of the following impacts of climate change on Australia: warming of 0.4° C – 0.7° C since 1950, with more heatwaves, fewer frosts, more rain in the north-west, less rain in the south-east, more intense droughts, and a sea level rise of about 70 mm.[4] For Australia as a whole, 2009 was the second hottest year since high-quality temperature records started in 1910 and August 2009, the warmest August on record; and 2005 was the hottest year, 1.09° C hotter than the standard mean for the period 1961–1990.[5] In 2010, the Bureau of Meteorology (BOM) reported that the decade of 2001–2010, the hottest on record, was 0.52° C above the average; and that Australia's climate had steadily warmed over the previous 60 years, with very few cool years occurring in the last three decades.[6] Australia was then ravaged by one of the most severe La Niña events in recorded history, in late 2010 early 2011, aggravated by record-high sea-surface temperature, very likely related to global warming. BOM analyst David Jones observed: 'The last year (of extreme weather events) has been really extreme, but in the Australian context the really major story is La Niña.'[7] Droughts and flooding rains are becoming more severe.

At the end of this hot decade, CSIRO and the Bureau of Meteorology (BOM) produced a report based on peer-reviewed data on temperature, rainfall, sea level, ocean acidification, and CO_2 and methane levels in the atmosphere, over the past five decades. Since 1960 the overall mean temperature has increased by 0.7° C and some areas have experienced warming since 1960 of up to 0.4° C per decade, resulting in total warming over the five decades of 1.5 to 2° C. The number of days with record hot temperatures has increased each decade over the past five and there have been fewer record cold days each decade. Since 1993 sea levels have risen 7–10 mm per year in the north and west and 1.5–3 mm in the south and east; and sea surface temperatures have increased by about 0.4° C in the past 50 years. Total rainfall has been relatively stable but its geographic distribution has changed significantly over the past 50 years, increasing in the north and decreasing in the south-west and south-east, including all the major population centres. 'Our observations clearly demonstrate that climate change is real', it concluded.[8]

As this report intimates, regions experience climate change differently. Altered rainfall distribution is especially significant. By 1996, climate change was already causing a rise in the incidence and distribution of Ross River virus, dengue fever (endemic in parts of tropical Australia, particularly Queensland) and asthma.[9] Already, as Lauren Rickards and Karlie Tucker's research on agriculture points out, many farming families in southern Australia have been forced to seek alternative forms of income.[10]

Asa Wahlquist notes the particularly negative impact of climate change on south-western Australia, which created a crisis situation for Perth's water supply:

> Perth's rainfall has had two step changes. Between 1911 and 1974 the rainfall averaged 1251 millimetres. It dropped to 1073 millimetres between 1975 and 1996, then dropped further to 997 millimetres between 1997 and 2005. By 2005, Perth was receiving 20 per cent less rain than it had for most of the previous century.[11]

Perth became the first major Australian city to commission and construct a desalination plant. Operating since November 2006 at Kwinana, 40 kilometres south of Perth, this desalination plant supplies an average of 130 million litres of desalinated water daily.

Precipitation in Victoria has decreased 13% over the past decade compared to the average during 1961–1990. In March 2007 Melbourne Water figures showed that the Thomson Dam, which provides more than 60% of Melbourne's water supply, reached a record low of 20%; and the total capacity of the nine reservoirs was only 33.5% of capacity.[12] In an effort to counteract water shortage, in 2007 the Victorian Labor government ordered the construction of a desalination plant in Wonthaggi in south Gippsland. Power for the desalination plant was supposed to come from a wind farm to be constructed by AGL Limited in Glenthompson in western Victoria.[13] This incredibly expensive and ecologically questionable project is nearing completion. Dam levels fell even further until spring 2010 when the effects of La Niña reversed the downward trend.[14] Though eased since this La Niña event, water restrictions in the 'Garden State' have become an accepted fact of life. Victorians now monitor water levels in dams whose storages they once took for granted; the *Age* newspaper front page now provides this information each day.

In Victoria, daily maximum temperatures have increased by 0.8° C since 1950.[15] According to Tony Kevin: 'A 30-degree day in southern Australia used to be considered hot, but people now see this as a relatively average midsummer's day. In the public mind, 40 degrees is becoming the new 30 degrees.'[16] In Melbourne, since complete annual records began in 1856 when the average maximum temperature was 19°C, there have been 6 years in which the average temperature was over 21° C. Five of these have been since 1999, with 2007 recording 21.8.[17] Hotter even than 2005, 2009 was the hottest year ever in the southern half of the continent.[18] On 18 March 2010, Melbourne experienced its hundredth consecutive day with a maximum temperature over 20° C, for the first time since partial records commenced in 1840.

The devastating 'Black Saturday' bushfires of 7 February 2009 in Victoria, sparked by unusually hot weather and dry conditions, concentrated national attention on the unprecedented high temperatures that preceded the disaster. David Karoly, a renowned climate scientist at the University of Melbourne and a lead author for Working Group 2 of the IPCC, linked the Victorian bushfires with climate change, when interviewed on *Lateline* on 9 February 2009:

> The records were broken by a large amount and you cannot explain that just by natural variability ... What we are seeing now is that the chances of these sorts of extreme fire

> weather situations are occurring much more rapidly in the past 10 years due to climate change.

Accumulated meteorological evidence about recent climate trends in south-eastern Australia substantiates Karoly's assertions. On 9 February 2009, the BOM stated:

> An exceptional heatwave affected south-eastern Australia during late January and early February 2009. The most extreme conditions occurred in northern and eastern Tasmania, most of Victoria and adjacent border areas of New South Wales, and southern South Australia, with many records set both for high day and night time temperatures as well as for the duration of extreme heat.[19]

Tasmania experienced seven of the eight highest temperatures on record in early 2009.[20]

In January 2009 Melbourne experienced the 'heatwave of the century', in which temperatures exceeded 43° C on three consecutive days. According to Rob Adams, Director of Design and Culture at the City of Melbourne:

> Despite its 2008 ranking as *The Economist*'s second most livable city in the world, Melbourne proved most unlivable for the majority of January in 2009. The city's inability to cope with heat led not only to the tragic loss of life, but potentially hundreds of millions of dollars of lost income, productivity, and property damage.[21]

The Victorian Department of Health Services reported that this heat wave resulted in an estimated 374 excess deaths in Victoria.[22]

On Saturday 7 February, the temperature hit 46.4° C in Melbourne, an all-time high for the city, and surrounding areas recorded even higher temperatures. Victoria experienced a record state temperature high of 48.8° C. The bushfires of this Black Saturday resulted in 173 deaths and the destruction of hundreds of homes. Badly affected communities, such as Marysville in the Yarra Ranges, are still rebuilding.

Abnormal heat returned in the spring of 2009. Central and south-eastern Australia experienced a prolonged spring heatwave in which November constituted the hottest on record for many places: the mean monthly temperature for NSW was 4.61° C above normal, the largest monthly anomaly ever recorded for an Australian state; the Victorian mean temperature anomaly of 4.36° C above normal also broke the previous record; Tasmania experienced a mean temperature 2.43° C above normal. In the same month, Adelaide experienced eight consecutive days above 35° C, breaking the previous November record of four such days in a row.[23] Barry Brook, a University of Adelaide climate scientist, indicated that Adelaide's late spring heat wave very likely resulted from climate change.[24] The extreme heat continued. Referring to the summer of 2009–2010, Royal Australasian College of Physicians Spokesperson Robyn McDermott observed: 'In Adelaide last summer, for the second summer in a row, we had a heatwave with two weeks above 35 degrees. By the second week we started having daily health tolls updates on the radio.'[25] Between May 2009 and April 2010 both Victoria and Tasmania experienced the warmest 12-month period ever recorded. [26]

Flooding rains then replaced drought to plague much of Australia, particularly Queensland and to a lesser degree NSW and Victoria, during late 2010 and early 2011. This book was finished under climate-change induced adverse circumstances. Verity Burgmann's house was flooded on 4 February 2011. Many of the research materials for this book, in a wooden filing cabinet, became water-logged and unreadable. In addition, all the flooring in the house had to be replaced, distracting greatly from book writing. Meanwhile, Sydney and the Hunter region of NSW endured a record-setting heat wave in early 2011. The BOM reported:

> Maximum temperatures were well above average across NSW between January 30th and February 6th, with the largest anomalies persisting along the central coast … Records for consecutive days above thresholds were set at many locations, with Sydney Observatory Hill recording seven consecutive days above 30° C between the 31st and the 6th with 41.5° C recorded on the 5th (the highest daily maximum temperature recorded during the hot spell at

> Sydney Observatory Hill). This is the longest hot spell recorded in the city in 152 years of records, beating the previous record by two days.[27]

The *Critical Decade Report* published by the Climate Commission Secretariat in May 2011 emphasised that the climate is indeed changing and humanity is almost certainly the primary cause; failing to take sufficient action today entails potentially huge risks to our economy, society and way of life into the future; and this is the critical decade for action.[28]

Predictions and Implications of Climate Change for Australia

The 2007 IPCC Report predicted substantial financial losses for Australia due to droughts, floods, fire and tropical cyclones. Water shortage could become an especially serious problem. This IPCC Report predicts that the annual stream-flow in the Murray–Darling Basin, which accounts for about 70% of irrigated crops and pastureland in Australia, is likely to fall 10–25% by 2050 and 16–48% by 2100. Furthermore, there exists a 50% probability that by 2020 the average salinity of the lower Murray River will exceed the threshold for acceptable drinking and irrigation water. Water shortages will adversely affect various industries and ordinary household activities, such as cooking, drinking, washing, bathing and gardening. The 2007 IPCC Report predicted a decline in agricultural and forestry productivity by 2030 in the south-east.[29] At the same time, as a Department of Climate Change report indicated in 2009, tropical cyclones will probably become more severe.[30]

Most recently, the CSIRO and Bureau of Meteorology 2010 *State of the Climate* report calculated that Australian average temperatures are projected to rise by 0.6 to 1.5° C by 2030; and, if global greenhouse gas emissions continue at current levels, warming is projected to be in the range of 2.2 to 5.0° C by 2070. Compared to the period 1981–2000, decreases in rainfall are likely in the decades to come in southern areas during winter, in southern and eastern areas during spring, and in south-west Western Australia during autumn. 'An increase in the number of dry days is expected across the country, but it is likely that there will be an increase in intense rainfall events in many areas.'[31]

More farming areas in southern regions will become too dry to be productive, directly and indirectly affecting the livelihoods of many.[32] At the same time, rising sea levels could create millions of climate refugees in the Asia–Pacific region; and Australia should play its part in accommodating these people. The Australian Greenhouse Office has published research by Barrie Pittock on the 'dislocations of populations' that will occur in Australia's vicinity as sea levels rise.[33]

Apart from adverse effects on employment and the obvious increased dangers of spectacularly bad bushfires and serious water shortage from decreased rainfall in the southern half of the continent, climate change threatens people's health, and will have a catastrophic effect on biodiversity. While Australia's land surface accounts for only 5.7% of the earth's land surface, Australia sustains 7–10% of the world's species, most of which are unique to the country.[34]

The 2007 IPCC Report predicted serious biodiversity loss by 2020 in various 'ecologically rich sites' in Australia.[35] CSIRO foresees significant loss in biodiversity by 2020 in the Great Barrier Reef (which already is undergoing massive coral bleaching), the Kakadu wetlands in the Northern Territory, south-western Australia, the Australian Alps, and the sub-Antarctic islands.[36] In the protected rainforests of northern Queensland, climate change threatens more than seventy species of birds and mammals.[37] According to Lindenmayer:

> Elevated levels of carbon dioxide in the atmosphere could alter the chemistry of plant leaves, such as the content of nitrogen (protein) and other plant compounds. Alarmingly, this could change the palatability and nutritional values of food plants for many leaf-eating animals, from the iconic koala to the lesser-known marsupials such as the greater glider, and leaf-eating invertebrates.[38]

At the base of the food chain in the Southern Ocean, which absorbs 40% of 25% of CO_2 generated by humans around the world, ocean acidification will decrease the ability of marine plants and animals to form shells, because the CO_2 concentration in 2009 of 386 ppm is much higher than the natural range of 170 to 300 ppm that has existed for at least the past 800 000 years and possibly the past 20 million years.[39]

The projected effects on human beings are devastating. The Royal Australasian College of Physicians has termed climate change the 'biggest global health threat of the twenty-first century'.[40] There are numerous potential health impacts resulting from climate change in Australia, in addition to fatalities and injuries from severe weather events, such as floods, storms, cyclones and bushfires. Increasing temperatures will increase: vector-borne infectious diseases (dengue fever and Ross River virus); aeroallergens (spores, pollens) potentially exacerbating asthma and other allergic respiratory diseases; food-borne infections and diseases, such as salmonella; water-borne infectious diseases and heath risks from poor water quality; poor nutrition due to rising food prices caused by diminished food production; and air pollution, for example, from bushfire smoke.[41] Rising temperatures will bring about an increasing number of heatwaves, which will result in more instances of heat stress, causing mental disturbances and disorientation, as well as severe physiological effects.[42] A chilling Table in the Garnaut Review indicates, for the 'no-mitigation case', a large national increase in temperature-related deaths, mainly from heart attack and strokes, in the second half of the century.[43]

Fictional Images of a Climate-Changed Australia

The portrayals of how climate change will affect Australia provided by the IPCC, CSIRO, BOM and other agencies are deeply concerning. Yet statistical predictions generally present overall outcomes, without much attention to the social distribution of the adverse effects of global warming, and none whatever to the political consequences. They are also abstract and impersonal. For us, as humans, to grasp fully the import of the scientific information provided in such accounts, imagination is needed: a sociological imagination or perhaps even a literary one. Fiction might constitute the best means for envisioning what Australia might look like and how Australians like us might behave if climate change were not effectively mitigated around the world.

Melbourne and its surrounding environs served as the backdrop for Nevil Shute's 1954 classic novel *On the Beach* that portrays the annihilation of the last affluent human beings on the planet in the wake of full-scale nuclear war. The novel was made into a film starring Ava Gardner and Fred Astaire as the principal Australians and Gregory Peck as the American protagonist. When asked her opinion about Melbourne, Gardner reportedly said it was 'the perfect place to make a film about the end of the world'.[44]

Australia's first climate change novel was also set in Melbourne, a city largely submerged under rising sea-levels in the middle decades of the twenty-first century. George Turner's dystopian novel, *The Sea and Summer*, was published in 1987 around that moment when public agitation about the greenhouse effect first became marked: Turner was speaking to this mood of the late 1980s, which constitutes a seriously lost political opportunity in mitigation initiatives. *The Sea and Summer* is an extraordinarily well-crafted and gripping novel that received international awards and critical acclaim but has not received the popular attention it deserves. Perhaps its fate was sealed by the fact that the novel's dystopia was simply too close to impending reality, too confronting because the nightmare world it described seems more inevitable than nuclear annihilation did in the 1950s. Turner's novel was written with expressly political intent. 'We talk of leaving a better world to our children', he wrote in the 'Postscript', 'but in fact do little more than rub along with day-to-day problems and hope that the long-range catastrophes will never happen'. *The Sea and Summer*, he explained, 'is about the possible cost of complacency'.[45]

The Sea and Summer depicts a world in the 2040s which exceeds ten billion people and where winter has disappeared as a season for most parts of the planet, as a result of a global temperature 4.5° C higher than in 1990. The global financial and exchange systems have collapsed. Australia as a nation–state has, under pressure from world powers, given up its northern third to accommodate climate refugees from Asia.[46] In Victoria, winter lasts only a few weeks and the climate fluctuates between drought and torrential storms and floods.[47] The Yarra River has risen over its banks due to rising tides, coastal roads have vanished, and the lower floors of many tenements are abandoned because of the water below.[48] Shortages of all sorts have become a normal part of social life; and a sharpened class divide therefore exists between 'the Sweet' and 'the Swill', with the former having claimed the higher ground and the latter forced to reside in decaying high-rise towers, surrounded by water, in locations such as Elwood and Newport. A social class called 'Fringers' has emerged—neither Sweet nor Swill—but 'seen by the Swill as fake Sweet and by the Sweet as Swill tainted'.[49] The Australian state has responded to the social chaos by becoming a totalitarian system with Police Intelligence Officers seeking to keep the Swill under control in the high-rise tower blocks. It has implemented a coupon system for obtaining goods and services under which the Sweet are able to obtain sufficient supplies and some of them to live in relative luxury. Fringers have barely enough, while life for the Swill is a struggle for mere existence.[50] Disturbingly convincing characters inhabit the starkly differentiated social layers—and live out the consequences of their class locations.

In 1992, American historian and novelist W Warren Wagar spoke to the greenhouse anxieties of that moment, and again, Melbourne is a chosen location for the novel's conclusion. Wagar's enthralling *A Short History of the Future* gives an account of the history of humanity into the twenty-first and twenty-second centuries. The period of 2001 to 2032 is characterised by an economic boom 'hinged in part on advances in the production and cheapness of energy'.[51] Various technological innovations, including the creation of expensive fusion nuclear power plants, enable reliance on fossil fuels to drop to 71% by 2030. Yet still the capitalist treadmill of production and consumption continues to ensure excessive emission of greenhouse gases. Thus, by 2040, the atmosphere contains 555 ppm of carbon dioxide and even more alarming increases in methane, and CFCs due to the burning of fossil and biomass fuels, fertiliser use, and the decay of organic matter in rice paddies.[52] The average global temperature increases 4.2° C between 1980 and 2040, which results in melting of the polar ice caps and glaciers, rising sea levels, heavy flooding in some regions and declining food production. These developments culminate in the Catastrophe of 2044, marked by the outbreak of worldwide nuclear war followed by a sudden shift to arctic cold. In the aftermath of the war, the power centre of the world moves to countries south of the 25th parallel. Under the leadership of the World Party founded by a New Zealander, Melbourne becomes the capital of a bureaucratic, technocratic socialist-oriented world government. However, failure of this new government to bring about world peace results in the development of an anarchistic global system committed to sustainable living, mysticism and small-scale technology.

In an otherwise non-fiction book entitled *Crunch Time*, Tony Kevin includes a chapter 'Southern Australia, 2060: drowning cities in a parched land' that is dystopian about the physical impacts of climate change but utopian about the ultimate manner of adaptation of Australian society. Average temperatures have risen 4° C globally, 10° C in the polar regions and 6° C in temperate regions. Thirty years ago world leaders had decided to stop burning fossil fuels, so CO_2 emissions are holding at 550 ppm but temperatures are continuing upwards. The sea level has risen one metre since 2010 and is still rising at about five centimetres annually. The permafrost in Siberia and arctic North America has been rapidly melting. Societies around the world have adopted autarkic solutions and the Australian state has peacefully ceded governance of the northern half of the country, making it available—as in Turner's novel—to climate refugees from Asia.

Australians have in large numbers turned their backs on the seashore; city centres are emptying out and turning into derelict tower-scapes.

> Even in the big cities, sewerage and stormwater drainage systems, having lost one metre of their design head, no longer work as they were designed. There are dangerous overflows from backed up drains into low-lying streets at times of high tides and heavy rains. For a few years, cities built expensive, stronger drainage-outflow pumping systems, trying to cope. They also contemplated building high dykes around vulnerable low areas on the Netherlands model, until the awful, complete uncertainty of any future sea level-rise predictions sank home to government planners.[53]

The regions west of the Great Dividing Range have become completely arid and uninhabitable, with remote highland and tablelands being more densely populated. The southern Australian landscape had been transformed into one of 'dissolving coastlines, parched tablelands, and a huge, virtually waterless inland desert'.[54]

The climate change crisis has in the first instance exacerbated class divisions:

> The old economic and social order, with its comforting certainties of jobs, asset wealth, pensions, and property rights under the rule of law, unravelled as the seas rose and increasing numbers of desperate, dispossessed, and unemployed people moved inland in search of new ways to keep alive. There were breakdowns of law and order, localised descents into bloodshed and anarchy, as local property-owning communities fought back, resisting invasions of desperate refugees from the drowning cities who demanded living space. Well-resourced fundamentalist and millenialist communities tried to seal themselves off at gunpoint from the growing disorder that surrounded them, in vain bids to create little sectarian utopias.[55]

Australians gradually found new sustainable ways of living in this harsher world. 'There had been a sharp public reaction in the 2040s against market capitalism and the associated resources-wasting consumerism—a collective social judgement that these excesses had been the major drivers of humanity's loss of its former climate security.'[56]

Australia adopted a 'regulated market-economy'[57] with an emphasis on local economic activity, a greater sense of social responsibility and a more vegetarian diet. People finally turned to large-scale reliance on solar, wind, and geothermal energy, all part of a national grid. Private cars virtually vanished and airplane travel was abolished; and railways enjoyed a revival. The new human settlements were whitewashed, sunbaked, mud-brick villages, of compact, terraced, two-storey thick-walled houses with tiled roofs and slab floors, which used abundant local clay soils. They made sense in the hot, dry climate. So the Australian countryside began to look like rural Spain or Greece. Arcadian themes pervade Kevin's description:

> Already, some of these new villages are starting to become settled and fairly sustainable communities, offering some quality of life and hope for a more secure future. Cellars are filling up with food stores, such as sweet potatoes, carrots, pumpkins, grains, wines, dried fruits, nuts, olives and olive oil. For the children being born now, those too young to remember the trauma and grief of the great evacuations from the drowning cities, life offers modest joys again. There are harvest festivals and dancing, local musicians, and story-tellers. The recent past is already fading into mythology, like stories of old Atlantis.[58]

The 'regulated market-economy' is overseen by a directorate, which is quick to intervene against business malpractice, hoarding or profiteering from scarcities. 'Nobody thinks about economic growth any more in the old way. Now, it is all about sustainability, trying to keep population and lifestyle in equilibrium with scarce natural resources.'[59] Kevin concludes the description of his green Keynesian semi-utopia with the observation that one thing is clear, that 'humanity will not make these kinds of environmental blunders again: it has finally left its childhood behind. It now understands, from the lessons of terrible experience, that it has to live hereafter within the physical limits of its world.'[60]

Melbourne writer Sean McMullen, whose day job is with the BOM, published 'The Precedent' in the international journal *Fantasy & Science Fiction.* Precise location is unspecified, but one character is a 'bushman' and some action occurs 'in the desert, a thousand miles from the coast'.[61] It is 2035 and three degrees warmer than now. People are responding to the climate crisis by riots, which 'caught the authorities by surprise', and in other ways more conducive to survival, such as 'Going Lateral', a movement that began on the internet. The lateralists worked out that they could actually live better by detaching themselves from the economic systems of derivatives, leverage, optionality and toxic assets. In just months, the lateralists' ranks swelled from thousands, to millions, to hundreds of millions. This generated a crisis in confidence that triggered the biggest financial collapse in history, and very soon the people who had formerly worked at generating meaningless wealth were out looking for real jobs. 'By then it was too late because the climate was severely screwed, there were famines in Western democracies, the trillions of dollars based on derivatives and options were fast becoming meaningless, and economic growth was considered as healthy as cancer.'[62]

People born before the year 2000 are known as 'tippers' and are held responsible by the 'victim' generations, those born after 2000, for pushing the global climate past the tipping point. A World Audit, similar in process to the Nuremburg trials, judges people for serious crimes, such as squandering, display and greed. Depending on the severity of each crime, all these carry the death penalty. Denial, the most serious crime, carries a penalty of life underground in the old mines, the only food being the flesh of your fellow deniers.

The trials tell just how ordinary are the 'tippers'. One was a mushroom hunter in his previous life.

> 'How does a mushroom hunter get denial, squandering and greed?' asked Chaz.
>
> 'I drove an SUV to reach the best spots.'
>
> Denial, because he said he loved the wilderness yet drove an SUV. Squandering merely because he drove an SUV. Greed, because he took from nature without giving back. Death, death, and death. Having spoken, he looked more relaxed, perhaps because he had left something of himself in our memories. Shepherded by the executioner, he walked out onto the tipping plank. The gallows were built of timber even though the old mine site was littered with steel pipes and beams. This was climatic symbolism. The Auditor General stood waiting.
>
> 'James Francis Harrington, you have been found guilty of denial, squandering, and greed,' she declared. 'For this you are sentenced to death by merciful means. As you did take from the Earth, so now you must give what remains to you back to the Earth. This is my tally, the

> Twenty-Fifth day of March, 2035. Wardens, reclaim his carbon'.[63]

The punishments grotesquely fit the crimes, as the narrator reveals:

> I saw a fit-looking man of about fifty being forced to the ground by the wardens. With the skill of much practice they spread-eagled him on his back and chained him to a wooden frame in the shape of an X. All the while he was shouting about his right of appeal, and calling for a proper lawyer.
>
> 'Your turn will come. You'll pay for this!' he screamed. 'This is a concentration camp.'
>
> 'You helped run the concentration camp called the global economy,' replied a warden. 'You kept Mother Earth in there until she was a living skeleton.'
>
> They took two large glass panels and clamped them over him in a tent shape. We were level with them as they fitted a pair of glass triangles over the ends.
>
> A greenhouse. For causing the greenhouse effect, death by greenhouse. It was a hideous way to die, roasted slowly in a glass oven. The unlucky ones lasted to evening, got the respite of night, then had to face a second day.[64]

Australia's Role in Climate Change

Australia has one of the highest per capita ecological footprints in the world. Ecological footprint, which correlates closely with energy use, is calculated by measuring the consumption of energy, biomass (such as food and fibre), building material, water, and other resources and converting this into land area, expressed as hectares per capita.[65] Calculations of each country's ecological footprint, with and without marine products, have been made. Australia sits easily in the largest five in each case, along with New Zealand, USA, Canada and Iceland. The world average without marine products is 1.8 and with marine products 2.3, which places the Australian figures of 7.4 and 8.1 respectively into perspective.[66]

Above all, Australia has the dubious distinction of being one of the highest per capita greenhouse gas emitters in the world. Ben McNeil describes Australia as having a 'carbon obesity epidemic'. Accounting for different levels of economic output, Australia is 'the most carbon obese nation in the developed world', even more carbon obese than the USA.[67] Only Bahrain, Bolivia, Brunei, Kuwait and Qatar are higher in terms of emissions per capita. Bolivia aside, these countries are oil-producers and their small populations maintain on average very energy-intensive economies and lifestyles.[68] Australia's per capita emissions are the highest of any OECD country, followed by Luxembourg, Canada, the USA and New Zealand, easily outstripping the USA.[69]

Australia also currently ranks as the thirteenth largest national producer of greenhouse gas emissions globally, despite a relatively small population of about 22 million.[70] Over 50% of Australia's emissions are produced by just fifty companies. The top ten polluters are Macquarie Generation, Delta Electricity, Great Energy Alliance Corporation, International Power (Australia), Alcoa Australia, Rio Tinto, CS Energy, TRUenergy, Eraring Energy and Electricity Generation Corporation.[71]

The IPCC specifies four sources of emissions: energy, industrial processes, agriculture and waste. The energy sector is broken down into fuel combustion and fugitive emissions. The latter come from coal mining and handling and decommissioned mines, oil and natural gas production, processing and distribution. In addition, countries must account for land use, land use change and forestry activities, namely deforestation, afforestation and reforestation.[72] In 2008, Australia's emissions were an estimated 553 Mt CO_2e. The breakdown by sector in terms of Mt CO_2e was: Energy—fuel combustion 377; Energy—fugitive emissions 39; Industrial processes 32; Waste 15; Agriculture 91.[73]

Long-term statistics reveal that Australia has managed to reduce its emissions per capita over the period 1990 to 2007, but only by 32.3 to 28.6 tonnes CO_2e.[74] Its net emissions increased between 1990 and 2008, while governments were fully aware of the need to reduce emissions.[75] There were significant variations between sectors. Emissions from land use change and waste declined 53.90%

and 11.40% respectively. However, emissions from agriculture increased 3.8%, industrial processes 17.70%, fugitive emissions 18.10%, transport 27.40%, stationary energy 47.30%.[76] The *Greenhouse Indicatory Report* released by the Climate Group on 13 January 2009 found that, weighed up against 1990 levels, NSW's emissions in 2008 had risen by 30%, Victoria's by 32% and Queensland's by a whopping 116%.[77]

Another significant measurement is the carbon intensity of the economy, expressed as emissions per dollar of GDP. Due to factors such as rising GDP figures, a decline in manufacturing and a rise in service industries, carbon intensity has declined from 1990–2007 by 36.4%, from 0.90 to 0.57 kg CO_2e.[78] Nonetheless, McNeil observes:

> All developed countries have considerably lower carbon intensity than Australia. Australian factories, on average while making the same product, produce up to three times more carbon dioxide than factories in Europe, nearly double those in Japan, 70 per cent more than New Zealand factories and 25 per cent more than US factories.[79]

Energy efficiency in Australia has improved, but is still low compared to other OECD countries.[80] Most significantly, energy usage per person has increased between 1976 and 2006. This is a tendency in developed countries, but the increase is more marked in Australia than in most other countries. Moreover, some developed countries have managed to decrease energy usage per person in these three decades. Australia lags behind. Its total primary energy supply tonnes of oil equivalent per person increased from 4.43 in 1976 to 5.93 in 2006, whereas Germany's decreased from 4.36 to 4.24, Denmark's from 3.87 to 3.73 and the USA's from 8.13 to 7.76.[81]

So emissions rise. The Climate Group's *Greenhouse Indicatory Report* in 2009 found that emissions in Queensland, NSW and Victoria in 2008 had increased by 1.3% on 2007 levels, whereas worldwide emissions increased 0.5% , according to the World Meteorological Organisation. Furthermore, this report only measured emissions from energy generation, discounting emissions from agriculture, the second largest contributor to Australia's overall emissions.[82] According to the Department of Climate Change 2009 report, Australia is increasing its output of greenhouse gases by 2% per annum.[83] The Australian Greenhouse Office, the government agency responsible for tracking emissions and promoting reduction targets, estimates that Australia will reach 127% of its 1990 emissions level by 2020.[84]

Why are Australia's Greenhouse Gas Emissions So High?

Compared with other advanced capitalist countries, there are distinctive features of the Australian economy that explain its high emissions. Australia's poor performance is mainly because of its reliance on abundant coal supplies for electricity generation and its low overall energy efficiency compared to other OECD countries. This section examines in particular the roles of the coal, automobile, aluminium smelting, agriculture and forestry industries—and Australians' enthusiasm for high levels of consumption—in contributing to high per capita emissions.

Dependence on Fossil Fuels for Energy

In his observation that 'governments have fanned a concept of national identity built on notions of economic success and personal status measured by material acquisition', Tony Cutcliffe adds: 'In Australia, this prosperity has its roots and stability entrenched in unbridled consumption of cheap energy provided by fossil fuels.'[85] Coal, oil, and natural gas provide most of Australia's energy. In 2005–2006, Australia generated 84% of its electricity from steaming coal (59% from black coal and 25% from brown coal), 9% from gas, 7% from hydroelectricity, and negligible amounts from wind or solar sources.[86] In terms of the percentage of electricity generated from coal, Australia is in third place internationally. Poland derives 96% of its electricity from coal, South Africa 92%, Australia 78%, China 75%, India 75%, Czech Republic 73%, Germany 53%, and the USA 52%.[87] Moreover, in contrast to steelmaking, which relies only on black coal, Australian electricity generation relies on both black coal and the even dirtier brown coal still in abundance in the Latrobe Valley in Victoria, which creates even more emissions per unit than black coal.[88] Yet the NSW and Victorian governments have plans to build new coal-fired power plants.[89]

Mining

Apart from heavy reliance on coal for power generation, the prominence of mining within the Australian economy adds significantly to emissions. Mining accounts for about 5.6% of Australia's GDP; in 1950 it accounted for 2.6%. Minerals comprise about 35% of the country's exports. In 2001–2002, the top Australian exports were coal, crude oil, iron ore, wheat, aluminium, meat, bauxite and alumina. Yet the mining sector employs only about 129 000 workers or a mere 1.3% of the total labour force.[90] It exploits natural resources much more than it utilises labour.

James Goodman and David Worth maintain the mining boom has been a 'curse' in disguise in that it has: sharpened socio-economic and community divisions; contributed to political conflict; and resulted in 'ecological mal-development' with serious environmental consequences.[91] Mining produces direct emissions due to electricity usage and also heavy fugitive emissions. Also, since most mines are located in remote areas, many operate on a fly-in/fly-out basis, with miners flying from their homes for two weeks of intense labour and then home for one week of respite, contributing undetermined emissions in this long-distance commuting.

Despite its relatively small population, Australia is the world's fourth largest coal producer, after China, USA and India. With three-quarters of Australia's black-coal production exported, Australia is the world's largest national coal-exporter. Australian coal, exported to more than thirty-five countries, provides 30% of global coal exports, contributing about 500 million tonnes each year to global emissions.[92] Nearly half of Australia's coal exports go to Japan; Europe is the second biggest purchaser.[93] Only about 2% goes to China and Australia provides China with less than 1% of its coal, most of it going to steel mills.[94]

Most of Australia's coal is mined in the Bowen Basin in Queensland and Hunter Valley in New South Wales, but there are also coal mines in Western Australia south of Perth and lower grade brown coal is mined in Victoria. Nine terminals at seven ports stretching along the east coast of Australia serve as the coal exporting sites, with Newcastle and Gladstone the largest, both competing with Richards Bay in South Africa for the distinction of world's largest coal port.[95]

Fred Pearce reports of Gladstone: 'Some 68 million tonnes a year comes here by rail from mines in the Queensland interior. Burning that coal ultimately puts into the air about 200 million tonnes of carbon dioxide a year.'[96]

And this industry keeps expanding. In April 2008 the federal government gave the go-ahead for a multi-billion dollar expansion of Gladstone, which will result in another 84 million tonnes of coal being exported annually. Despite his insistence that climate change was the great moral challenge of this generation, former Prime Minister Rudd announced on 12 December 2008 that $580 million of public money will be allocated to the expansion of coal facilities in the Hunter Valley and the Port of Newcastle, predicted to increase coal exports from Newcastle from 97 to 200 million tonnes annually.[97] The NSW government has given Newcastle Coal Infrastructure Group, a consortium of six mining companies, permission to construct a new $900 billion coal port. Ironically, in anticipation of rising sea levels, the NSW government has also issued new planning guidelines to coastal councils such as Newcastle, increasing the height of the new coal port by several metres. According to NSW Department of Planning estimates, the 66 million tonnes of coal that could be handled by the new coal port 'would release 174 million tonnes of greenhouse gas when burnt in overseas power stations'.[98]

Bauxite mining for aluminium production is also significant. For example, Rio Tinto, the world's largest mining corporation, has a license to strip-mine 2600 square kilometres at Weipa on the Gulf of Carpentaria, the traditional homeland of the Alngith people.[99] Ten per cent of the world's bauxite comes from this area and is shipped to Gladstone. The conversion of bauxite into aluminium requires lots of electricity. Indeed, aluminium smelting accounts for about 13% of Australia's electricity consumption, most of which comes from coal-fired power plants.[100] As a result of this, in contrast to most other countries which rely on hydroelectric power for smelting, Australia's aluminium industry has the highest greenhouse intensity in the world, accounting for around 6% of Australia's total emissions.[101]

Transport and the Domination of Motor Cars

Australia—like the USA and Canada—is a car-oriented society. Automotive technology, according to Brian Ladd, is politically constructed: 'powerful interests have made their influence felt in lasting ways, notably in the construction of roads'.[102] In Australia this influence was glaringly apparent when Australian cities, except Melbourne, got rid of trams and trolley buses between 1950 and 1970.[103] The motor vehicle lobby easily persuaded the Liberal Party and eventually also convinced the Labor Party to promote road construction rather than improve public transport.[104]

By the 1990s, more than 80% of daily journeys in Melbourne were made by car, and more than 50% of households had two or more cars.[105] By 1996, the percentage of public transport use as a share of total transport use was 12.6 in Sydney, 7.7 in Melbourne, 5.5 in Brisbane and 4.5 in Perth.[106] The situation has deteriorated further since. By 2006, Australia had 659 motor vehicles per 1000 population, fourth-highest after the USA (755), New Zealand (735) and Italy (666); and Australia remains one of the leading countries in the world in terms of passenger-kilometres travelled by private car per person.[107] In *Shocking the Suburbs* (2006) and *Unsettling Suburbia* (2008), Griffith University researchers Jago Dodson and Neil Sipe found that automobile dependence of the least wealthy rose between 2001 and 2006 and that there was a strong relationship between car dependency, low incomes and mortgage stress in the outer suburbs of Australian cities.[108]

Compared to most other developed countries, Australia is especially laggard in train usage, and train usage per person in 2005 was half that of 1970. In 2005, passenger-kilometres by train per person were 3062 for Japan, 2178 for Switzerland, 1257 for France, 1032 for Austria, 986 for Sweden, 908 for Germany, 717 for the UK, 692 for Norway, 570 for Australia, 436 for Ireland and a paltry 29 for the United States.[109] Rail is also woefully underused for freight purposes in Australia. By 2000, rail accounted for only 20% of the land freight between Sydney, Melbourne and Brisbane, significantly increasing emissions.[110] According to Diesendorf, in terms of energy use and emissions, the most efficient freight transport mode is sea (4.2 tonne-km/MJ), followed by rail (3.6 tonne-km/MJ) and then road (0.37 tonne-km/MJ), where MJ represents a megajoule.[111]

Agriculture and Forestry

Despite its extensive aridity, Australia has had a strong agricultural sector and exports about two-thirds of its total agricultural production. It is the world's second leading exporter of cotton, beef and mutton.[112] This important part of the Australian economy is, by world standards, a poor performer environmentally, because a large proportion of Australia's agricultural production involves grazing of sheep and cattle. Australia's per capita emissions arising from agriculture reflect Australia's small population and its significant agricultural sector but it is still alarming that they are six times the world average, more than four times the OECD average, and the third highest in the OECD.[113]

Agriculture accounts for around 16% of Australia's emissions, exceeded only by stationary energy. According to Rickards and Tucker:

> Agricultural emissions come from a range of farming activities, including methane from rice fields, application of nitrogen fertiliser, burning of savannas and crop residues and, where it is done, land clearing. By far the largest contribution, however, comes from manure management and digestion in sheep and cattle (approximately 70% of Australian agricultural emissions and 11% of net national emissions). It is livestock emissions that make Australia the third-largest agricultural emissions per capita country in the world.[114]

The beef cattle industry in particular contributes significantly to Australia's emissions. In Queensland alone, in 2004, methane emissions from enteric fermentation were calculated at 1.03 million tonnes, the equivalent of 22.8 million tonnes of CO_2 or 14% of all greenhouse emissions in that state.[115] Methane emissions from cattle and sheep account for about 10% of total national emissions, though this percentage declined recently with the fall in animal numbers due to the combined effects of extensive drought and the fall in the price of wool with the global recession.[116]

Land clearing is another significant factor. Up until 1990–2000, Australia had the sixth highest annual rate of land clearing in the world and was the only country in the top twenty land-clearing nations with a developed first-world economy; more than 60 million trees were cleared in Queensland alone between 1997 and 2003, eliminating an important source of carbon sequestration.[117] Governments, both federal and state, have not placed sufficient emphasis on forestry protection as a means to reduce emissions. A team of ANU scientists asserts that the National Carbon Accounting System underestimates the carbon-carrying capacity of natural forests with high biomass stocks. The researchers argue that green carbon accounting and protection of natural forests in all nations should become part of a comprehensive approach to solving the climate change problem.[118]

The 'Shop Till You Drop' Culture of Consumption

Australia has more shopping centres per capita than any other Western nation and Australians spend more time in them than any other place outside home and work.[119] Fuelling Australia's high consumption pattern is the large average size of residences. Dwelling units in Australia, like in the United States, are becoming bigger and bigger. In the mid-1950s, the average size of new houses was about 115 square metres, that is, half the size of houses today. Between 1985 and 2000 the average floor area of new houses increased by 31%, from 170 square metres to 221 square metres, and the size of apartments increased by 25%, to an average of 139 square metres.[120] Needless to say, these larger dwelling units generally require more and more energy to heat or cool; and entice their occupants to fill them with more and more consumer items, such as additional computers, wide-screen television sets, huge barbecues, and so on.

Arguably, the high acquisition rate of consumer durables then encourages the purchase of even larger residences or the building of extensions, in an ecologically irresponsible vicious circle. This culture of consumption ensures that Australia, like other developed countries, contributes to emissions elsewhere by importing consumer products manufactured overseas, in particular in China. In 2004, China emitted 4732 million tonnes of CO_2, 1109 million of which were from exports. Tao Wang and Jim Watson argue the industrial world is becoming ever more reliant on imports from China, effectively 'exporting carbon' to it.[121] As Monica Oliphant observes, 'many countries are "outsourcing" their greenhouse-gas emissions and not accounting for them in their national totals'.[122]

There are of course tremendous class-based variations in

consumption. A wealthy person with multiple residences, such as a holiday McMansion on the Great Ocean Road or Gold Coast, in addition to a spacious principal residence, leaves a much greater ecological footprint than a flat dweller or a homeless person. Climate justice issues are clearly raised by this culture of consumption. Not only do some Australians consume considerably more than others, but Australians as a whole, compared with people in developing countries, are contributing disproportionately to global warming. Social scientist Ted Trainer has long argued that developed societies such as Australia exceed sustainable levels of production and consumption.[123] He points out that Australians 'can only live as affluently as we do because we are taking and using up most of the scarce resources and preventing most of the world's people from acquiring anything like a fair share'.[124]

* * *

Australia's high per capita greenhouse gas emissions have been encouraged by decisions made for many decades by governments at all levels. In Part I we explore this nexus between corporations and the state to explain why state action to reduce emissions is inadequate to meet the challenge of both climate change and the climate justice issues associated with the problem of a warmer world and an even more sunburnt Australia.

Notes

1 Flannery, *An Explorer's Notebook*, p. 54.
2 Global Humanitarian Forum, p. 66.
3 Wahlquist, p. 38.
4 Hennessey et al, p. 509.
5 Bureau of Meteorology, *Annual Australian Climate Statement 2009*; Lindenmayer, p. 59.
6 Bureau of Meteorology, *Annual Climate Summary 2010*, p. v.
7 Tippet and Russell, p. 9.
8 CSIRO and BOM, *State of the Climate*, pp. 1–5.
9 Curson.
10 Rickards and Tucker, p. 89.
11 Wahlquist, p. 62.
12 *Age*, 10 March 2007. http://www.theage.com.au/articles/2007/03/09/1173166985969.html# (viewed 22 September 2010).
13 A Morton, 'A glimpse of the future'; *Weekly Times Now*, 5 August 2010.

14 http://www.melbournewater.com.au/content/water_storages/water_report/zoom_graph.asp (viewed 22 September 2010).
15 Environment Victoria, p. 8.
16 Kevin, p. 123.
17 http://www.bom.gov.au/jsp/ncc/cdio/weatherData/av?p_nccObsCode=36&p_display_type=dataFile&p_startYear=&p_stn_num=086071 (viewed 27 April 2010).
18 Bureau of Meteorology, *Annual Australian Climate Statement 2009*; Lindenmayer, p. 59.
19 BOM, 'Special Climate Statement 17'.
20 Department of Climate Change, 'Fact Sheet'.
21 Adams, 'Transforming Australian Cities', p. 121.
22 A Morton, 'Climate change biggest world-health threat'.
23 BOM, 'Special Climate Statement 19', p. 2; Bureau of Meteorology, press release, 3 May 2010.
24 Bradstreet, p. 3; http://www.bravenewclimate.com (viewed 3 February 2009).
25 Quoted in A Morton, 'Climate change biggest world-health threat'.
26 BOM, 'Record warmth'.
27 Bureau of Meteorology, 'Special climate statement 27'.
28 http://www.greenlivingpedia.org/ (viewed 25 October 2011).
29 Hennessey et al, pp. 509, 516–17.
30 'Climate change in Australia: Science Update 2009, Issue One'. http://www.climatechangeaustralia.gov.au (viewed 29 June 2010).
31 CSIRO and BOM, *State of the Climate*, p. 5.
32 Rickards and Tucker, p. 89.
33 Pittock (ed.), *Climate Change*, esp. pp. 7, 41.
34 Lindenmayer, p. 2.
35 Hennessey et al, p. 509.
36 CSIRO, 'Climate change in Australia', p. 9.
37 Lindenmayer, p. 60.
38 ibid., p. 61.
39 CSIRO and BOM, *State of the Climate*, pp. 1–5.
40 A Morton, 'Climate change biggest world-health threat'.
41 Garnaut, *Climate Change Review: Final Report*, p. 139.
42 A Morton, 'Climate change biggest world-health threat'.
43 Garnaut, *Climate Change Review: Final Report*, pp. 140–1.
44 Milner, '*On the Beach*', p. 191.
45 Turner, p. 318.
46 ibid., p. 29.
47 ibid., p. 153.
48 ibid., p. 20.
49 ibid., p. 157.
50 ibid., p. 20.
51 Wagar, p. 63.
52 ibid., p. 67.
53 Kevin, p. 252.

54 ibid., p. 256.
55 ibid., p. 259.
56 ibid., p. 265.
57 ibid., p. 265.
58 ibid., p. 262.
59 ibid., p. 266.
60 ibid., p. 268.
61 McMullen, pp. 249–50.
62 ibid., p. 235.
63 ibid., pp. 232–3.
64 ibid., pp. 243–4.
65 Wackernagel and Rees.
66 Lindenmayer, p. 57.
67 McNeil, *The Clean Industrial Revolution*, p. 45.
68 Garnaut, *Climate Change Review: Final Report*, p. 153.
69 ibid., p. 154.
70 Environment Victoria, p. 10.
71 Climate Institute, *Action Plan on Pollution and Climate Change*, p. 8.
72 IPCC, *Fourth Assessment Report*.
73 Department of Climate Change, *Australia's National Greenhouse Accounts*, p. 1.
74 Garnaut, *Climate Change Review: Final Report*, p. 154.
75 Department of Climate Change, *National Greenhouse Gas Inventory 2009*. http://ageis.climatechange.gov.au/Chart_KP.aspx?OD_ID=21308211426&TypeID=2 (viewed 9 September 2011).
76 Brohe et al, p. 201.
77 Butler, 'Australia—world's biggest climate vandal', p. 4.
78 Department of Climate Change, *Australian National Greenhouse Accounts. National Inventory by Economic Sector 2007*, p. 3.
79 McNeil, *The Clean Industrial Revolution*, p. 48.
80 Hamilton, *Scorcher*, p. 43.
81 Tiffen and Gittens, p. 158.
82 Butler, 'Australia—world's biggest climate vandal', p. 4.
83 'Climate Change in Australia: Science Update 2009, Issue One'. http://www.climatechangeaustralia.gov.au (viewed 29 June 2010).
84 Hickson, p. 345.
85 Cutcliffe, p. 247.
86 Kevin, p. 141.
87 Diesendorf, *Greenhouse Solutions*, p. 215.
88 Davies and Morton, p. 2.
89 *Solidarity*, no. 17, August 2009, p. 5.
90 Pearse, 'Quarry Vision', p. 12.
91 Goodman and Worth, pp. 201–207.
92 Diesendorf, *Greenhouse Solutions*, p. 2; Goodman and Worth, p. 216; Kevin, p. 166.
93 McNeil, *The Clean Industrial Revolution*, p. 57; Diesendorf, *Greenhouse Solutions*, p. 215.

94 McNeil, *The Clean Industrial Revolution*, p. 57; Pearse, 'Quarry vision', p. 13; Goodman and Worth, p. 216.
95 'Coal mining in Australia—overview'. http://www.mbendi.com (viewed 29 June 2010).
96 Pearce, *Confessions of an Eco-Sinner*, p. 193.
97 Butler, 'Australia—world's biggest climate vandal'.
98 Cubby, 'Coal group coy about port exposure to rising seas'.
99 Pearce, *Confessions of an Eco-Sinner*, p. 192.
100 Diesendorf, *Greenhouse Solutions*, p. 6.
101 Turton, p. vii.
102 Ladd, p. 8.
103 Ponting, p. 337.
104 Davison, p. 125.
105 ibid., p. 261.
106 Newman, p. 104.
107 Tiffen and Gittens, p. 210.
108 Quoted in Matthews, 'How do we fix public transport?'.
109 Tiffen and Gittens, p. 210.
110 Diesendorf, *Greenhouse Solutions*, p. 203.
111 ibid., p. 203.
112 Rickards and Tucker, p. 85.
113 Garnaut, *Climate Change Review: Final Report*, p. 162.
114 Rickards and Tucker, p. 93.
115 McAlpine et al, p. 25.
116 Garnaut, *The Garnaut Review 2011*, pp. 141, 149.
117 Lindenmayer, p. 42.
118 Mackey et al.
119 Davis, *The Land of Plenty*, p. 138.
120 Hamilton and Denniss, p. 20.
121 Wang and Watson, pp. 88–9.
122 Oliphant, p. 322.
123 Trainer, *Saving the Environment*, p. 8.
124 ibid., p. 10.

Part I
Climate Politics 'From Above': Governments and Corporations

3
The Public and the Politicians: the Political Effects of the Greenhouse Effect 1980–2007

It is often observed that politicians produce a lot of hot air. In the case of global warming, Australian parliaments have made a significant contribution. This chapter charts the history of public concern about climate change in Australia from the 1980s until the final days of the Howard government—and the largely ineffectual and inadequate responses of governments, both Labor and Coalition.

The Greening of the 1980s and the Labor Governments 1983–1996

Australian scientists and governments were quick to engage with the issue of global warming. In 1980 the Australian Academy of Science held a conference that examined 20 years of measurements indicating an increase in CO_2 levels and how they might contribute to climate change.[1] In November 1981 the federal government's Office of National Assessments prepared a confidential report for the Fraser Coalition government, 'Fossil Fuels and the Greenhouse Effect', which presented the global warming problem simply and accurately. It predicted that, by the turn of the century, concern about greenhouse gases could culminate in pressure to restrict the use of fossil fuels, with adverse consequences for Australian coal exports; and that an anti-fossil fuel lobby could emerge, perhaps generated by public alarm about rising sea levels. The date of this prescient document is significant, for it proves that the federal government was alerted by its primary intelligence-gathering agency to the problem of global warming much earlier than previously thought.[2] In 1987, CSIRO held a conference called 'Greenhouse '87', which examined scenarios based on the most likely changes in climate, with a range of experts analysing the impacts of these possible changes on the physical, biological and social environment. The conference attracted considerable media interest, mainly concentrated on the worst possible consequences of climate change.[3]

Extensive newspaper coverage of scientific information about the greenhouse effect saw a dramatic increase in environmental awareness—reflected in opinion polls, election outcomes, marketing pitches, politicians' rhetoric and growth of environmental organisations.[4] In May 1989, a Saulwick Poll found that 87% of respondents agreed that: 'The threat to our environment is real and must be treated seriously.'[5] Corporations changed their advertising to pander to 'green consumerism'.[6] Politicians responded to public opinion with platitudes. As an *Age* editorial on 12 August 1989 observed: 'Hardly a day goes by without a Government or Opposition spokesman at either the state or federal level expressing concern over air pollution, forest clearance, soil degradation, the ozone layer or the greenhouse effect.' Rhetoric proved much stronger than action.

The late 1980s and early 1990s was a crucial moment in climate change politics: the dangerous effects of greenhouse emissions had been publicly recognised and there was extensive public concern, yet political will to reduce emissions was absent. Perversely, the power of fossil fuel industries escalated, because they responded to the potential threat of the greenhouse debate by going on the political offensive. At the same time, the influence of climate and atmospheric scientists at CSIRO, who had expressed concern about climate change, was diminished in the process of a general restructuring of CSIRO undertaken to 'push' scientists into collaborative arrangements with corporations.[7]

So, although the Hawke governments recognised growing public anxiety about global warming and capitalised politically by appearing to respond to this concern, mitigation initiatives were piecemeal and sporadic. The Commission for the Future, a government agency created in 1986, sponsored 'Greenhouse '88', a series of conferences in nine cities in 1988.[8] In conjunction with the Australian Consumers' Association, the government issued each household with a booklet on how to use energy more wisely and efficiently, because, as Hawke's introduction explained: 'the emergence of "The Greenhouse Effect" as a global environmental issue, and the need for energy conservation and efficiency, mean that nowadays we all should "do something about it", and help reduce this accelerated global warming and its potentially drastic effects'.[9]

Jim Falk and Andrew Brownlow's *The Greenhouse Challenge: What Is To Be Done?*, confirmed in 1989 that significant segments of Australian society, ranging from the prime minister to environmental groups, were aware of the greenhouse effect. The issue was significant at the March 1990 federal election, which saw another Hawke Labor government returned, thanks to green preferences.[10] Clive Bean and John Warhurst affirm 'the centrality of environmental politics in the 1990 election' and consider this election marked 'a turning point in the nature of political discourse ... with the new politics coming to the fore'.[11]

Late in 1990 Australia adopted a watered-down version of the Toronto Target that resulted from an international greenhouse conference in 1988, which called for a 20% reduction in emissions on 1988 levels by 2005. Environment Minister Graham Richardson, ever attentive to opinion polls, had the previous year proposed such a target to Cabinet, but it had been rejected after opposition from ministers with economic and resource portfolios. Now they accepted it, on condition it would not adversely affect the economy.[12] Some state governments discussed adopting the target but, in realising that the 'adoption of such a target could have been threatening to Australia's energy-intensive industries and the electricity supply industry', they quickly backed off.[13] So the target adopted in October 1990 by the state and federal governments was a modified one, which stipulated that: 'Australia should not implement response measures that would have net adverse economic impacts nationally or on Australia's trade competitiveness, in the absence of similar action by major greenhouse gas producing countries'. This modification of the

Toronto Target later provided John Howard with an excuse not to ratify the Kyoto Protocol.[14]

After becoming Prime Minister in 1991, Paul Keating actively tried to minimise federal involvement in environment issues.[15] Keating did not bother to attend the Rio Earth Summit, which led to the creation of the UNFCCC, though his government did sign it. However, in 1992 his government did formulate a National Greenhouse Response Strategy under which the National Greenhouse Advisory Panel (NGAP) was established—a broadly based body comprising representatives of scientific, forestry, rural, industrial, conservation and administrative interests, to act as a source of advice to government on greenhouse issues, including progress with implementation, review and development of national greenhouse measures.[16] In 1995, Keating threatened industry with a carbon tax, which prompted the latter to accept the voluntary Greenhouse Challenge Program.[17]

Throughout Keating's prime ministership, polls continued to reveal that environmental concern was widespread and people favoured environmental protection over development, when these two options were presented as incompatible. The environment, according to Elim Papadakis, had become 'a permanent feature on the political agenda'.[18] Survey data also indicated that, despite mythology to the contrary, working-class people were as concerned about green issues as middle-class people: tertiary educated people were more agitated about logging but people with no post-secondary schooling were more anxious about pollution and greenhouse emissions than were people with university degrees.[19] The widespread concern about emissions was expressed by increasing numbers of people attempting, in their daily routines, to tread more lightly on the planet.

Yet the intensity of greenhouse anxiety that marked the late 1980s had passed. Commentators during the 1990s sometimes referred to the 'ebbing' of the green tide. However, according to Jan Pakulski and Stephen Crook, the apparent downturn was due to 'routinisation': the acceptance by mainstream society of the importance of environmental concerns and the partial accommodation by governments to this concern.[20] Pakulski and Crook elaborate the dangers involved:

> Routinisation leads to political-ideological fragmentation, de-mobilisation, and a shift from a 'moralistic-normative' idiom to an 'interest' idiom. Protests, demonstrations and other value-laden mobilisations give way to more conventional lobbying and voting. Protest groups and movements transform into parties and quangos, and the polycephaly of charismatic mobilisers gives way to hierarchies of officials and bureaucrats who live 'for and off' politics.[21]

Though routinisation was indicative of past successes, it was also one of the means by which the greenhouse challenge was managed by neo-liberal forces most resistant to change. Spokespeople representing the major green organisations were routinely consulted and therefore distracted by top-level negotiations, the public profile of the movement was reduced and the movement weakened as time, energy and resources were diverted from grassroots activities and campaigning. At the same time, routinisation placated opinion: people were persuaded that governments cared about the threat of global warming and were responding to the situation.

Corporations responded by greenwashing their public images and by counterattack against the global warming message from scientific sources. From the early 1990s, according to Cam Walker, companies engaged in concerted attempts to reclaim public sympathy and support, involving a range of tactics, such as front groups, PR campaigns, legal cases and 'dirty tricks'.[22] Various authors, notably Bob Burton, have tracked these campaigns.[23] However, most of the anti-green message was coming directly from industry sources or closely aligned right-wing think-tanks.[24]

The dominance of neo-liberal ideas and policies ensured the anti-green message was received sympathetically.[25] The opportunity to respond responsibly to scientific information and public concern about global warming was wasted. Policies enacted during the Keating governments 1991–1996 reversed rather than advanced the timid Hawke government initiatives of 1983–1991. Governments were only really concerned to appear concerned, to placate public opinion, while doing nothing to confront the real problems. Assessing the record of governments in making hard environmental decisions opposed by the major resource industries, Hutton and Connors concluded in 1999 that Labor governments had 'mediocre to bad records' and Coalition governments had been 'disastrous'.[26]

The Howard Governments 1996–2007

The Howard governments were closely associated with the fossil fuel industries' 'carbon lobby', also known as the 'greenhouse mafia'; and John Howard as Prime Minister afforded considerable comfort to climate change deniers in both his policies and pronouncements. So Howard government policies ensured that the 1990s continued to be a wasted decade, while climate change was becoming an ever more serious issue for Australia.[27] The incoming Howard government accepted the Greenhouse Challenge Program established by Keating, because it was voluntary; and on 20 November 1997 Howard announced the creation of the Australian Greenhouse Office.[28] However, this Office relied largely upon advice from the Australian Bureau of Agricultural and Resource Economics (ABARE), subject to heavy greenhouse mafia lobbying.[29]

This first Howard government also ceased funding Macquarie University's Climate Change Centre, resulting in its closure; abolished the Energy Research and Development Corporation, the body responsible for researching and developing renewable energy sources, resulting in the loss of numerous researchers to other countries; cut spending to energy-efficient programs by 50%; refused to make retention of native bushland a condition of grants from the Natural Heritage Trust; and continued to massively subsidise the cost of diesel fuel to mining companies.[30]

Emblematic of Howard government climate change policies was refusal to ratify the Kyoto Protocol and persistent criticism of the Protocol, Howard declaring it was 'not in Australia's interests' to ratify it.[31] The Howard government had played an invidious role in its negotiations in 1997. Environment Minister Senator Robert Hill had negotiated the adoption of 'the Australia clause', which included emissions from land clearing (deforestation) in the base year, thereby easing the degree of any agreed reduction subsequently for Australia.[32] Then, at the Sixth Conference of Parties to the UNFCCC—held in The Hague in November 2000—Australia aligned itself with the USA in fighting hard to avoid emissions reductions. Amongst other concessions, Australia wanted unlimited carbon trading—that is, buying credit for reductions other countries make; to count 'sinks' (forests and grasslands to absorb carbon) as being effectively a reduction in emissions; and little punishment for countries that fail to meet Kyoto targets.[33] In preparation for this convention, ABARE had charged $50,000 for any organisation that wished to join the policy defining committee, effectively promoting business input and excluding NGOs.[34] In 2004, the Senate Environment, Communications, Information Technology and Arts Committee held an inquiry as to whether the government should ratify the Kyoto Protocol. The Committee, dominated by Coalition Senators, solicited submissions from business and environmental groups and input from scientists and public servants. As anticipated, it recommended against ratification, with Labor, Democrats and Greens dissenting.[35]

One of the world's first mandatory trading schemes occurred during Howard's reign, but in a different sphere of government. This was the NSW Labor government's 2003–2012 Greenhouse Gas Reduction Scheme, which has reduced emissions by 90 million tonnes. State and Territory governments established a working group in 2004 to consider the design of a national ETS, which recommended a start date of 2010 and that emissions should be reduced by 60% on 1990 levels by 2050. Kate Crowley considers this action important for signalling that the inaction of the Howard government would not be tolerated by these governments. This was a time when ETSs were being established internationally: in the European Union from 2005; in Japan from 2005; in New Zealand from 2008; in various states of the USA from 2008; and in Canada from 2010. Crowley argues the policy failure of the Howard government in relation to climate change occurred because Howard was 'shackled to self-proclaimed climate change scepticism for over a decade'.[36]

In 2005 the Howard government joined the Asia Pacific Partnership on Clean Development and Climate (AP6) initiated by US President George W Bush. China, Japan, India and South Korea also belong to AP6, whose member countries account for about half the world's emissions.[37] AP6 held its inaugural ministerial meeting in Sydney in January 2006 and stressed voluntary mitigation strategies, such as sharing technologies for renewable and reduced-carbon fuel sources. According to Christoff and Eckersley:

> The AP6 focuses only on future growth in emissions and ignores historical responsibility for past emissions and the advantages that have accrued to developed countries as a result of these past emissions. The AP6 also ignores the IPCC's warnings that developing countries are expected to suffer much greater harm than developed countries from climate change.[38]

Clive Hamilton claims the Howard governments 'worked diligently to control public debate over greenhouse, including making extensive efforts to deny the existence, or the seriousness, of climate change'. For example, when the Science, Engineering and Innovation Council urged the government to take a more proactive stance on climate change, Howard changed the membership of the group.[39] CSIRO scientists were discouraged from speaking on topics such as emissions trading, emissions targets and climate refugees or from joining the Wentworth Group of Concerned Scientists or the Australian Climate Group.[40] A 2006 article in *Nature* maintains that during the Howard government various CSIRO scientists functioned in a 'culture of fear' in speaking out on climate change issues.[41] Dr Graeme Pearman's story is discussed in Chapter 8. Another former CSIRO scientist, biofuels researcher Barney Foran, spoke in February 2006 about his experience: 'Just recently in August I was back at work late in the afternoon, took a call from our corporate centre saying, "Barney, how are ya?"—"Okay mate." He said, "Barney, we've just had a call from the Prime Minister's Department. They'd really appreciate it if you didn't say anything about ethanol." '[42]

The Howard governments were careful to avoid input from 'green' business interests. For example, Howard did not simply abolish the Energy Research and Development Corporation, responsible for researching and developing renewable energy sources, but established instead the Lower Emissions Technology Advisory Group, representing various 'brown' corporations, which encouraged the Howard governments to retard investment in renewable energy sources in its recommendations on climate policy.[43] In 2002 the Howard government decided not to provide funding for a proposed Cooperative Research Centre for renewable energy, while resourcing new coal-fired power plants at a total cost of over $60 billion.[44]

Tim Flannery maintains that three fallacious premises underpinned the Howard governments' climate policies: renewable sources of energy such as wind and solar could not provide base-load electricity and thus had only a limited role to play; funding and developing new technologies without imposing carbon taxes or trading schemes can avert catastrophic climate change; and funding of 'clean coal' initiatives was sufficient to develop the technology in time to avert climate chaos.[45] The Howard governments' especially poor record on renewables was stressed by Kevin Rudd during the 2007 election campaign. Rudd stated on the *7.30 Report* on 30 October that when Mr Howard won office in 1996, renewable energy represented 10.5% of Australia's total electricity production. 'Roll the clock on 11 years it's shrunk to 9.5%. That's solar, wind and all the others put together and that's despite the fact that in the intervening decade we've had this mounting evidence on climate change.'[46]

Hamilton ascribed a 'state of denial' about climate change on the part of the Howard government, despite opinion polls indicating 'widespread concern about the issue'.[47] Howard, normally sensitive to opinion polls, did not respond adequately. This was especially true in relation to his chosen techno-fix for the climate change problem. When he visited the United States in 2006, the Bush administration urged him to establish closer ties with the nuclear industry.[48] So, having previously sided with the climate change deniers of the world, Howard began to give greater credence in his public utterances to the need to mitigate climate change, in order to press the claims of nuclear energy. Howard formed a task force to advise the government on the contribution nuclear energy could allegedly play in terms of mitigating climate change, to which he appointed Dr Ziggy Switkowski, head of the Australian Nuclear Science and Technology Organisation, a vigorous and outspoken proponent of Australia's further involvement in a nuclear power industry.[49] Craig Dilworth observes: 'The idea of replacing millions of tons of fossil fuels, every year, by nuclear energy, means "solving" the fuel problem by creating an environmental and ecological problem of a monstrous magnitude.'[50] Apart from the dangers posed by nuclear waste materials, the fact that uranium is not a renewable resource, and possible proliferation of nuclear weapons, construction of nuclear power plants creates significant emissions.

The shift in Australian consciousness also prompted Howard in December 2006 to commission a Task Group on Emissions Trading, which recommended that Australia implement an ETS by 2011, setting a low emissions price (a somewhat arbitrarily determined cost allowing companies to emit by paying governments a fee) with deeper emissions cuts being required later on.[51] However, his responses were generally deemed too little too late, in the case of the ETS, or, in his championing of nuclear energy, considered by most people to be no solution at all. Howard was losing political ground on the increasingly significant issue of climate change.

Many Australians had previously accepted Howard's reassurances that climate science was indefinite about the seriousness of the global warming problem and whether the sources were anthropogenic. However, during the last year or so of his prime ministership, mainstream attitudes underwent a seismic shift as authoritative scientific pronouncements that human-induced climate change was a considerable problem, altered the thinking of large numbers of people. The discrepancy between Australian public opinion and Howard's equivocal attitudes to climate change undermined him politically.

The long-running drought in much of the continent made Australians more responsive to information about global warming, but there were not just local climatic reasons for increased anxiety. During 2006 Al Gore's book and film *An Inconvenient Truth* popularised the message that anthropogenic climate change was real. Gore visited Australia for the release of his film in September 2006. Howard agreed to talk to Gore on the phone but announced he did not 'take policy advice from films'.[52] Climate change was shaping up as a wedge issue for the Coalition. To add to the Gore effect, the Stern Review was discussed extensively in Australian media in late 2006.[53] The *Sydney Morning Herald* reported on 1 November: 'Climate change leapt to the top of the political agenda yesterday after the release of an alarming report by a former World Bank economist, Sir Nicholas Stern, warning of global depression should the problem not be addressed within 10 years.'[54] The public was now inclined to agree—and judge Howard's prevarication as costly and dangerous. Geoff Boucher and Matthew Sharpe observed that by late 2006, following *An Inconvenient Truth* and the Stern Review, and the drought deepening and water restrictions affecting Australians everywhere, Howard could no longer claim to be speaking for the silent majority on the issue.[55]

Opinion polls in 2007 confirmed that concern about climate change had risen dramatically. In July 2003, only 38% of respondents had felt that 'climate change is extremely important for Australia'; by July 2006, 68% believed 'global warming is a serious and pressing problem and that we should begin taking steps now even if this involves significant cost'; and by January 2007, 78% of respondents were very concerned or extremely concerned about climate change. ARG polling early in 2007 discovered people's knowledge about climate change predominantly came from media sources but most people also referred to their own experiences, such as flowers in the garden blooming out of season and the extent of the drought, to support their view that climate change was real.[56]

Of interest is the fact that these polls between 2003 and 2007 also showed a rising number of people pointing the finger at government. In 2003 a poll undertaken by TNS for GreenPower asked respondents who was mainly responsible for solving climate change: 68% had singled out the government. In October 2006, Newspoll found 81% said the government should be doing more to tackle climate change.[57] Early in 2007, the Climate Institute commissioned qualitative and quantitative research by the Australian Research Group (ARG) on attitudes to climate change and climate change solutions.[58] The report summary observed:

> concern about climate change is at an all time high and the vast majority of people no longer doubt that it is real or that it is caused by greenhouse gases created by human activity ... climate change now ranks as more important to people than a wide range of issues including housing affordability and national security. While many people are still unfamiliar with the detailed science of climate change, this does not detract from their passion to deal with it.

This summary added: 'A common theme in the research is that people are looking for leadership. They accept there may be a price to pay and they are hungry for decisive action. They are also keen to know more about the problem, and importantly they want to be able to take action which will make a real difference.'[59]

Support for clean energy solutions like solar and wind was very strong and there was a view that Australia's abundant sunshine was not being put to good use. People also felt strongly about cutting energy waste. February 2007 ARG polling found 91% support for more solar panels, 82% for more wind turbines, and 78% for reducing overall energy consumption. Also in February 2007, Newspoll found 77% of respondents willing to pay more for renewable or other alternative sources of energy to help reduce emissions. March 2007 ARG polling found strong support for legal targets to reduce greenhouse pollution.[60]

This ARG poll of 1000 people in March 2007 asked respondents whether Australia should lead the way on climate change solutions or wait and see what the rest of the world did. Compared with only 9% who thought we should wait, 81% agreed Australia should lead the world. In this poll too, 87% agreed 'We must act now to protect the world for our kids' and only 3% disagreed.[61] This 2007 *Climate of the Nation* report concluded: 'People now expect concerted action on the part of their political leaders and they want a way to be able to make a direct and useful contribution themselves.'[62]

Al Gore, now armed with the Nobel Peace Prize, returned to Australia shortly before the 2007 election—inconvenient timing for the Howard government.[63] Liberal MPs declined invitations to events starring Gore.[64] Rudd met with Gore, reiterating his promise to ratify Kyoto and break with US policy. Gore was photographed with Rudd and a raft of Labor luminaries, including Maxine McKew, candidate against Howard in Bennelong. *Sydney Morning Herald* Environment Editor Marian Wilkinson predicted that Gore's passionate call for Howard to ratify the Kyoto Protocol would 'resonate throughout the election campaign'.[65]

2007: 'The World's First Climate Change Election'

The public mood had altered so much during the last year or two of Howard's reign that climate movement activists rather than Howard became more representative of mainstream opinion. As with the issue of the Vietnam War in 1972, Labor was to be the electoral beneficiary of this radical shift in opinion. Indeed, the Australian federal election of 2007 has been described by political commentators and political scientists internationally as 'the world's first climate change election'.[66] Chris Rootes queries the significance of climate change in the election outcome,[67] but evidence for its importance can be found in the distinction voters made between the climate change policies of the two major parties and the emphasis voters put on the issue as a factor in influencing their votes.

With Howard responding inadequately to heightened public concern about climate change, the Labor Opposition under Kevin Rudd moved to own the issue—and succeeded. Rudd positioned himself as the alternative Prime Minister who took much more seriously the threat of global warming, indicating a Labor government would ratify the Kyoto Protocol, fast-forward the development of an ETS to 2010 and implement other mitigation policies. In his first speech in parliament as Leader of the Opposition, early in December 2006, he stated:

> If we do not take sustainability and climate change seriously, what will happen to the future of the market economy? Sustainability is a core Labor value ... If we do not rescue this planet from itself in terms of the damage being done to it by unrestrained market capitalism, let me tell: you the entire market system ultimately will fragment.[68]

It seemed a particularly promising omen for a new Labor government's action on climate change that in December 2006 Rudd had appointed Peter Garrett, a well-known and long-time environmental campaigner, as Shadow Minister for Climate Change, Environment & Heritage. Rudd did not support Garrett when he indicated that a 'Labor government would be willing to commit to a new emission target, even if countries like China and India did not sign up' and obliged him to clarify that Labor would require binding commitments from both developed and developing countries in a post 2012 agreement.[69] Nonetheless, the impression remained that Garrett's presence on the ALP frontbench indicated honourable intentions towards climate change mitigation.

Tackling global warming was a stance Rudd was as keen to cultivate as fiscal responsibility. On the first day of the 2007 Parliament early in February, Rudd emphasised Labor's priorities regarding the future: 'to build long-term prosperity and take action on climate change and water'.[70] When the IPCC released its Fourth Report, Rudd immediately announced a climate change summit for late March, declaring: 'Climate change is an environmental challenge. It is an economic challenge. And it directly affects Australia's long-term sustainable prosperity. It is in the national interest that we work together to develop the best national response to the climate change challenge.'[71]

By this stage, opinion polling conducted by ARG confirmed a perceived gap between the major parties on climate change that made the issue a difficult one for the Coalition. Moreover, ARG focus group and poll research conducted in February 2007 showed that the leading issues of concern were water shortages, climate change, cost of housing and workplace relations, ahead of terrorism and national security. A poll of 1200 people conducted in mid-February 2007 revealed that 48% thought Rudd was serious about dealing with climate change compared with only 17% who thought Howard was. Similarly, 49% said Garrett would make a better Minister for the Environment and Climate Change; only 18% preferred the actual incumbent, Malcolm Turnbull. In contrast to very high support for solar and wind as renewable sources of power, only about a third supported Howard's preference for 'nuclear reactors for electricity'.[72]

In March 2007, ARG explicitly asked 1000 people whether a clear plan to address climate change would influence their vote. Sixty-six per cent said they would be more likely to vote for a party with a strong, clear plan to tackle climate change. Forty-two per cent of respondents who said they voted for the Coalition last election listed the government's failure to deal with climate change as amongst the top three reasons for deciding to change their vote away from the Coalition at the election. Additional focus group research discovered swinging voters were interested in hearing about solutions that 'offer an answer for our children' and were concerned about the 'legacy' we leave our children.[73]

Rudd and the Opposition continued to speak to this anxiety. The National Climate Change Summit was held on 31 March in Canberra, attended by premiers, scientists, environmental and industry groups, as well as mining, banking and insurance company representatives. It was in his opening remarks to this summit that Rudd famously announced: 'Climate change is the great moral challenge of our generation.' Rudd said he wasn't proposing to release a grand Labor climate change policy initiative from this summit. Rather:

> My intention is to harness the best brains and talent available in the country to get our response and the nation's response to climate change as right as possible. To do that we have to begin by fashioning, shaping and encouraging a national political and policy consensus on climate change. Our job is to listen. But subsequent to that, our job is then to act.[74]

On 10 May 2007, Rudd delivered his Reply-to-the-Budget, in which he stressed that Australia's long-term economic future required action on climate change: 'The science is in. Climate change is a reality. It's happening now ... The core economic reality is that there is an economic cost to Australia if we fail to act on climate change ... the economic cost of not acting will be far greater than the cost of taking early and responsible action.' He reiterated aspects of Labor's recently released ten-point plan on climate change, including: ratifying the Kyoto Protocol; reducing greenhouse gas emissions by 60% by 2050; establishing a $500 million national clean coal fund; boosting the use of renewable energy through enabling Australians to access low-interest loans to help install energy-efficient measures such as solar panels.[75]

With the election called for 24 November 2007, on 29 October Labor announced its climate change proposals: ratify the Kyoto Protocol, Rudd to attend the December climate change summit in Bali, introduce an emissions trading scheme to reduce emissions by 60% of 2000 levels by 2050, and introduce policy mandating a renewable energy target of 20% by 2020.[76] Throughout the campaign, Labor depicted itself as the party committed to serious action on climate change and the Coalition as part of the problem, maintaining that Howard's refusal to ratify the Kyoto Protocol was being used by developing countries to walk away from their obligations. Shadow Water Minister Anthony Albanese described the Howard government as 'all over the shop when it comes to climate change', whereas Labor's position was 'very clear' and a Labor government would be 'part of the global effort' to tackle the problem.[77] Coalition advertisements confirmed Labor's greater commitment to climate change mitigation by specifically targeting Garrett as one of Labor's 'fanatics, extremists and learners', making much of his remark that Labor green policies appeared cautious but 'once we get in, we'll just change it all'.[78] This advertisement probably won Labor votes.

During the campaign, Rudd presented Howard as 'locked into the past, old thinking and old thinking generally on climate change'.[79] He emphasised that Labor's renewable energy target would reverse the decline in renewable energy under the Howard government and 'create a new vibrant, clean, green energy industry in Australia'. It was more stringent, he stressed, than the clean energy target announced by Howard the previous month, which was only 15% and included power from 'clean coal'-fired power stations or nuclear reactors, whereas Labor's target would be met entirely from renewable sources such as solar, wind, wave and geothermal. Turnbull, as Environment Minister, criticised Labor's plan, saying it would cost businesses and households two billion dollars.[80] Rudd told the *Australian Financial Review* in mid-November that tackling global warming would be his 'top priority' if he became Prime Minister.[81] Rootes is incorrect that Labor did not highlight climate change during the campaign.[82]

As the election loomed, ARG, on behalf of the Climate Institute, examined the perceptions and attitudes of voters in nine key marginal seats in NSW (Bennelong, Parramatta, Lindsay), Queensland (Petrie, Bowman, Bonner) and South Australia (Wakefield, Makin, Kingston). The poll was conducted online using a representative sample of 877 voters from 2 to 5 November 2007.[83] These marginal electorate respondents indicated levels of concern about climate change similar to the nationwide pattern; and this was true too amongst the respondents who indicated they were undecided voters.

More than 90% of voters surveyed wanted to see climate change policies given attention during the election campaign.[84] There was a clear preference (66%) for the government to move quickly to renewable energy alternatives to create jobs in clean energy industries.[85]

On reduction targets, 75% preferred Australia to commit to targets to set an example rather than wait until developing countries did so.[86] More than seven out of ten (73%) stated climate change would have either a 'very strong' (34%) or 'strong' (39%) influence on their vote.[87] On the Kyoto Protocol, 56% of undecided voters believed Australia should sign and only 16% not sign.[88] Amongst undecided voters, climate change was a 'strong' or 'very strong' influence for 63%.[89] A large proportion said the parties were the same (30%) but the ALP led the Coalition by 35% to 15% as the party better at dealing with climate change.[90] A clear majority indicated they would be more likely to vote for a party that would: set energy efficiency targets (72%); ensure all new electricity generation comes from clean energy (70%); set greenhouse pollution reduction targets within 5 years (71%); and ratify the Kyoto Protocol (54%).[91]

Four days before the election, the Climate Institute assessed the parties' climate policies against criteria of clean energy, international leadership and reversing greenhouse pollution within 5 years, giving the ALP a score of 60%, compared with 30% for the Coalition, and 90% for the Greens.[92] The night before the election, the *Guardian* observed: 'Australia goes to the polls tomorrow in what is arguably a milestone in 21st-century history: the world's first climate-change election.' It had shaped up as such because Howard had 'poured scorn on the idea of global warming', while the trees were dying, the crops failing and the rivers drying up; whereas 'Rudd has been much bolder on climate change, making it a defining point of difference'. The *Guardian* quoted Lynton Crosby, the pollster and campaign strategist who helped Howard win four elections in a row: 'For the first time in 25 years in this country, the environment is an important voting issue.'[93] The Monday after the election, Malcolm Turnbull stated publicly that the Coalition had been damaged by the non-ratification of the Kyoto Protocol.[94]

Was the 2007 election the world's first climate change election? Certainly, as we have seen, public opinion polls before the election, both across the nation and in marginal electorates, indicated three salient matters: there were serious and all-time high levels of concern about climate change; the ALP was identified with a better response to the problem; and there were Coalition or Coalition-leaning voters prepared to switch their vote away from the Coalition because of the climate change issue. In a tight race, it was not surprising that victory was accorded the party that seemed most committed to tackling the problem in earnest.

Rootes argues that industrial relations was more important to Labor voters than climate change and that, if there was an issue that determined party preference, it was most likely Howard's unpopular WorkChoices legislation that Labor promised to repeal.[95] Industrial relations was indeed the other big issue of the campaign, and one felt very keenly by Labor voters and which kept them in the Labor fold, but global warming was more decisive in persuading many who had voted Coalition in 2004 to switch to the ALP in 2007. Rootes' analysis overlooks the fact that more progressive industrial relations was intrinsically a Labor issue, whereas climate change mitigation was 'up for grabs' by either party and was therefore a more significant potential vote-changer.

Closer examination of polls is revealing in this regard. An Exit Poll analysis commissioned by Australian Policy Online phoned 7825 phone numbers on the afternoon of the election, resulting in 744 completed surveys. It found that the issue with the greatest impact among Labor voters was industrial relations (73%), followed by global warming (71%). However, among the sub-group of voters who changed their vote compared to last time, global warming was the decisive issue, followed a long way behind by industrial relations; and 73% of these vote-changers switched to voting for the ALP. The Australian Election Study (AES) poll, a much larger sample, likewise showed that 74% of the vote-changers had shifted to the ALP and that global warming was foremost among their concerns, with this and industrial relations well ahead of any other issues.[96]

Political scientists largely agree that climate change was 'perhaps the most decisive issue of the campaign'.[97] Christine Jackman's study of the ALP campaign observed that, given heightened levels of public concern about climate change, accompanied by on-going drought and stringent water restrictions in many Australian cities and towns, Rudd was able to capitalise on climate change as 'powerfully totemic of the future' and present himself as a more progressive politician on this serious problem than Howard.[98] Clive Bean and Ian McAllister's analysis of the AES study concluded that 'Labor's largest advantage was on the issue of global warming (with a

38% differential in its favour) followed closely by the environment more broadly'.[99] Green organisations were in no doubt. Greenpeace declared the election result a 'comprehensive thumping' for the conservative parties, and 'one of the things that hurt them most was their (non) position on climate change'.[100] Climate Institute chief executive John Connor cited the exit poll by Auspoll that showed climate change was a 'critical issue for voters' and summed up the significance of the election result: 'We congratulate the new Prime Minister Kevin Rudd who has a clear mandate, responsibility and, it would seem, intent to combat dangerous climate change and reverse rising greenhouse pollution.'[101]

Notes

1 Staples, p. 113.
2 Hamilton, 'Climate Change and its Implications for Australia', pp. 9–10.
3 Lowe, 'The greenhouse effect', p. 318.
4 Burgmann, *Power and Protest*, pp. 217–19; Burgmann, *Power, Profit and Protest*, p. 189.
5 *Age*, 16 May 1989, p. 18.
6 For example, *Weekend Australian*, 8–9 July1989, p. 3; *Age*, Good Weekend, 19 August 1989, p. 26; *Choice*, January 1990, p. 12.
7 Lawrence, p. 97.
8 Lowe, 'The greenhouse effect', p. 318.
9 Australian Government and Australian Consumers' Association.
10 *Age*, 29 March 1990, p. 17.
11 Bean and Warhurst, p. xv.
12 Staples, p. 112.
13 Bunting, p. 291.
14 Staples, p. 112.
15 Staples, p. 116.
16 AGO, National Greenhouse Strategy, p. 4.
17 Hamilton, *Scorcher*, p. 97.
18 Papadakis, 'The Environment', pp. 201–4.
19 Tranter and Pakulski, pp. 68–9.
20 Pakulski and Crook, p. 19.
21 ibid., p. 12.
22 Walker, 'Time to Choose Sides', p. 24.
23 Burton, 'Nice Names'.
24 Hutton and Connors, p. 258.
25 Toyne, p. 17.
26 Hutton and Connors, pp. 264–6.
27 Yencken and Wilkinson, pp. 18, 23–5.
28 Hamilton, *Scorcher*, p. 97.
29 Firth, p. 310.

30 Doyle, p. 178; Flannery, *An Explorer's Notebook*, p. 248; Hutton and Connors, p. 246.
31 Singer, p. 2.
32 Brohe, Eyre and Howarth, p. 200.
33 Settle, p. 26.
34 Doyle, p. 178; Hutton and Connors, p. 246.
35 Vromen, Gelber and Gauja, p. 319.
36 Crowley, 'Climate Policy Failure', pp. 5–6, 1
37 Lowe, *Living in the Hothouse*, p. 178.
38 Christoff and Eckersley, 'The Kyoto Protocol', p. 38.
39 Hamilton, *Scorcher*, pp. 13, 124.
40 Davis, *The Land of Plenty*, p. 194; Enting, p. 102.
41 Dennis, pp. 896–7.
42 'Scientists make CSIRO censorship allegations', AM—Monday, 13 February 2006, 08:15, Reporter: David Mark. http://www.abc.net.au/am/content/2006/s1568322.htm (viewed 12 October 2011).
43 Flannery, *An Explorer's Notebook*, p. 248; Davis, *The Land of Plenty*, p. 193.
44 Lowe, 'Sunshine, Social Factors, Solar Energy and Sustainability', p.124.
45 Flannery, *An Explorer's Notebook*, pp. 243–4.
46 http://www.abc.net.au/7.30/content/2007/s2076131.htm (viewed 14 July 2010).
47 Hamilton, 'The social psychology of climate change', p. 187.
48 Lowe, 'Reaction time', p. 124.
49 Hickson, p. 249; Lowe, 'Reaction time', p. 75.
50 Dilworth, p. 416.
51 Murphy, 'Carbon trading by 2012'.
52 Topsfield, p. 9; Minchin, p. 7.
53 Frew and Peatling; Kleinman and Topsfield; ABC Newsonline, 30 October 2006, http://www.abc.net.au/news/newitems/200610/s1776304.htm (viewed 14 July 2010).
54 Coorey and Peatling.
55 Boucher and Sharpe, p. 219.
56 Climate Institute, *Climate of the Nation*, p. 4.
57 ibid., p. 8.
58 ibid., pp. 2, 4
59 ibid., p. 3.
60 ibid., pp. 3, 13, 15.
61 ibid., p. 7.
62 ibid., p. 17.
63 Minchin.
64 Hawthorne and Ziffer.
65 Wilkinson, 'Maybe the PM … '.
66 Glover.
67 Rootes.
68 http://webdiary.com.au/cms/?q=node/1771 (viewed 29 July 2010).
69 Jackman, p. 198; AAP.

70 Quoted in Brohe et al, p. 204.
71 *Sydney Morning Herald*, 4 February 2007, p. 1.
72 Climate Institute, *Climate of the Nation*, pp. 3, 6, 9, 13–14.
73 ibid., pp. 3, 10–11, 13.
74 *Age*, 31 March 2007, p. 1.
75 Rudd, 'Budget Reply Speech 2007'.
76 Goot, pp. 103–4; Colebatch.
77 AAP.
78 Glover.
79 *7.30 Report*, 30 October 2007. http://www.abc.net.au/7.30/content/2007/s2076131.htm (viewed 14 July 2010).
80 Davis and Wilkinson.
81 Popular News.
82 Rootes, p. 475.
83 ARG, pp. 3, 6.
84 ibid., pp. 4, 18.
85 ibid., pp. 5, 22–3.
86 ibid., pp. 4, 16.
87 ibid.,, pp. 3, 7.
88 ibid.,, pp. 4, 14.
89 ibid., pp. 3, 7–8.
90 ibid., pp. 3, 9.
91 ibid., pp. 3, 10.
92 Climate Institute, 'World's first climate changed election?'.
93 Glover.
94 Greenpeace, 'What the Oz election outcome means'.
95 Rootes, p. 475.
96 Watson and Browne, pp. 4–6.
97 Brohe et al, p. 199.
98 Jackman, pp. 107–108.
99 Bean and McAllister, 'The tale of the rabbit-less hat', p. 12.
100 Greenpeace, 'What the Oz Election Outcome Means'.
101 Climate Institute, 'Climate Change Top Tier Issue'.

4
Carbon Pollution Reduction and Carbon Pricing: The Rudd and Gillard Governments

There was a clear mandate for the Rudd government to enact significant legislation to reduce emissions, even if it was costly in the short-term and other countries did not make similar commitments. Even Rootes, who questions whether climate change decided the election, concedes it marked a clear break with the climate policies of the Howard government and ushered in 'a new era in which Australia promises to play a leading role in global environmental politics'.[1] It was widely understood that delay would only add to the cost of mitigation, because it would necessitate deeper emissions cuts, as the Stern Review stressed. The Department of the Environment stated that 'the cost of avoiding dangerous climate change is much lower than even the most conservative estimates of inaction'.[2]

The Rudd government started on a promising note. Rudd's first official act as Prime Minister was to sign the instrument of ratification of the Kyoto Protocol, to demonstrate, he said, 'my government's commitment to tackling climate change'.[3] He created a new Ministry of Climate Change, with Senator Penny Wong as Minister. In addition, Shadow Environment Minister Peter Garrett, with all his green credentials, assumed this portfolio in government.[4] Yet the new government also contained mining industry enthusiasts, including Martin Ferguson, who became Energy and Resources Minister. And, at the United Nations climate change conference in Bali in December, Rudd refrained from committing Australia to any reduction target.[5] The pressure exerted on the Rudd and Gillard governments by the 'carbon lobby' is discussed in the next chapter; here we tell the unadorned story of these governments' attempts to enact mitigation policies.

Waiting for Garnaut

Before election to government, the Labor Opposition and Labor state governments commissioned a study on mitigation and adaptation strategies, headed by leading economist Professor Ross Garnaut. Garnaut, who has remained economic adviser to the Labor governments on climate change, is an ANU Professor of Economics, a board member of nearly a dozen companies, organisations, and academic journals; and the chairperson of four of them, including Lihir Gold Limited with a dubious environmental record in PNG.[6] During the 2007 election, Labor announced it would base its mitigation policies on recommendations made in this anticipated Garnaut Review. In December 2007, new Deputy Prime Minister Julia Gillard stated that Garnaut would take into account a just-published PriceWaterhouseCoopers report that Australia could reduce its emissions by 30% by 2020 at no net cost to the economy.[7]

Repeatedly during his reviewing period, in town meetings and media interviews, Garnaut referred to climate change as a 'diabolical' matter which was 'harder than any other issue of high importance that has come before our polity in living memory'.[8] An interim Garnaut review was released in February and a draft review in June. The long-awaited final Garnaut Review was released on 30 September 2008.

It stated that growing emissions risked 'dangerous climate change' and argued that 'the costs of action are less than the costs of inaction'. It proposed two sets of targets: one as part of a global strategy to limit greenhouse gases at 450 CO_2e ppm in which Australia commits itself to a 25% reduction of 2000 levels by 2020 and 90% by 2050; and one that accepts a limit of 550 CO_2e ppm 'more realistic in terms of what other countries are likely to do', thereby having

Australia aim for a 10% reduction of 2000 levels by 2020 and an 80% reduction by 2050.[9] It called for a 'cap and trade' ETS but argued a simple carbon tax would be preferable, if the government 'felt unable to resist pressure on the political process for ad hoc and overly generous assistance' to polluting industries. It specifically recommended against free allocation of permits and urged that any support for emissions-intensive trade-exposed (EITE) industries be phased out once other countries priced carbon. This needed to be handled well, otherwise it would lead to 'rent seeking behaviour rather than the pursuit of low-emissions production processes'.[10]

Garnaut conceded that developed countries have contributed a disproportionate share of past cumulative emissions and developing countries had a right to object to a 'per capita approach' for avoiding that issue of 'historic responsibility'.[11] He recommended that developed countries provide funding for developing countries to adapt to climate change.[12] Garnaut also proposed a system of 'green credits' that would assist with 'funding of investments in energy efficiency in housing, household appliances, and transport'.[13] In terms of transport, he recommended: electric cars; decarbonisation of electricity production; improvements in vehicle technology; greater reliance on trains within and between cities; and improved technologies in aviation, shipping, rail and ancillary road transport.[14] He suggested Australia should shift from sheep and cattle production to 'lower-emissions forms of meat', such as chicken, fish, pork and kangaroos.[15]

Guy Pearse, a former adviser to Howard but a strong mitigation advocate, criticised Garnaut for not prioritising wind and solar energy development, backing use of ETS revenue for 'clean coal' projects, and supporting international carbon credits to help Australia meet its target and thereby endorsing 'almost every loophole required for Australia to export its emissions obligations and for our worst polluting sectors to offload theirs'. In particular, Pearse was concerned that Garnaut accepted the notion that the Australian economy required continuing growth, ignoring the impact growth might have on emissions.[16] Yet the Garnaut Review, in its firm reiteration of the climate science consensus, was nonetheless a clarion call to the Rudd government to meet the moral and diabolical policy challenge. 'There are times in the history of humanity when fateful decisions are made', Garnaut concluded. 'On a balance of probabilities, the failure of our generation on climate change mitigation would lead to consequences that would haunt humanity until the end of time.'[17]

The Carbon Pollution Reduction Scheme

On 15 December 2008 the government released its White Paper, delineating its Carbon Pollution Reduction Scheme (CPRS). It fell considerably short of adopting most of Garnaut's recommendations, while admitting the seriousness of global warming to Australia:

> As a hot and dry country, we have more to lose than any other developed nation if the world fails to reduce the carbon pollution that causes climate change ... In the past 12 years, we have experienced eleven of the hottest years since records began and temperatures are projected to continue to rise over the next century. In Australia, temperatures are expected to rise by around five degrees by the end of the century.[18]

This White Paper emphasised, early and often, that mitigation would not threaten economic growth. Australia can 'reduce carbon pollution while continuing to grow our economy and jobs'.[19] Unmitigated climate change posed 'a significant threat to Australia's economic security' so early action was essential but 'the government has sought to get the balance right: to secure Australian jobs today while at the same time moving to a low pollution economy that will deliver growth and jobs of the future'.[20]

The CPRS would be a 'cap and trade' scheme from 1 July 2010 that would set a limit on total annual emissions; the government would announce the caps for at least 5 years in advance; businesses would then be required to obtain permits for every tonne emitted and would be permitted to purchase an unlimited number of permits from offshore, such as in tree planting projects. It would cover a thousand companies; emissions from stationary energy, transport, industrial processes, waste and 'fugitive emissions', such as in oil and gas production; agricultural practices, including deforestation, were to be excluded from the scheme until at least 2015. The White Paper also proposed other mitigation strategies, including an expanded Renewable Energy Target (RET) investment in renewable energy sources, carbon capture and sequestration (CCS), and improved energy efficiency.[21]

Although the market would, in theory, dictate the price of carbon, some companies would be granted free emissions permits, especially those in EITE industries. The CPRS provided $7.4 billion compensation in the first 2 years, with a further $2.25 billion designated primarily for coal-fired power stations. Coal-fired electric power plants would have free permits valued at almost $3.9 billion in the first 5 years of the CPRS. Brown coal-fired plants would receive the biggest share of this assistance. Ninety per cent of permits required by 'emissions-intensive' activities, such as aluminium smelting, were to be provided free.[22]

There was considerable criticism of this aspect of the scheme; an *Age* article by Richard Denniss on 17 December 2008 was headlined 'Households pay as big polluters cash in on climate change'. The government justified the compensation payments on the grounds they were necessary to avoid EITE industries moving offshore, resulting in 'carbon leakage'. Pearse argued the chance of that happening on any significant scale was virtually nil.[23] Garnaut was critical of the 5 years' notice needed for withdrawing compensation to EITE industries as other countries adopted carbon pricing and was contemptuous of the compensation offered for coal-fired power generators, saying there was no public policy justification for $3.9 billion in unconditional payments in relation to hypothetical future loss of asset value. 'Never in the history of Australian public finance has so much been given without public policy purpose, by so many, to so few.'[24]

Apart from overly generous treatment of polluters, the minimal extent of the 'cap' was the other main yardstick by which the scheme was judged. Again, Garnaut was critical here, not just because the reduction target was too low but because it ruled out in advance any greater reductions.[25] Despite the government's recognition of the seriousness of climate change and its pledge to 'reduce our carbon pollution by 60% of 2000 levels by 2050', for the short term it was only committing itself by 2020 to 'reduce Australia's carbon pollution by up to 15% below 2000 levels in the context of a global agreement where major economies agree to substantially restrain carbon pollution and advanced economies take on reductions comparable to Australia' and an 'unconditional 5% reduction in carbon pollution below 2000 levels by 2020'.[26] Treasury economic modelling indicated that a 5% reduction in emissions by 2020 would probably result in an initial carbon price of less than $25 per tonne of CO_2.[27]

While the government acknowledged its targets were more modest than those the EU had proposed for 2020, it insisted its target range translated into a 34–41% reduction in per capita emissions, given that population was projected to grow by around 45% over the 1990–2020 period.[28] Rudd explicitly rejected Garnaut's proposed 2050 targets of 80–90% reductions, saying he would have to seek a new election mandate for any proposed 2050 targets above 60%, which had been his campaign promise in the 2007 election.[29] If the Garnaut Review was as significant and authoritative as the government claimed, why were its recommended targets not adopted in the proposed CPRS? Rudd appeared to be bending to the 'carbon lobby', the unfortunate possibility Garnaut had anticipated.[30]

The Greens criticised the CPRS for its modest targets and overly generous assistance to polluters. However, the Greens did not at this stage hold the balance of power in the Senate, so were unable to negotiate a more ambitious scheme. The government needed the support either of the Coalition parties or the Greens plus Senators Fielding and Xenophon, both of whom opposed the CPRS. Fielding was insisting that global warming had stopped and was primarily due to solar activity.[31] He attended a Heartland Institute conference on climate change in the USA in June 2009, supported by a think-tank that had received $1 625 000 from Exxon-Mobil between 1998 and 2005.[32] On his return, he presented data to Wong and her scientific advisers, which claimed that global average temperatures had not risen over the previous decade.[33] Liberal leader Brendan Nelson stated on television that 'it would be irresponsible to endanger our primary industries by introducing an emissions trading scheme as soon as 2010', because resources accounted for 37% of Australia's economy, so an ETS should not be implemented until at least 2011.[34] Senator Barnaby Joyce stated the National Party would oppose the CPRS, regardless of the Liberal Party's position.[35] Passage through the Senate was looking unlikely for the CPRS legislation that the government announced on 10 March 2009, especially when the Liberals started to question whether Australia's emissions were globally significant enough to warrant any action at all.[36]

The Rudd government announced a modified CPRS proposal on 4 May 2009; and introduced this legislation into the House of Representatives on 14 May 2009. It negotiated the modifications directly with the Coalition, to secure passage through the Senate. By this stage, Turnbull had replaced Nelson as Liberal leader, which made bipartisan negotiations simpler, given Turnbull's public commitment to mitigation and an ETS in particular. The government pursued this strategy rather than negotiating a stronger scheme with the Greens and, if necessary, using defeat in the Senate as a double-dissolution election trigger. Carol Johnson argues Labor acted in keeping with its long-term fear that antagonising business would result in investment flight and subsequent job losses and possibly a concerted business campaign against it, but also to win broad initial support for the scheme, with the intention of strengthening it incrementally over a period of years as business and public opinion was won over to tougher measures.[37]

Apart from delaying implementation by a year, which the Coalition had always favoured, the modified CPRS proposed: an increase from 15 to 25% emissions reduction if a global agreement was attained (which was unlikely); the carbon price would be $10 per tonne for the first year then market-based; even more free trading permits for EITE industries, which increased the proportion of free permits to 95% for large EITE industries and from 60–70% for other eligible industries; a doubling of assistance to the coal-mining industry; granting manufacturing and mining companies another $1.5 billion to offset power price increases; providing $200 million in 2009–10 for business energy efficiency projects and a further $50 million for a business energy efficiency trust to provide loan money for energy-saving programs. It was described as 'boosting smoke and mirrors rather than cutting emissions'.[38]

The legislation introduced into parliament on 14 May 2009 was scheduled to reach the Senate on 13 August 2009. Because it had negotiated this CPRS with the Coalition, the government was expecting it to pass. However, despite Turnbull's attempts to negotiate further amendments to assuage Liberal concerns, the Liberal

Party opposed the legislation, so the bill was defeated in the Senate on 13 August and again on 22 October 2009.

However, parliament did pass the RET legislation on 20 August 2009, designed to obtain 20% of the country's energy from renewable sources by 2020. (It was presented as an expansion of the Mandatory Renewable Energy Target implemented by the Howard government but discontinued in 2006 due to carbon lobby pressure.) The RET creates a guaranteed market for additional renewable energy deployment using a mechanism of tradeable Renewable Energy Certificates (RECs), the demand for which is created by a legal obligation on parties who buy wholesale electricity, the retailers and large users of electricity. The supply of RECs will be created by renewable energy power stations using wind, large-scale solar, hydroelectric and potentially geothermal power; as well as small-generation units, including small-scale solar panels, small wind turbines and micro hydro systems, solar and heat pump water heaters.[39] This bill originally had been part of the CPRS but, in the wake of its defeat, the government separated the two bills; the RET legislation, unlike the CPRS, enjoyed bipartisan support.[40]

The government responded to the defeat of the CPRS by announcing it would reintroduce the legislation before the Copenhagen conference in December, and called on the Opposition to undertake negotiations in 'good faith', to which Turnbull responded.[41] The negotiations secured even more support for polluting industries but $5.67 billion less compensation to households. Agriculture was to be excluded indefinitely. Electricity generators would get an additional $4 billion in free permits over 10 years and coal-fired generators could get further free permits if they increased their energy efficiency. Coal-mining companies were granted an additional $750 million over 5 years. EITE industries would have the 5-year deadline to reduce the rate of free permits removed. There would be an additional $1.1 billion over 2 years to compensate mining and manufacturing companies for electricity price increases.[42]

The shadow cabinet accepted the outcome, but on 24 November the Liberal party room greeted it with uproar. On 1 December Abbott deposed Turnbull as Liberal leader. In the context of climate change politics, the change of leadership on the Opposition benches was highly significant, as Nick Economou observes: 'Turnbull got his party to the brink of providing bipartisan support for the Rudd government's ETS, only for conservatives within his own party to mobilise against him in favour of Tony Abbott, whose inclination on the matter of CPRS bills was to reject them again.'[43]

Abbott argued the CPRS constituted a 'great big new tax'.[44] He offered an 'environment policy' rather than a 'tax policy' that proposed to subsidise companies for voluntarily cutting emissions and to cut 140 million tonnes of CO_2 by 2020 by 'direct action on soil carbons' and tree planting. The Minerals Council of Australia welcomed the Coalition's policy for providing 'incentive to reduce greenhouse gas emissions without negatively impacting on jobs, investment, exports and growth'.[45] Abbott's ascendancy to the Liberal leadership both expressed and reinforced the power of climate deniers within Coalition ranks; and influenced Coalition voters' attitudes to global warming. Gary Morgan found in January 2010 that 51% of Coalition supporters (up 5%) felt concerns about global warming were exaggerated, compared to 22% of Labor supporters.[46]

The Rudd government had repeatedly weakened its ETS to secure Opposition support, yet these accommodations still did not secure its passage. The government did not use this defeat of major legislation twice in the Senate to seek a double-dissolution election immediately with the aim of resubmitting the bill to a new parliament. Instead, Rudd indicated plans to introduce yet another version of the scheme about three months later. Rudd and Wong then attended the Copenhagen summit—notwithstanding the *Age*'s observation that Wong had 'notched up $120 000 of plane travel and hotel bills in 20 months, delivering speeches and attending conferences and meetings'.[47] Wong met with jeers and chants when she spoke. Rudd termed the Copenhagen Accord a 'big step forward', though Indian Environment Minister Jairam Ramesh denounced Australia's attempts to shift more of the burden for emissions cuts onto poor countries, telling media on 16 December that Australia was 'the Ayatollah of the single track'.[48] In contrast to events about to play out on the Australian stage, reduction pledges under the Copenhagen Accord by China, Indonesia, Brazil and South Africa were strong enough to satisfy the Rudd government's criteria for Australia to commit to a 15% reduction.[49]

The 'Great Moral Challenge' Unmet

Dennis Woodward argues that, long before the CPRS received its second Senate rejection, it had departed from Garnaut's recommendations 'and been successively watered down to appease the vested interests of polluting industries to the extent that its usefulness in combating climate change was doubtful'. The Rudd government's approach was determined more by considerations of achieving domestic political advantage than meeting the great moral challenge. Yet, when it resorted to negotiating with the Opposition, it achieved not only the defeat of the legislation but also the triumph of climate change deniers within the Liberal Party, declining support for climate change mitigation 'and ultimately Rudd's political demise'.[50]

The government made much of the fact that, by 2009, opinion polls indicated the public had become less concerned than in 2007 about tackling climate change. In particular, the Global Financial Crisis (GFC) was distracting attention from climate change when it came to ranking issues as 'very important', with job protection and strengthening the economy increasing in significance to respondents. Yet even at this comparative low point in climate change concern, Lowy Institute 2009 polling found it was still considered important to 85% of respondents.[51] From 2006 to 2009 there was only a 5 point drop from 92 to 87% of respondents saying global warming had to be addressed, though the proportion who preferred cost to be taken into consideration had increased from 24 to 39%; and the 13% who agreed 'Until we are sure that global warming is really a problem, we should not take any steps that would have economic costs' represented a 6 point increase from 7% in 2006.[52]

Although there was a decline in levels of concern about global warming, well over four-fifths of Australians still considered it important.[53] The majority of Australians still accepted the need to change their economic lifestyle to deal with climate change, were themselves prepared to pay extra to this end, and believed the government was not doing enough to tackle climate change.[54] What is true is that the GFC had reordered people's immediate priorities. Also, the appearance of activity on the part of the Rudd government probably eased anxieties about the issue or possibly bored people with the tortured parliamentary manoeuvrings. Alternatively, the Solar Rebate, Green Loans and the badly botched Housing Insulation subsidy scheme—good ideas administered badly—contributed to public scepticism about mitigation policies.[55] More significantly, with the Rudd government threatening to tackle seriously the issue of Australia's high per capita emissions, climate change deniers busied themselves with attempts to confuse and confound the scientific messages and public understanding of these.

Associated with this offensive was wilful misrepresentation of opinion polls. For example, a Lowy Institute poll 12–28 July 2008 asked 1001 respondents how much extra per month they were prepared to pay on their electricity bills to help fight climate change. Only 21% would not pay anything extra; the remainder prepared to pay varying additional amounts. However, under the heading 'Climate fight only worth $10 to Aussies', the *Australian* reported that 32% of people favoured paying only $10 extra, without mentioning how much extra other percentage groupings had been prepared to pay.[56] The *Australian* ought to have reported that another 20% were willing to pay $11–20 extra and another 19% were willing to pay $21 or more. In other words, 71% were prepared to pay at least $10 per month extra, especially younger people.[57]

For the majority of Australians still deeply concerned about climate change, disappointment at the miniscule nature of the guaranteed reductions turned to serious discontent when the projected CPRS came to naught. In early 2010 the government had attempted to reach a compromise with the Greens, Senator Xenophon and dissident Liberal Senators over the Greens' proposal for an interim carbon tax, but to no avail.[58] On 27 April 2010 Rudd announced the CPRS would be shelved, saving $2.5 billion, and reviewed again in 2012. Rudd's rationale was extraordinary: 'The opposition decided to backflip on its historical commitment to bring in a CPRS and there has been slow progress in the realisation of global action on climate change', then added that his government's reduction targets remained unchanged. 'Climate change remains a fundamental economic environmental and moral challenge for all Australians and for all people of the world.'[59] Rudd's final decision was in contempt not only of the mandate received at the 2007 election but also of the opinion polls that continued to confirm support for unilateral action. A Lowy Institute poll released the day Rudd made this announcement found that 72% agreed 'Australia should take action to reduce its carbon emissions before a global agreement is reached'.[60] Rudd's failure to meet 'the great moral challenge of our times' triggered a dramatic slide in his approval rating. On 16–18 April Newspoll had found 50% of respondents were satisfied with the way Rudd was doing his job as Prime Minister; on 2 May, after the ETS backdown, this figure had plummeted to 39%. The same polls showed a similar downward slide for voting intention, with Labor moving from 43 to 35% of the primary vote and 54 to 49% of the two-party preferred vote. Rudd's approval rating had plummeted in the fastest collapse of support for a PM in the twenty-year history of Newspoll and one of the two sharpest drops in the 40 years of Nielsen polls.[61] On 1 June 2010 two-thirds of Auspoll respondents could not distinguish between the two major parties on climate change politics.[62] Labor had thrown away this advantage it clearly had, underestimating how determined Australians were for genuine and urgent action to mitigate climate change.[63]

The Greens and a wide array of climate action groups had mobilised around a critique of the CPRS as seriously flawed. However, the government's decision to retreat from its ETS in the absence of Coalition support instead of negotiating with the Greens to present a more far-reaching scheme and, if necessary, accepting defeat in the Senate as a double-dissolution trigger, was in clear defiance of the wishes of the Australian people at what was rightly dubbed the world's first climate change election: the Labor government was not merely given permission to reduce emissions immediately and considerably but instructed by the Australian people that it must not fail to do so. Journalists instantly and stridently condemned the 'backflip' for treating such a clear electoral mandate with contempt.[64]

Moreover, a two-year study of attitudes towards ETSs, conducted by the Centre for the Study of Choice at the University of Technology, Sydney, found in July 2010 that the majority of 7000 randomly selected people wanted to see a more ambitious and more immediate ETS than the one abandoned by the government, whether or not the USA and China took similar steps.[65] It concluded: 'Their choices are remarkably stable: Start a plan NOW & do not wait for other countries'; 'Redistribute revenues—reduce taxes, give to poor & seniors';

'Invest some revenues in R&D'; 'Exempt transport for some period of time'; 'Do not exempt energy-intensive sectors'; and 'Set relatively high targets for cuts in emissions for 2020'.[66]

If one accepts the evidence that retreat from the CPRS was not based on electoral considerations, why did the government back down? It is always helpful to inquire 'cui bono?' Who benefitted from the shelving of the scheme? The accounting and law firms—'the carbon-trading lobby'—that stood to profit from the introduction of the scheme were disappointed. Spokespeople for 'green corporations' said the decision would hamper investment in renewable energy projects and low-carbon technologies. On the other hand, the backflip was welcomed by brown corporations. The circumstantial evidence implicating the 'greenhouse mafia' is strengthened when considering its role in breaking Rudd's prime ministership when he did not cave in to its pressures to abandon or reduce the Resource Super Profits Tax (RSPT). The *Guardian* reported of events down under that Rudd was 'driven from power' when he tried to impose this tax on mining industry super-profits. 'Commodity conglomerates tilted him out of office with a lavishly funded publicity campaign that claimed he was threatening Australia's prosperity.'[67] Having betrayed the Australian public on climate change mitigation and suffering the opinion poll consequences, it was a greatly weakened Rudd and a nervous Labor caucus who faced this greenhouse mafia assault, resulting in Julia Gillard becoming Prime Minister in June in a set of circumstances that appeared something of a coup d'état. Arguably, Rudd, after Howard and Turnbull, was the third person to lose leadership of a major political party due to climate politics.[68]

The Gillard Government

As Prime Minister, Gillard moved swiftly to reduce the RSPT tax rate from 40% to 22%, add extra forms of compensation and discursively conceal the very existence of 'super-profits' by renaming it the Mineral Resource Rent Tax. Moreover, the tax was only to be applied to iron ore and coal companies and only those who earn annual profits above $50 million.[69] Another early initiative of Gillard, as she reiterated Rudd's instance that climate change constitutes a serious threat to Australia and the world, was to send Trade Minister Simon Crean to broker a deal to export Victorian brown coal to Vietnam.[70]

Soon after the 2010 election was called, on 23 July 2010, Gillard proposed the formation of a Citizens Assembly of 150 average Australians, one from each electorate, to help build consensus for an ETS. She stated her government could not take action on climate change until a 'community consensus' on the issue had been achieved. This assembly never eventuated, because the proposal was widely derided.[71] The public was inclined to the view that the government should just get on with climate change mitigation and stop convening talk-shops. Labor also proposed: a $2000 rebate to get old cars off the road and set mandatory emissions standards for new vehicles; $1 billion for energy network upgrades to help connect renewables to the grid and new emissions standards for coal power plants; and $46 million to help farmers create carbon credits through tree planting and better farming practices.[72]

Throughout the election campaign, Gillard referred to 'the breakdown of the domestic consensus' to tackle climate change and the need to rebuild consensus to do so. Bipartisan support had not been forthcoming over the CPRS, but that was a different matter. Conflating 'domestic consensus' and 'bipartisan support' distracted attention from its betrayal of the mandate—or instruction—it had received at the world's first climate change election in 2007. The Coalition also downplayed climate politics during the 2010 election, simply promising a $3.2 billion 'direct action' climate change fund to store carbon in soil, plant trees and reduce power plant emissions.[73] The Morgan Poll released on 29 July 2010 found both major parties had lost ground on environmental issues and the Greens, whose much stronger climate policies are discussed in Chapter 6, were seen as the best party for 'managing global warming and climate change' (46%, up 5%) compared with the Coalition (19%, down 4%) and ALP (17%, down 4%). Morgan also predicted 'The Federal Election will be closer than anyone realises.'[74]

It was. The historically close numbers eventually permitted Gillard to continue as Prime Minister of a minority Labor government with promised support on confidence and supply motions from the Greens (Adam Bandt) and three Independents (Andrew Wilkie, Tony Windsor and Rob Oakeshott) in the House of Representatives. The Greens, also holding the balance of power in the Senate, were able to extract assurances of action on mitigation, a stance reinforced by the three Independents. Tony Windsor, for instance, insisted Greg Combet, the new Minister for Climate Change, immediately implement an Australian Productivity Commission study into carbon prices in other countries, including the USA, UK, Germany, China, Japan and India.[75]

In early February 2011 the Gillard government announced its plan to introduce a carbon price in the following July, to be replaced in 3–5 years by an emissions trading scheme. The Greens aimed to strengthen the proposed legislation by working within the Multi-Party Climate Change Committee, whose members were Julia Gillard, Wayne Swan, Greg Combet, Independents Tony Windsor and Rob Oakeshott, and Greens Senators Bob Brown and Christine Milne. Advisors to the committee were Patricia Faulkner, a partner at KPMG, which specialises in facilitating privatisations and public-private partnerships; Rob Sims, a former economic advisor to the Hawke government; and Ross Garnaut.[76]

However, the move towards carbon pricing was not necessarily good news for renewable sources of energy, because the government intimated it was more interested in promoting both CCS and gas-fired power stations to replace coal-fired ones. On 13 September 2010 Combet told the *Australian*: 'The coal industry is a very vibrant industry with a strong future. What you've got to do is look to how we can achieve in the longer term things like carbon capture and storage for coal-fired power stations'.[77] On 9 March 2011, he stated: 'For base-load electricity generation it will be a gas-fired electricity that we see emerge, and for that investment to be committed, we need a carbon price in the economy.'[78]

The Garnaut Climate Change Review team commissioned research to investigate attitudes to climate change and to what extent these had changed since 2008. In March 2011 it reported that: most Australians believe the climate is changing, though belief in its anthropogenic drivers had waned in recent years, matching trends in other Western countries; most Australians believe Australia should take action on climate change without waiting for global consensus; and there was no clear consensus of what policy actions Australians prefer, such as a carbon price or an ETS.[79] Income had no noticeable effect on beliefs, though one poll had found wealthier people were less inclined to support mitigation. People with a university degree were more likely to be concerned about climate change (83%) than those with no post-secondary qualifications (69%). Location appeared significant, with 58% of urban dwellers and only 27% of primary producers agreeing that human activity was the primary cause of climate change. However, this might reflect political persuasions.[80] The research found the most significant demographic factors were politics and gender: 'support for left-wing political beliefs is associated with greater belief in human-induced climate change' and 'Women consistently were more likely to believe in human causes of climate change and are more likely to be concerned by it'.[81]

Gillard was reluctant to refer to carbon pricing as a tax, because she had asserted during the election campaign there would be no carbon tax under her government. In an interview with John Laws on Radio 2SM on 28 February 2011, she attempted an explanation: 'I have agreed that there will be a fixed price period before we get to a full market based pricing scheme. That is effectively like a tax, I'm happy to say that ...'. She phrased it more judiciously on ABC radio's *AM* program on 24 June 2011 when she insisted the carbon price mechanism was not a tax but a carbon trading scheme that was starting with a fixed price on carbon and added: 'We are going to get there through a path I didn't expect during the election campaign, of a fixed price period for around three to five years.'[82]

Predictably, the Coalition accused the Gillard government of imposing 'a great big new tax' on the Australian people. Abbott had for a while backed away from his controversial statement that 'climate change is crap', but on 23 March 2011 he again exhibited his contrarian predilections when he told an audience in Perth: 'Whether carbon dioxide is quite the environmental villain that some people make it out to be is not yet proven.'[83] The same day, in Canberra, Abbott spoke at a rally against the carbon price, pictured in front of a banner reading 'Ju-Liar Bob Brown's Bitch'. On the platform beside him were Barnaby Joyce and Deputy Opposition leader Julie Bishop. Also in attendance was Pauline Hanson, the former One Nation party leader.[84]

On 31 May 2011 a follow-up Garnaut Review was released. Garnaut surveyed alarming new information about rising temperatures internationally and how 'rising sea levels continue to track the upper levels of modelling'. He regretted that the Copenhagen agreed targets were voluntary and called for 'concerted unilateral action' in which countries examine and encourage each other's efforts. In contrast to the 'inaction in Australia', he pointed to other countries' mitigation policies, including China's pledge to reduce its carbon intensity by 40–45% between 2005 and 2020 and bipartisan efforts in Germany, the UK, France and South Korea to achieve carbon pricing. Australia was on track to meet its Kyoto Protocol target, because emissions were projected to reach only 106% of 1990 levels in 2008–2012; but emissions were projected to increase by 24% from 2000 levels by 2020 in the absence of further policy action. He stated that carbon pricing would drive significant emissions reductions in the electricity sector and generate huge amounts of revenue, about a fifth of that collected through the GST.[85] Speaking at the launch of his book, Garnaut recommended a starting carbon price of $26, raising $11.5 billion in the first year; that households should get 55% of that revenue via compensation, rising to 65% by 2021–22, but households earning over $80 000 per annum should not receive any compensation; and that industry should receive less, sliding from 35% to as low as 20% by 2011. Garnaut also took a swipe at Graham Bradley, the chairperson of the Business Council of Australia, for allegedly favouring 'narrow business interests above the many benefits that add up to the national interest'.[86]

Also in late May, an International Energy Agency report revealed that humanity had emitted some 30.6 billion tonnes of CO_2 in 2010, 5% higher than the previous year. The Gillard government offered in the upcoming legislation to boost its 2050 emissions reduction target from 60% to 80% of 2000 levels in exchange for compromises by the Greens on compensation for households, heavy industry and the coal-fired power industry.[87] The following week, Wayne Swan tried to reassure the Australian public that a carbon price would not have a severe impact on the Australian economy by citing a Treasury modelling report indicating that the real national income would grow under a carbon price of $20 per tonne, at an average annual rate of about 1.1% until 2050.[88]

On 10 July 2011, dubbed Carbon Sunday, the Gillard government announced the details of its implementation of a carbon tax on 1 July 2012, to transition into an ETS on 1 July 2015. The new Climate Change Authority to oversee the scheme would be headed by Bernie

Fraser, former governor of the Reserve Bank. In May 2014 this body will set emissions caps for 2015–2019, and each year by 30 June set a cap 5 years ahead. The cap, which limits the amount of emissions, effectively sets the carbon price. It is designed to result in a 5% cut in emissions, based on 2000 levels, by 2020, and an 80% cut on those levels by 2050. Some 500 big polluters will pay a tax to be set at $23 for every tonne of emissions, rising to $24.15 in 2013–2014 and $25.40 in 2014–2015. Agricultural emissions are exempted. It is an extremely complex scheme that includes numerous subsidies for both industry and low- and middle-income households, and free permits for some industries. The government estimates the carbon tax will increase costs for the average household by 0.7% or $9.90 a week and that the tax will result in a short-term 0.7% increase in inflation. The scheme compensates low-income households by lifting the tax-free threshold from $6000 to $18 000 in 2012 and then to $19 400 in 2015–2016 and grants a tax cut to all households below $80 000. The government projects that two out of three households will have their costs fully offset and that nine out of ten households will receive some compensation.[89]

Compensation to industry was exceptionally generous. It allocates $9.2 billion over three years to support EITE industries; heavy polluting industries such as steel, aluminium, zinc, pulp and paper will get 'free permits' covering up to 94.5% of their emissions; $1.2 billion is allocated for a 'clean technology program' to help manufacturers; $1.3 billion will assist affected coal miners find alternative employment. To encourage renewable energy, $10 billion is set aside to set up the Clean Energy Finance Corporation, which will invest in renewable energy projects; and $3.2 billion is allocated for research and development of renewable energy projects. The scheme intends that 20% of Australia's energy will come from renewable sources, such as wind, solar and geothermal, by 2020; and large-scale renewable energy will provide 40% of electricity by 2050. The dirtiest generators will be paid to close, cutting 2000 MW of coal-fired power by 2020. (Of a national total of $5.4 billion compensation to electricity generators, $5.22 billion or 97% is projected to go to Victorian generators.) Gas-fired electricity will increase by more than 200% by 2050. Other funding initiatives include: $429 million for a Carbon Farming Initiative for reforestation and revegetation schemes that would allow farmers to earn money by selling carbon credits to big polluters; $947 million over six years for a biodiversity fund to protect Australian native species from climate change; $40 million over four years to promote energy efficiency measures among small businesses and community groups.[90]

Tony Abbott referred to it as a scheme for redistributing wealth that smacked of 'socialism masquerading as environmentalism'.[91] He reiterated Coalition policy that emissions should be cut by planting numerous trees, shifting from chemical to organic fertiliser, and improving energy efficiency; and continued to support the coal-fired power industry. 'I think there is no reason why we shouldn't build base-load power stations using coal', he stated. 'You can't run a modern economy without a lot of power. Now and for the foreseeable future, the most cost-effective way of producing power is going to be coal.'[92] Malcolm Turnbull confirmed that Liberal Party policy accepted the scientific consensus that global warming is real and regretted there had been a very effective campaign against climate science by those opposed to cutting emissions, and this had affected the carbon tax debate. Rejecting climate science, he said, was like ignoring the advice of your doctor to give up smoking on the basis that somebody down the pub told you their uncle Ernie had lived to 95 and smoked like a train all his life.[93]

The week after Carbon Sunday, an Age/Nielsen poll indicated Labor's primary vote had plummeted to 26%, the lowest on record, and was trailing the Coalition 39 to 61% in the two-party preferred vote. Gillard's approval rating fell to 34%, the lowest since Paul Keating's low of the same percentage in March 1995.[94] On 19 July a Morgan Poll found 52% agreed Australia should not have a carbon tax until China and the USA had a similar tax, while 47% disagreed. The number of Australians worried about global warming was 46%, down 21% since April 2006.[95] When the legislation passed the House of Representatives 74–72 on 12 October 2011, protesters in the public gallery yelled 'no mandate' and 'democracy is dead'; and Abbott made 'a pledge in blood' to repeal the carbon tax as the 'first order of business' for an incoming Coalition government.[96] Early in November, a Morgan poll found that only 45% of electors supported Abbott overturning the tax if elected compared to 48% who were opposed, but a clear majority of electors (57%) still opposed the carbon tax, despite the legislation passing federal parliament, compared to only 37% that support it.[97]

Robyn Eckersley and Peter Christoff, University of Melbourne climate policy experts, blamed 'the brutal debate and weak public support' for the softness of the scheme, arguing that Australian lobby groups and conservative politicians had 'beaten up the carbon tax as a radical and painful measure' and Australia's media bore some responsibility 'for perpetrating this deception'. They described the scheme as 'an astute package of measures that will alter Australia's carbon economy in the best traditions of ecological tax reform' but pointed out how inadequate it was to cope with the extent of the global warming problem and redress Australia's position as the world's highest per capita emitter, because the short-term reduction target was among the weakest of all industrialised and major industrialising nations. The real price of this 'modest carbon tax' would be paid, they concluded, 'by our children and by countless future generations'.[98]

Notes

1 Rootes, pp. 479–80.
2 Jowit and Wintour.
3 'Ratifying the Kyoto Protocol: media statement', 3 December 2007. http://.www.alp.org.au/media/1207/mspm030.php (viewed 5 December 2007).
4 Marr, p. 54.
5 Hamilton, 'Rudd at Bali and beyond'.
6 'Who is Ross Garnaut?', 25 February 2008. http://newmatilda.com (viewed 20 August 2008).
7 Kevin, p. 15.
8 Quoted in Hickson, p. 505.
9 Garnaut, *Climate Change Review: Final Report*, pp. 279–80.
10 ibid., pp. 315–17, 332, 297.
11 ibid., p. 202.
12 ibid., p. 223.
13 ibid., p. 385.
14 ibid., pp. 503–519, 525.
15 ibid., p. 547.
16 Pearse, 'Quarry vision', pp. 3, 59–60.
17 Garnaut, *Climate Change Review: Final Report*, p. 597. Also available in the Synopsis of the Review. http://garnautreview.org.au/synopsis.htm (viewed 1 June 2010)
18 Australian Government, *CPRS—White Paper*, p. iii.
19 ibid., p. iv.

20 ibid., pp. xv–xvi.
21 ibid., p. xxiii.
22 ibid., pp. xxiv–xxvi; Diesendorf , *Climate Action*, p. 56.
23 Pearse, 'Quarry Vision', p. 55.
24 Ross Garnaut, 'Australia counts itself out', *Age*, 20 December 2008, quoted in Woodward, p. 5.
25 ibid.
26 Australian Government, *CPRS—White Paper*, pp. iv, xix.
27 Diesendorf, *Climate Action*, p. 56.
28 Australian Government, *CPRS—White Paper*, pp. xix–xx.
29 Kevin, p. 24.
30 Garnaut, *Climate Change Review: Final Report*, p. 317.
31 Morton, 'Fielding's climate mission'.
32 Glickson, 'Noah's ark revisited'.
33 Kevin, p. 182.
34 ABC News, 23 July 2008; Pearse, 'Quarry vision', p. 1.
35 *Age*, 24 February 2009, p. 1.
36 *Age*, 17 April 2009, p. 1.
37 C Johnson, 'The CPRS', p. 1.
38 *Age*, 5 May 2009, p. 1. See also White, 'Rudd gambles with the planet'.
39 Koh, pp. 31–2.
40 Butler, 'The renewable energy scheme'.
41 *Age*, 20 November 2009, p. 1; *Age*, 22 November 2009, p. 1; Woodward, p. 7.
42 Woodward, pp. 7–8.
43 Economou, p. 6.
44 *Australian*, 2 December 2009, p. 1.
45 Mitchell Hooke, CEO, MCA, media release, 2 February 2010.
46 http://www.roymorgan.com/news/polls/2010/4458 (viewed 1 February 2010).
47 Gordon, 'Carbon offset plans'.
48 *Australian*, 17 December 2009; Butler, 'Australia's dirty role'.
49 Manning, 'One climate policy'.
50 Woodward, pp. 1, 9.
51 Hanson, *Australia and the World*, 2009, pp. 21–22.
52 ibid., p. 27.
53 ibid., pp. 21–22.
54 Pietsch and McAllister, p. 232.
55 For discussion of these, see Economou, p. 7; Crowley, 'Climate Policy Failure', p. 10.
56 *Australian*, 29 September 2008. http://www.theaustralian.com.au/news/nation/climate-fight-only-worth-10-to-aussies/story-e6frg6nf-1111117611839 (viewed 2 June 2010).
57 Hanson, *Australia and the World*, 2008, p. 13.
58 *Age*, 23 February 2010, p. 1.
59 *Age*, 28 April 2010, p. 1.
60 Grattan and Arup, 'Rudd in ETS backflip'.

61 Hartcher, 'How Rudd lost his nerve'.
62 Davis, 'A "pox on both their houses" '.
63 Pearse, 'Quarry vision', p. 74.
64 For example, Grattan and Arup, 'Rudd in ETS backflip'; Morton and Arup; Carney; Grattan, 'PM's climate credibility'.
65 Cubby.
66 Louviere et al.
67 *Guardian*, 7 September 2010. http://www.guardian.co.uk/environment/2010/sep/07/Australia-green (viewed 10 September 2010).
68 ABC, 'The carbon war'.
69 McAuley, p. 20.
70 Butler, 'Gillard: same old spin on climate'.
71 Economou, p. 8.
72 *Age*, 21 August 2010, p. 3.
73 ibid.
74 http://www.roymorgan.com/news/polls.2010/4544 (viewed 1 August 2010).
75 I Bailey, p. 33.
76 Dall.
77 Quoted in Courtice.
78 Quoted in Moore.
79 Leviston et al, p. 27.
80 ibid., pp. 7–8.
81 ibid., p. 6.
82 Quoted in Butler, 'Carbon trading promise'.
83 Quoted in *Green Left Weekly*, 23 March 2011, p. 3.
84 Butler, 'Climate crisis'.
85 Garnaut, *The Garnaut Review 2011*, pp. xi, xiii, xv, xviii, 5, 26, 87, 149.
86 Arup, 'We're not pissants'.
87 Arup and Morton, 'Emissions surge to record level'.
88 Grattan and Arup, 'Carbon pricing impact "minimal" '.
89 *Australian*, 11 July 2011, p. 6.
90 *Australian*, 11 July 2011, p. 6; S Maher, 'Gillard's clean energy crusade'; S Maher, 'Coal and steel'; Gordon and Arup.
91 Quoted in Franklin.
92 Quoted in Morton and Schneiders.
93 Turnbull, pp. 4–5.
94 *Age*, 18 July 2011, p. 1.
95 http://www.roymorgan.com/news/polls/2011/4686 (viewed 19 July 2011).
96 Grattan and Wroe.
97 http://www.roymorgan.com/news/polls/2011/4710 (viewed 6 November 2011).
98 Eckersley and Christoff, 'A soft start on carbon'.

5
Corporations and the State

Awareness of the need to reduce greenhouse gas emissions has been on the policy agenda of federal and state governments for quarter of a century. Yet, both Coalition and Labor governments have encouraged the exploitation of Australia's coal and natural gas supplies, both for export and domestic purposes, rather than stimulating investment in and development of renewable energy sources. At state level, governments have also privatised electricity and gas provision in many jurisdictions, which has also adversely affected energy efficiency and reversed earlier trends towards renewable energy sources, such as the pioneering developments in solar energy during the 1970s. In this chapter we explore how the power over governments wielded by brown corporations has retarded efforts to reduce emissions and disadvantaged green corporations.

The Greenhouse-Gas-Producing Double Standard

Governments have achieved adverse ecological effects in philosophically contradictory ways. On the one hand, there has been both refusal to intervene in the market to restrict emissions by direct regulation and also withdrawal from the market via privatisation of public assets, including electricity. On the other hand, there has been enthusiastic intervention in the market in support of emissions-intensive industries. It has been a sad case of 'now you see the free market, now you don't'.

Since the mid-1980s, Australian governments have professed commitment to neo-liberal notions that government should withdraw as much as possible from the economy to allow free-market forces to reign[1]—while pouring public subsidies into private corporations. The Labor governments from 1983 to 1996 engaged in what Tom Conley describes as a 'love affair' with neo-liberalism.[2] Australia became a world trendsetter in privatisation programs, with massive transfers of publicly owned assets to private ownership.[3] These Labor governments were truly 'Moloch's Little Mate'.[4] The Coalition governments between 1996 and 2007 continued this political trajectory even more forcibly. This was less surprising than Labor's 'love affair' with neo-liberalism, because the Liberal Party is the main party of urban capital. More surprising was that, despite the devastating impacts on rural and regional Australia of free-market forces, the National Party became neo-liberalised, more representative of mining and agribusiness than pastoralists, farmers and small business people in country towns upon whom it had historically drawn support.[5]

Australian governments have traditionally subsidised the costs to private enterprise of large-scale development projects and worked in close concert with corporations, including mining companies. As Bob Fagan and Dick Bryan observed in 1991: 'Large state government investments in electricity generation, and ports and harbour developments were often a subsidy to the private sector to encourage mineral exports.'[6] When governments were not committed to free-market ideology but instead touted the advantages of government sponsorship of free enterprise in developing the nation, such subsidies were politically unremarkable. There was no inconsistency at play. Not so for the neo-liberal moment from the mid-1980s. In this context, state subsidies to private companies indicated a double standard and often a greenhouse gas-producing one at that.

Since the mid-1980s, despite vigorous—but hyprocritical—commitment to free-market principles, Labor and Coalition governments at federal and state levels have subsidised the most polluting industries in the country, promoting especially the exploitation of Australia's coal and natural gas supplies. 'The push was on', Andrea Bunting observes, 'to entice energy-intensive industry and encourage greater fossil fuel exports.'[7] In 2000–2001, for example, Diesendorf calculates that governments provided $6.54 billion to subsidise the production and consumption of fossil fuels. This included $750 million, possibly as much as $1.2 billion, in tax benefits for salary packaging of motor vehicles; $400 million in automotive industry support; $153 million in research and development support for fossil fuels; $2000 million in road and car-parking subsidies; $410 million for electricity supply subsidies to the aluminium industry; and various other examples.[8] In 2009, Pearse published his influential essay, 'Quarry Vision':

> Mining, metals, and energy industries enjoy billions of dollars in tax breaks, fuel excise rebates, cheap electricity contracts, royalty holidays and infrastructure ... In Victoria ... state and federal governments are jointly spending over $150 million to help build a new brown-coal-fired power station. In Queensland, the state government is subsidising 'clean coal' and is also spending $5.4 billion on a Coal Transport Infrastructure Investment Program. In New South Wales, half a billion dollars in federal money is being spent in the Hunter Valley alone to improve railway infrastructure to increase coal exports.[9]

In May 2009 the Victorian Government announced that Latrobe Valley brown coal, not wind turbines, would provide electricity to the desalination plant in Wonthaggi.[10]

Even while preparing its CPRS legislation, the Rudd government was pouring huge subsidies into emissions-intensive enterprises. Apart from budgeting $200 million to increase the capacity of Eraring coal-fired power plant in NSW and assisting with the construction of two new coal- or gas-fired power plants at Bayswater and Mt Piper and the retrofitting of the existing coal-fired power plant at Munmorah,[11] it directed significant funds away from renewable energy initiatives to develop 'clean coal' technology. CCS is an unproven technology and unlikely to be available commercially until after 2020.[12] Results to date are unimpressive, judging by the Hazelwood CCS pilot project. This $10 million CCS installation captures 25 tonnes of CO_2 a day, only 0.05% of Hazelwood's emissions.[13] The Victorian, NSW and Queensland governments also provide big funding and policy support for CCS. The Victorian government, with the dirtiest coal, has made CCS a central feature of its future energy development strategy.[14]

In September 2008 the Rudd government established a Global Carbon Capture and Storage Institute, with $100 million, to promote research into CCS, with a view to creating twenty commercial plants by 2020. Ironically, the Institute estimated the technology would not be commercially viable until emissions cost $60 a tonne and it would increase electricity prices by up to 78%.[15] Then, in November 2008, Martin Ferguson announced the formation of the National Low Emissions Coal Initiative to be funded through the National Low Emissions Coal Fund outlined in the CPRS. Also in the CPRS was a Coal Sector Adjustment package of $250 million over 5 years to coal-mining operators to clean up their coal, as well as a further $500 million over 5 years direct assistance to 'gassy coal mines' to help them explore 'abatement opportunities', amounts far in excess of that allocated the Renewable Energy Fund.[16] In 2009, the Rudd government's $2.4 billion for research on CCS outstripped the $1.5 billion it committed to build new solar power plants.[17]

The coal-mining industry spends only one thousandth of its revenue on CCS research. According to Hamilton:

> Big coal thinks it's cheaper to lobby governments to provide public funding, with the result that governments are investing heavily in research into 'clean coal' while the fossil fuel industries spend up on PR campaigns and political lobbying to keep the public funds flowing and quash any thoughts of the need to phase out coal.[18]

Sometimes the PR campaigns are subsidised by the state. For example, the Australian Coal Association devised a COAL21 National Action Plan, supported by the Rudd government, to promote 'clean coal' technologies, which asserts that Australia needs to improve power-station efficiency and geosequestration to retain coal as an energy resource.[19]

Tony Kevin describes Rudd's promotion of CCS as 'an expensive public facade, hiding the stark policy contradiction between the government's rhetorical commitment to reducing Australia's greenhouse-gas emissions and its real commitment to supporting the Australian coal industry's unimpeded growth'.[20] Kenneth Davidson describes CCS as 'a fig leaf to maintain business as usual for as long as possible'.[21] And the money poured into clean coal technology is money not spent on developing renewable sources of energy.

Aluminium smelting operations are also highly subsidised by state governments, at the same time as they profess commitment to both free-market principles and climate change mitigation. Estimates of the public subsidies that the Queensland, New South Wales, Victorian, and South Australian state governments give the aluminium industry have varied between $210 million and $400 million per annum.[22] Also, the Victorian government built a transmission line from the coal-fields in the east across the state to feed power to the smelter at Portland in the west and permitted Alcoa to use it free of charge; and its subsidies for the Portland and Point Henry smelters will continue until 2016, then replaced by a contract between Alcoa and Loy Yang Power that locks in coal-fired power for smelting for another 20 years.[23]

In stark contrast to the billions of public money poured into emissions-intensive industries, much smaller amounts are directed towards development of renewable energy sources—despite expressed concern about climate change.[24] In 2011 an audit of Victoria's renewable energy targets found 3.9% of Victoria's energy was generated by renewable sources of energy, only 0.3% more than in 2002 when the government announced a 10% target to be achieved by 2010. Given this failure, it was relatively simple for Ted Baillieu's new Coalition government to gut the renewable targets completely.[25] In August 2011, the Baillieu government drastically reduced Victoria's feed-in tariff rates, for households with solar panels to sell electricity back to the grid, from 60 cents per kilowatt hours to 25 cent per kilowatt hours.

Federal government activity on renewables is similarly unimpressive. Prior to the 2007 election, Labor promised the creation of a $500 million Renewable Energy Fund. Apart from some funding for geothermal drilling, the Rudd government did not allocate funds for renewable energy in its first budget but instead allocated $500 million to the coal industry from a Clean Coal Fund.[26] The Rudd government also backtracked on promising funding for solar thermal and solar photovoltaic electricity research. In May 2009, the Rudd government launched a Solar Flagships program that would allocate $1.6 billion for the construction of four power plants providing for a total of 1000 megawatts of solar power, but the Clean Energy Council argued the funding was insufficient.[27]

The Rudd government did grant an $8000 rebate for solar panels that was to be effective until 30 June 2009. However, it axed the scheme on 9 June 2009, blaming a cost blow-out from $150 to $750 million annually due to its popularity.[28] The effect was unsettling for renewable energy companies. Solar Shop Australia claimed on 30 July 2009 that it might have to lay off 100 sales and marketing people due to the subsequent crash in the rooftop solar panel market. Managing director Adrian Ferraretto said sales had declined from 450 a week to fewer than ten. The government replaced the original scheme with one with lower rebates but not means-tested and including small businesses.[29] But, in September 2009, Solar Systems a company that had planned to build Australia's first large-scale solar power plant near Mildura, providing 1000 renewable energy jobs and powering 45 000 homes, went into receivership due to inability to raise loans. In December 2009, AGL Energy, Australia's largest renewable energy company, claimed it might have to curtail its plan to construct an $800 million wind farm in western Victoria due to insufficient funding from the Rudd government.[30]

At the same time as governments intervene to subsidise the most polluting industries—in contrast to the niggardly support for the development of renewable energy sources—they withdraw from the economy in ways that also have detrimental effects. So, when governments practise the free-market principles they persistently preach, this also encourages emissions. A good example here is government retreat from electricity provision, which has discouraged energy efficiency and harmed development of solar energy, which seemed a promising Australian initiative during the 1970s and 1980s.

From the mid-1990s, electricity utilities underwent corporatisation, which in some cases included privatisation.[31] Privatisation of electricity supply discourages use of domestic solar panels in particular, because it reduces demand on the grid and affects the profits of these powerful corporate electricity suppliers. Victoria was a spectacular case in point. In 1994 the State Electricity Commission's energy-saving unit was closed down by the Kennett government, because it wanted to encourage electricity consumption to 'fatten' the industry for privatisation.[32]

At precisely the moment in history when greater government controls were needed to reduce electricity consumption and invest public money into renewable sources of electricity generation, private sector interests secured greater involvement in electricity generation—with obviously little interest in discouraging either electricity consumption or reducing the provision of electricity generated by the still plentiful supplies of black and brown coal. Had Victorian electricity supply remained in public hands, subsequent Victorian government attempts to increase the proportion of power generated from renewable sources might have been more successful. Privatisation of electricity has retarded the increase of solar energy as a proportion of overall energy consumption, a ludicrous trend in Australia's climate.

Where electricity remained in public hands, governments were able, if willing, to force reductions in emissions—as initiatives under the Carr Labor government in NSW demonstrated. In contrast to the Kennett government's privatisation of electricity supply that saw Victoria move backwards in terms of energy efficiency, NSW became the leader in energy-saving industry, with its retention of publicly owned electricity provision and the setting up of its Sustainable Energy Development Authority. By 1999 this Authority found that the growth rate of the sustainable energy industry was over 25% per annum, higher than that of other major industries in NSW, such as tourism and information technology.[33] The Carr government then established one of the first mandatory ETSs in the world, the Greenhouse Gas Reduction Scheme, under the Electricity Supply Act in 2003.[34] It set a per capita CO_2 cap from electricity at 5% below 1990 levels, which had saved about 40 megatonnes of CO_2 in total by 2008.[35] In 2004 Carr committed NSW 'to further reducing greenhouse emissions' and established a state Greenhouse Office.[36] However, in late 2011 the NSW Labor government sold its electricity assets for a paltry $5.3 billion, an amount slated to be diminished with associated costs, including about $1.5 billion in government funds to buy a new coal mine north-east of Lithgow to ensure a cheap coal supply for the new private owners.[37]

There is a clear contradiction in the way that governments affirm neo-liberal principles when they privatise utilities and refuse to regulate emissions directly, but deny these principles by gross interventionist favouritism towards resource industries. This double standard has greatly increased greenhouse gas emissions.

The Carbon Lobby

The clue to this double standard so manifestly at play is the power of the 'greenhouse mafia' or 'carbon lobby': a small cabal of lobbyists, including CEOs from the oil, coal, cement, aluminium, mining, and electricity industries, who have open access to politicians in both government and opposition.[38] Garnaut explains they operate by avoiding analytical discussion of issues and by relying on private briefings with ministers and senior political personnel.[39] Their most effective strategy is to conflate their own economic interests with Australia's national interests. According to Pearse, they realised that: 'The best way to avoid regulation and attract government support was to confuse the national interests with their own, which would require strong relationships with both sides of politics, with bureaucracies, think-tanks, industry associations, scientific and economic agencies, and media commentators.'[40] Crucial to carbon lobby power is its links with the media, especially Rupert Murdoch's News Limited. Indicative of media bias, a Media Monitors study found climate denier Lord Monckton's 2010 speaking tour received 455 media mentions, whereas the coinciding tour by famous climate scientist James Hansen scored only twenty-one mentions.[41]

Organisations associated with the greenhouse mafia include the Australian Coal Association (ACA), the Minerals Council of Australia (MCA), the Business Council of Australia (BCA) and Australian Industry Greenhouse Network (AIGN). A particularly influential individual greenhouse mafioso is Hugh Morgan. Former President of the BCA, Managing Director of Western Mining Corporation (WMC) 1986–2003 and a director of Alcoa 1977–2001, Morgan claims emissions reduction targets violate Australian sovereignty. He influenced the Howard governments to take a go-slow policy on climate change and oppose the Kyoto Protocol.[42] Gina Rinehardt is certainly the leading greenhouse mafiasa, using her wealth as a mining mogul to become a media mogul as well, to use this power to counter what she describes as the 'global warming fear campaign'.[43]

The ACA represents the interests of black-coal producers in NSW and Queensland and by the 1990s had become a major donor to both the Liberal and Labor parties.[44] The MCA is a lobby group based in Canberra to encourage the expansion of the mining industry. It represents mining and mining processing corporations that produce more than 85% of Australia's minerals production.[45] The BCA consists of the CEOs of about 100 major corporations and, according to Pusey, was 'writing business national budgets sometimes almost line by line' in the 1980s.[46] When public alarm about the greenhouse effect began providing impetus to green politics in the late 1980s, the BCA attacked the green movement for encouraging 'emotion, hysteria and hidden agendas' that worked against sustainable growth.[47] According to Ian Lowe, the BCA acts as 'the voice of the large companies that might lose most from curtailment of emissions, such as those in the coal and aluminium industries, overlooking the interests of those firms and industrial sectors that could benefit from a government strategy promoting conservation and alternative energy resources'.[48]

The AIGN constitutes the 'organisational form of the greenhouse mafia'.[49] The AIGN claims on its website that it consults with 'government and other opinion leaders on the adoption of a principled national and international greenhouse policy within which effective, equitable, efficient and appropriate domestic greenhouse gas abatement actions are developed and implemented with sustainable development'. Its members comprise the heaviest polluters in the country. Its listing of 'industry association members' and 'individual business members' reads like a *Who's Who* of the Australian corporate sector, particularly the fossil fuels industry, including: the Australian Aluminium Council, the ACA, the Australian Petroleum and Exploration Association, the MCA, the National Generators Forum, Alcoa World Alumina-Australia, BP Australia, Caltex, Chevron Australia, ExxonMobil Australia, Origin Energy, Rio Tinto Australia, Santos, Shell Australia, Woodside Petroleum and Xstrata Coal

Australia.[50] Both Liberals and Labor have received millions since 1998 from organisations directly or indirectly represented by AIGN.[51]

The greenhouse mafia recognises that direct involvement in denying the connection between fossil fuels and global warming is risky and that a better public relations exercise is to contract out the propaganda work. With the huge funds at its disposal, it has established or strengthened existing neo-liberal think-tanks and front groups to espouse a contrarian stance in relation to climate change.[52] Most of these refuse to disclose their funding sources, though the Competitive Enterprise Institute is funded by Exxon Mobil in Australia.[53] These think-tanks are important in what Peter Ferguson describes as the 'discursive struggle' prosecuted by neo-liberal spokespeople 'to undermine the claims of environmentalists in order to legitimise a set of ecologically and socially destructive corporate practices'. The success of this latest 'culture war' battle partly accounts for Australia's recent poor record on environmental issues, particularly the Howard government's climate change denial.[54]

The Institute of Public Affairs, formed in 1942 by various industry groups and which played a key role in the founding of the Liberal Party,[55] has become involved in climate change denial. In 1991 it began a public campaign arguing there was no scientific consensus on the greenhouse effect and ozone depletion.[56] Its closest political links are with the Coalition, but it has also attracted prominent Labourites, such as Gary Johns, a former Labor government minister, who became one of its senior fellows.[57] Other neo-liberal think-tanks that have waged denialist propaganda are the Committee for Economic Development of Australia, the Centre for Independent Studies, the Sydney Institute, the Tasman Institute and the Australian Institute for Public Policy.[58]

A recent example of a denialist front group is the Lavoisier Group, established by Hugh Morgan and Ray Evans in 1999. It promotes the interests of the mining industry and has been very hostile to the Australian Greenhouse Office. Its board includes Ian Webber, a former director of WMC and Santos; former director of WMC Harold Clough, whose money has supported a wide range of right-wing organisations and is a current board member of the Institute of Public Affairs; and former BHP executive Bob Foster. Peter

Walsh, a former Labor Minister, serves as president of the group.[59] The Lavoisier Group publishes books such as *Thank God for Carbon* and *Nine Lies About Global Warming*.[60] According to Indymedia, Morgan launched the Lavoisier Group to 'legitimise his claims that climate change is the product of green extremists and "nazi propaganda" '.[61]

While these think-tanks and front groups provide supportive background noise, the greenhouse mafia's most effective tactic is direct lobbying, and it was there at the beginning of climate politics in Canberra. Joan Staples argues that the Labor cabinet decision in 1990 to accept a 20% cut in emissions on 1988 levels by 2005 only on condition that it did not adversely affect the economy indicated greenhouse mafia mobilisation.[62] Pearse agrees it began to mobilise in the early 1990s:

> Its members immediately recognised that any Australian commitment to absolute cuts in greenhouse-gas emissions was a problem. It threatened the place of cheap coal in Australian energy and trade policy, raised operating costs and could expose the fact that many commodities were produced with more emissions here than virtually anywhere else. These industries were under threat of losing their privileged position in Australia's political order, but they were well placed to defend it and they knew the issue would take decades to play out.[63]

Its influence and activity increased noticeably during these final years of the Hawke era, a momentum sustained during the Keating governments.

The Howard governments' climate change policies were largely dictated by the greenhouse mafia.[64] Pearse maintains Howard allowed his climate policy to be shaped by Australia's biggest polluters. Members of the greenhouse mafia were regularly consulted by government departments in the drafting of ministerial briefings on climate change; and the primary source of greenhouse economic advice came from ABARE, whose steering committee included the ACA, the Australian Aluminium Council, CRA, the BCA, Exxon Corporation, Mobil Australia and Texaco.[65] When the Howard government set up the Shergold Committee to investigate an ETS for Australia, a majority of its members were heads of companies that were big emitters or had a vested interest in avoiding punitive mechanisms to mitigate climate change; and there were no other interests of the Australian community represented on the committee.[66] Pearse observed that 'the carbon lobby was designing a carbon-trading scheme to suit themselves'.[67]

Greenhouse mafia influence did not decline dramatically after the defeat of the Howard government, though the carbon lobby did 'reinvent' itself suitably for dealing with a Labor government that had ratified the Kyoto Protocol: shifting away from outright climate change denial, attempting to incorporate mining industry unionists and stressing more than ever the alleged job losses attendant upon climate change mitigation.[68] Rudd appointed Sir Rod Eddington, who had served on the board of Rio Tinto, to serve as a key business advisor on climate policy; and many senior public servants who had been stanch supporters of Howard's climate policies retained their positions. Pearse notes: some shifted from the Department of Foreign Affairs and Trade to the new Department of Climate Change; some took jobs with the carbon-lobby Industry associations and economic consultancies, further bolstering the greenhouse mafia's position. 'Australia's most effective and notorious "iron triangle" of polluter, political and bureaucratic elites survived.'[69]

The Carbon Lobby, the CPRS and the Carbon Price

The form and fate of the CPRS was clearly influenced by the greenhouse mafia. The High Level Consultative Committee for the White Paper included representatives from the coal, oil, gas and uranium industries but none from renewable energy or energy efficiency operations.[70] Yet the greenhouse mafia is not necessarily representative of business interests in general. Corporations have varied, sometimes opposed, interests when it comes to climate change mitigation; business circles also sense that carbon pricing is an international trend, so the sooner it happens the better for investment decisions.

The business response to the White Paper was in general 'rather muted', according to Woodward, 'suggesting that they were happy with what they got but did not want to draw attention to the fact'. While the National Generators Forum argued the compensation package was inadequate, only the MCA was implacably opposed.[71] MCA CEO Mitchell Hooke stated the CPRS was overly ambitious, would risk 'severe economic contraction' and force EITE industries with the 'highest carbon costs in the world' to either 'scale down, shut down, or move offshore'.[72] By contrast, the Australian Industry Group (AIG), which mainly represents manufacturing capital and is not part of the greenhouse mafia, backed the CPRS but wanted two years' delay because of the GFC, a view echoed also by the Australian Chamber of Commerce and Industry (ACCI).[73]

The government responded to greenhouse mafia pressure making even more concessions to its interests when reformulating the CPRS. The AIG, BCA and Aluminium Council grudgingly approved the changes.[74] The AIGN pretended support-in-principle for emissions trading as a framework for 'least-cost abatement of greenhouse gases' but insisted the CPRS would not achieve this, particularly for EITE industries; and cautioned that 'Australia will lose existing and new jobs and investment to other countries for no global environmental gain'.[75] The MCA remained firmly opposed, while requesting yet more concessions and delays in implementation.[76] Of the final form, Garnaut commented 'the big emitters were given so much'[77] and concluded that 'a massive lobbying exercise by vested interests, unprecedented in the history of Australian public policy, has secured overly generous deals for business'.[78]

But still they lobbied, until the CPRS was abandoned. Kate Crowley's study of the Rudd government's backflip concludes that the most compelling explanation is the structure of the Australian economy and the influential lobbying of the industries likely to be affected. 'While popular and political commentators have long pointed to the constraining influence of a greenhouse mafia over Australian climate change policy, the influence became overtly apparent during the ETS debate.'[79]

When the CPRS was shelved in April 2010, business groups expressed nervousness about the uncertainty created. The CEO of the Electricity Supply Association of Australia, representing the country's biggest power generators, stated: 'The consequences will be quite dire if people can't make investment decisions.' Another senior energy industry executive even said the Greens' proposal of an interim carbon price of $23 per tonne, rising each year by 4% plus the consumer price index, was a 'perfectly valid' response that would stop coal-fired power stations being built and bring on a conversion to gas. The BCA also met with the Greens. It would not make public comment, but Greens Senator Christine Milne stated after the meeting that business should be concerned, because 'if a carbon price is not to come in in this way, then the logical next step is regulation'.[80] The AIG said industry remained committed to reducing emissions. Renewable energy companies were concerned, stressing that the uncertainty about the future of the scheme would hamper investment in renewable energy projects. A spokesman for Origin Energy said: 'Ongoing uncertainty risks delaying both the investment necessary to meet Australia's long-term base-load electricity needs and the investment in lower-carbon technology required to gradually reduce Australia's emissions.'[81]

The insatiable tendencies of the greenhouse mafia, evident during the CPRS negotiations, were revealed spectacularly by the removal of Rudd as Prime Minister shortly after he announced plans to legislate the Resource Super Profits Tax (RSPT). Mining companies immediately and expensively mounted a media scare campaign about job losses and economic downturn if the RSPT were to proceed. Mining magnates Clive Palmer and Andrew Forrest bragged that they had orchestrated ousting Rudd from his position as Prime Minister.[82] Treasury Secretary Ken Henry hinted that the defeat of the RSPT was the work of rent-seeking mining companies who, even as they announced an investment strike, were redoubling their efforts to exploit Australia's mineral riches.[83] The secretary of the union representing mine workers agrees with this assessment.[84]

Gillard's Mineral Resource Rent Tax did not pose the same threat to mining super-profits. In arriving at this new arrangement, Gillard had entered into negotiations with the three largest mining corporations: BHP Billiton, Rio Tinto and Xstrata. A Treasury report indicated that the reformulated mining tax resulted in a forfeiture of $60 billion compared with what the Rudd-proposed mining tax would have brought into government coffers.[85]

On 15 September 2010, Marius Kloppers, CEO of BHP, became front-page news when he called for a carbon price when addressing a business luncheon group in Sydney. His rationale was that such pricing was inevitable and the sooner it was done the better for the international competitiveness of interested Australian corporations.[86] The government was indeed preparing for such a policy, as part of the deal brokered with Greens MP Adam Bandt, whose support in the House of Representatives was crucial to the formation of the new Gillard government. The government and greenhouse mafia organisations appeared at this stage to be lobbying each other. For instance, while the BCA and the AIGN continued to oppose a price on carbon, Combet delivered an address at the AIGN/BCA Carbon Pricing Forum at Parliament House on 23 March 2011.[87]

Interested industry organisations continued to assert support for climate change mitigation while objecting to specific proposals. On 26 May 2011 the BCA submission on the proposed carbon pricing mechanism reiterated its opposition to the CPRS but asserted it was 'committed to a multifaceted approach to reducing global greenhouse gas emissions using a market-based mechanism such as an emissions trading scheme, as well as research and development of new technologies and improving Australia's adaptation capacities'. However, despite this declared commitment to an ETS, it rejected carbon pricing on the grounds that the 'introduction of any policy to price greenhouse gas emissions will mean additional costs that will flow through to all businesses and the consumers at a time when the Australian economy is experiencing uneven growth'.[88]

When the details of the carbon price were announced by Gillard on 'Carbon Sunday', 10 July 2011, the responses were predictable. The ACA placed a full-page advertisement in the *Age* maintaining that 'No other major coal exporter has a carbon tax. Not one,' listing Canada, South Africa, Indonesia, the US, Russia and Colombia. The ad went on to state: 'The Australian coal industry supports action on climate change. However, the Greens and Government already know the proposed carbon tax will cost thousands of jobs and harm exports.'[89] *Sydney Morning Herald* columnist Ross Gittens and Citibank questioned this.[90] The ACA also released a report entitled 'Impact of Proposed Carbon Tax on Black Coal' that asserted the proposed carbon price would force eight coal mines to close. (This could be regarded as proof of its effectiveness.) For the MCA, Hooke described the proposed carbon price as 'an exercise in revenue-churn futility, not a credible or effective climate change policy'. The BCA characterised it as a 'risky' scheme.[91] BCA CEO Jennifer

Westacott stated: 'The government has not considered alternative ways to introduce an emissions trading scheme that would pose less risk to the economy'.[92] AIG CEO Heather Ridout stated: 'Our initial evaluation of the federal government's climate change package is that large parts of industry, already under stress, will feel their concerns have not been addressed and their long-term competitiveness will be damaged.'[93]At ground level, a spokesperson for International Power, the operator of the Hazelwood plant, announced the company was open to discussion with the government about plans to shut down 2000 megawatts of coal-fired power generation by the end of the decade. This company recently purchased Hazelwood cheaply and is poised to profit considerably from its impending enforced shut down.[94]

Green Capitalism and the Carbon-Trading Lobby

In recent years, in Australia and around the world, many corporate elites have come to recognise the reality of climate change and that it is in the interests of capitalism to respond; and that failure to do so makes for poor public relations and contradicts claims of social responsibility on the part of corporations. All corporations have an interest in good public relations. Even greenhouse mafia companies have paid lipservice to 'sustainable development'. For Rio Tinto this provided 'business reputational benefits' that resulted in greater access to land, human and financial resources; BHP Billiton found it was 'about ensuring our business remains visible'.[95]

Some corporations have a real material interest in climate change mitigation; their support for emissions reduction is genuine. The rise of green capitalism demonstrates capitalism's ability to adapt to new circumstances: seizing climate change concern and mitigation as a business opportunity. Bunting notes that some corporations are looking 'to enhance their competitiveness through the innovation of environmental technologies', a change in thinking that has only recently started to permeate the Australian business community, but was embraced in some European countries much earlier.[96] In June 2008 a report from huge transnational financial institution HSBC showed shares in companies taking action to reduce emissions and develop products and services to cut pollution were 'outperforming the market'.[97] More cynically, on 8 July 2011 a group of New Zealand business leaders launched their 'Pure Advantage' campaign: 'Even if you don't believe in climate change, there's money to be made doing something about it. There's money in being green and we need to start turning Green Growth into wealth.'[98]

Green capitalism is characterised by sincere support for 'sustainable development' and the practical embrace of technological improvements that reduce emissions: renewable energy sources, such as solar energy, wind energy, geothermal energy, wave or tidal energy, and hydropower; technological advances, such as electric and hybrid cars; more energy-efficient appliances, such as the new form of light bulbs; and ecologically better-designed commercial and residential buildings. It looks favourably upon government schemes that enable green technologies to compete with brown technologies by pricing carbon and mandating energy-efficient building regulations and codes for products such as whitegoods. A price on carbon is considered essential by corporations engaged in the development of renewable energy sources, because financial institutions will not provide finance for ventures that cannot provide energy as cheaply as energy sourced from coal. Green capitalism requires a greener state, because the market does not spontaneously give cleaner, greener technologies a competitive edge.

It might be assumed that green capitalism invented renewable energy sources, given the importance of such technologies to the green capitalist agenda. In fact, advocates of 'alternative technology' have been with us a long time. Enthusiasm for renewable energy sources (and the construction of shelters using simple, locally available materials, such as mud) is associated with the radical hippy moment of the 1970s; and it is because renewable sources of energy are normally uncompetitive within capitalist societies that their use has until recently been associated with social marginality and political extremism. The early alternative technology advocates of the 1970s often promoted ideas subversive of capitalism, such as workers' control of industry. Many in this alternative technology movement 'were trying to create a better society', according to Bunting, but later sought to use alternative technology 'within the existing society'.[99]

Thanks to public concern and political pressures about climate change, alternative technology in the form of renewable sources of energy has become a business opportunity. So, although associated originally with the hippies, alternative technology advocates are nowadays more likely to have an MBA than live on a commune. Nonetheless, many climate activists feel clean energy entrepeneurs are insufficiently pushy. According to Pearse, some sections of ENGOs are frustrated with the lack of active, visible lobbying by the clean energy industry, because these ENGOs feel these businesses would be more likely to have political influence in the policy processes than would ENGOs.[100] According to Agnes Koh, this is unlikely: 'Comprising many small and medium enterprises, the renewable energy sector lacks the economic power needed to effectively participate in political processes for the furtherance or protection of their interests.'[101]

The peak organisations that sponsor renewable energy are respectable and well resourced, but considerably less powerful and wealthy than those associated with the greenhouse mafia. One is the Clean Energy Council, formed as a merger of the Australian Business Council for Sustainable Energy and the Australian Wind Energy Industry Association. Its members seek to provide Australia energy with lower emissions than conventional energy sources.[102]

The wind energy industry exemplifies the tensions and contradictions in the renewable energy sector. Climate campaigners and wind energy industry leaders are often allies in their support for wind energy. However, their motivations are different. As Bunting notes, 'the wind energy industry is principally motivated by commercial imperatives'. A second contradiction is that fossil fuel interests have been moving into the wind energy industry.[103] Many companies involved in wind and other forms of renewable energy still have an interest in fossil fuels and mining activities. 'And while energy sources may be renewable, the manufacture, usage and disposal of the energy conversion technologies may require the use of non-renewable materials, or may be environmentally damaging or energy-intensive.'[104]

Fierce competition in the automobile industry to develop the electric car also illustrates the problematic nature of green capitalism. The *Wall Street Journal* refers to 'the auto industry's high-stakes race to crack the market for electric cars'.[105] Arguably, such a technological improvement and the hopes invested in it encourage continued widespread use in the future of the private motor vehicle—and thereby distract attention and funding away from improved public transport. The electric car, far from being an ecological good, might have adverse effects if developed as the means to continue with transport-as-usual. For example, Better Place Australia is 'part of a global company dedicated to zero emissions driving' that is promoting the 'transition to electric vehicles (EVs) in Australia by providing the infrastructure and services that make it easy, affordable and attractive for motorists to adopt and drive electric vehicles.'[106]

Another dubious aspect of green capitalism is the obvious material interest of its close political allies: the 'carbon-trading lobby'. These are the businesses, such as accounting and law firms, which benefit from legislation that creates a trade in emissions permits. Emissions trading schemes elsewhere in the world have created billions of dollars in rights and assets.[107] Stuart Rosewarne shows how this 'ostensibly simple solution for addressing market failure is far from a costless exercise', involving as it does a complex institutional architecture to oversee trading, entailing substantial costs. 'The irony here is that the institutional transformation that is being proposed has resulted in more energy, and funds, being invested in establishing measures to expand the possibilities for polluters to trade their way to meeting designated emissions reduction targets.'[108]

In Australia, the carbon-trading lobby was poised to profit from the CPRS and will undoubtedly benefit from the intricacies for corporations of responding to the carbon price and the amount of accounting and auditing required. Koh describes the carbon-trading lobby as 'those who see a buck to be made in the price on carbon'. She bases her analysis on private communications from Pearse, Garnaut and Shadow Minister for Climate Change Greg Hunt. Hunt said of emissions trading: 'You've got the banking, trading and services sector. They're like, "Gosh we really want this, because we can make an awful lot of money trading permits".'[109]

However, as Koh stresses, the carbon-trading lobby's activity and influence 'pales in comparison to the carbon lobby's coordinated political participation'; and this comparative lack of influence prompts it to cooperate with ENGOs to work towards market-based mitigation policies.[110] Pearse informed her: 'It's basically the carbon-trading lobby that is funding the [climate policy, renewable energy, carbon-trading] conferences around Australia ... one is likely to find the same suspects, who are looking to make a dollar out of the carbon trade.'[111] Garnaut confirmed the carbon-trading lobby's relative lack of influence compared with the carbon lobby: 'Businesses that are operating now, and making profits now are much more powerful in the political processes than businesses which have an opportunity in the future ... The established interests always have a big advantage in the political process over those who will benefit from reform.'[112]

Whether or not such legislation succeeds in reducing emissions does not necessarily concern this carbon-trading lobby. Such businesses also benefit from other forms of environmental legislation, such as energy-efficiency regulations and codes. Again, their real effectiveness is not the important issue for such firms. Their primary concern—and their source of income—lies in assisting other corporations to comply as cheaply as possible with mitigation policies and supporting governments to implement and administer them.

Scholars of environmental politics locate green capitalism initiatives within the paradigm of 'ecological modernisation'. This school of thought constructs the climate change problem as one created by inefficiency, so proposes better technological and institutional design as the core of an effective response.[113] While innovations and improvements are helpful in climate change mitigation, it is the obsessive focus on them as proof of the ingenuity of capitalism that characterises ecological modernisation: 'a strong tendency towards hardware fetishism', in Bunting's words.[114] By emphasising such wondrous developments, ecological modernisation is able to concede that environmental problems are a by-product of global capitalism but also to reject any argument that a healthy planet requires transcendence or transformation of capitalism. For example, Michael Howes argues that using a strong ecological modernisation approach can overcome resistance to carbon trading by selling the change as a win for both business and the environment.[115]

An excellent example of an eco-modernist vision for Australia is articulated in Ben McNeil's *The Clean Industrial Revolution* (2009). McNeil, a senior research fellow at the Climate Change Research Centre at the University of New South Wales, argues that business as usual will not solve climate change, because businesses always try to maximise profit at the expense of other considerations; but with modifications, the Australian market economy could become part of the solution, because 'imposing a strong and broad carbon price in the Australian economy will drive new innovative climate-change-solving start-ups'.[116] With a carbon price in place, there are potentially tens of trillions of dollars to be made from investment and developments in a low-carbon economy; and once carbon emissions affect profits, capitalists will become 'accidental environmentalists'. So the 'new generation of climate-solvers must seek to harness, rather than hinder, markets and technologies to reduce greenhouse gas emissions'; and environmentalists must view economic growth as a positive feature, rather than necessarily a contributor to climate change, because 'cutting emissions and growing the economy are not independent of one another'.[117] Public funding for clean energy research and development can also contribute to the creation of a vibrant Australian economy, which would break its addition to coal and promote wind, solar and geothermal as part and parcel of the creation of a 'carbon-neutral energy supply', a win–win situation for both environment and business.[118]

The Climate Group, established in Australia in 2005, is another example of an eco-modernist approach to climate change. John Thwaites, Victoria's first Minister for Climate Change and Deputy Premier from 1999 to 2007, serves as Chairperson. Tim Flannery is Vice-Chairperson and other board members include Andrew Stock, a general manager with Origin Energy, and Jim Walker, a consultant on public and private sector responses to environmental issues.[119] The Climate Group's two key aims are 'Bringing business and government together' and 'Accelerating the uptake of new technologies'. Its website explains: 'We continue to forge partnerships at the highest levels. We work closely with Australian state governments, developing new policies and leadership that will fast-track the low-carbon economy, bringing long-term benefits to Australian businesses and the community.'[120] Its enthusiasm for new technologies includes CCS.[121]

Critics of ecological modernisation see much hype and not much substance in terms of mitigating climate change; ecological modernisation insists, in effect, that capitalism is capable of becoming sustainable by utilising the profit motive that is the cause of environmental problems. Craig Dilworth observes: 'The idea of an ecological capitalism, or a sustainable capitalism, has not even been coherently conceptualised, not to speak of becoming embodied in an institutional infrastructure.' The notion ultimately privileges profit-making and economic expansion over environmental sustainability, the aspects of capitalism that have caused climate change to commence. In other words, as climate justice activists argue: 'How can you expect the system that created the problem to solve it!'[122] At an even deeper level, Ariel Salleh argues that eco-modernist policies impose ecological, social and embodied debts at the periphery of capital; and it is time for climate activists to validate the leadership of communities in the global South who have pioneered low-carbon, nature attuned, eco-sufficient economies by rejecting the strategy of ecological modernisation in favour of the strategy of 'Living Well'.[123]

Crucially, ecological modernisation assumes that societies will develop 'green states' in which environmental regulation will produce a sustainable future.[124] However, green states might not be capable of realisation, because the state in capitalist society is always concerned that private enterprise prospers. Eckersley's arguments about 'Greening the Modern State' are ultimately pessimistic, because 'only the state can tackle the ecological myopia of the market', and yet 'any concerted attempt to regulate private investment and management decisions to a point where negative ecological externalities are eliminated (as distinct from merely ameliorated) would be deeply inimical to the interests of private capital and likely to lead to a capital strike or flight'. Governments, she concludes, simply cannot risk serious economic dislocation or a cessation of growth, as either of these is likely to bring about their demise. She argues that the 'contradictory imperatives'—to appease public concern over ecological degradation and maintain private capital accumulation—are not resolvable. 'They can only ever be managed, and usually only at the margins through incremental change.'[125] In *The Green State* in 2004, Eckersley again identifies the role of the state in promoting capital accumulation as one of the principal obstacles to the development of the green state, yet recommends the green movement embrace the *possibility* of a green state in order to move closer towards its vision of a socially just and ecologically sustainable society.[126]

The record so far invites pessimism about the extent to which the state in capitalist societies is able to be greened. States the world over have preferred market-based mitigation policies rather than direct regulation to reduce emissions—and there is inconclusive evidence at best about the effectiveness of market-based schemes. An ETS simply commodifies the 'external costs' of pollution and makes these costs tradeable. So industries, especially in developed countries, will buy the right to pollute, if profitable, instead of investing in greener solutions and technological improvements. In the EU in 2009 the ETS saw the price of carbon pollution valued at only seven euros a tonne.[127] There have also been problems with accountability and transparency. The EU's 'recycled' carbon permit scandal of March 2010 is a notable example of this, providing evidence of the potential for corruption in such schemes.[128]

Many in the climate movement accept the premises of ecological modernisation, placing their hopes in green capitalism facilitated by a green state, ignoring the problematic nature of any sort of capitalism in addressing not only climate justice issues but also those of simple environmental sustainability. While green capitalism seeks to incorporate 'externalities' such as greenhouse gas emissions into its economic calculations, it is oblivious to issues such as population growth driven by poverty and the reality of limited natural resources.

Moreover, green capitalism is almost entirely a phenomenon of the developed world, unable to have any real impact on the problem of ever-increasing industrial production in the developing world, particularly China and India—parts of the world where multinational corporations are relocating their most polluting operations, precisely because environmental protection laws are weakest. Arguably, green capitalism would contribute to this tendency to the extent that it succeeds in greening the state in the developed world. The point here is that capitalism is crafty, remorselessly seeking locations with the fewest restrictions on how commodities are produced—so even direct regulation might not work unless imposed globally. As a system that spreads itself to the furthest corners of the globe in the pursuit of profit, attempts to ecologically modernise global capitalism do little more than create niche marketing opportunities for a limited number of companies in developed economies. Climate justice activists therefore argue the need to challenge global capitalism as an entire system rather than simply green the state one is in.

Notes

1 Pusey, *Economic Rationalism in Canberra*; Pusey, *The Experience of Middle Australia*, p. 4; Kapferer, p. 145.
2 Conley, 'The domestic politics of globalisation', p.225.
3 Pusey, *The Experience of Middle Australia*, p. 164.
4 Burgmann, 'Moloch's Little Mate'.
5 Burgmann, *Power, Profit and Protest*, p. 324; S Pietsch, p. 29; Brett, pp. 48–53.
6 Fagan and Bryan, p. 27.
7 Bunting, p. 160.
8 Diesendorf, *Greenhouse Solutions*, pp. 289–92.
9 Pearse, 'Quarry Vision', p. 15.
10 Davidson, 'Editorial', p. 7.
11 Supple, 'Billions'.
12 Angel, p. 130.
13 Hickson, p. 66.
14 Kevin, p. 166.
15 Woodward, p. 4; Morton, 'Coming clean'.
16 Australian Government, *CPRS—White Paper*, p. xl–xlv; Kevin, p. 144.
17 Butler, 'The false hope'.
18 Hamilton, *Requiem*, p. 162.
19 Angel, p. 129; http://www.newgencoal.com.au (viewed 15 January 2010).
20 Kevin, p. 146.
21 Davidson, 'Editorial', p. 6.
22 Keane, p. 1; Davis, *The Land of Plenty*, p. 196; Hamilton and Turton, p. 2.
23 Diesendorf, *Climate Action*, p. 60; Keane. p. 2.
24 Victorian Government, p. 12.
25 Gordon, 'ALP energy targets'.
26 Diesendorf, *Climate Action*, p. 26.
27 Supple, 'Obsession'.
28 *Age*, 10 June 2009, p. 1; Arup and Morton, 'Garrett pulls plug'.
29 Morton and Arup, 'Solar panel companies'.
30 'Save Solar Systems Rally', poster, n.d. [2009]; Yeates.
31 Bunting, p. 24.
32 Davidson, 'Bush'.
33 Davidson, 'Bush'.
34 Brohe, Eyre and Howarth, p. 207.
35 Angel, p. 229.
36 Firth, p. 310.
37 'Duroyan. Inquiry exposes power rip-offs', *GLW*, 26 January 2011, p. 5.
38 Mackey et al.
39 Ross Garnaut, Interview with Agnes Koh, 19 April 2010, cited in Koh, p. 38.
40 Pearse, 'Quarry Vision', p. 31.
41 Benedek, p. 12.
42 Wiseman, p. 98; Hamilton, *Scorcher*, p. 133.
43 Benedek, pp. 3, 12.
44 Wikipedia, 'The Australian Coal Association' (viewed 7 October 2008).

45 Coates, p. 1.
46 Pusey, *The Experience of Middle Australia*, p. 11.
47 *Weekend Australian*, 23–24 September 1989, p. 12; BCA, 'Strategic Framework'.
48 Lowe, 'The greenhouse effect', p. 326.
49 Hamilton, *Scorcher*, p. 134.
50 http://www.aign.net.au (viewed 24 February 2010).
51 Pearse, 'Quarry vision', p. 38.
52 Pearse, 'Quarry vision', pp. 34–36.
53 Davis, *The Land of Plenty*, p. 191.
54 Ferguson, pp. 289, 293.
55 S Pietsch, p. 27.
56 Lowe, 'The greenhouse effect', p. 321.
57 Diesendorf, *Climate Action*, p. 37.
58 Pusey, *The Experience of Middle Australia*, p. 11.
59 Hamilton, *Scorcher*, pp. 139, 142.
60 Evans.
61 'Australia: climate change, corporations and PR', Indymedia Climate, 28 November 2005. http://www.climateimc.org (viewed 22 February 2009).
62 Staples, esp. p. 112.
63 Pearse, 'Quarry vision', p. 31.
64 Maddison and Hamilton, 'Non-government organisations', p. 84; Pearse, *High and Dry*; Hamilton, *Scorcher*.
65 Pearse, 'Quarry vision', p. 32; Firth, p. 310.
66 The members included: Xstrata's Peter Coates; the managing director of International Power, Tony Concannon; the director of Australian Pipeline Trust, Russell Higgins; the executive director of BHP Billiton, Chris Lynch; the chief of Alumina Limited, John Marlay; and National Australia Bank's John Stewart. (Ross Garnaut, personal communication to Agnes Koh, 19 April 2010, cited in Koh, pp. 36–7.)
67 Guy Pearse, Interview with Agnes Koh, 16 February 2010, quoted in Koh, p. 37.
68 Koh, pp. 38, 40.
69 Pearse, 'Quarry Vision', pp. 45, 48.
70 Diesendorf, *Climate Action*, p. 28.
71 Woodward, p. 5.
72 Hooke, 'Media release', p. 1.
73 *Age*, 27 February 2009, p. 1; 28 February 2009, p. 1; Breen and Thomas, p. 9.
74 *Age*, 5 May 2009, p. 1.
75 Quoted in Koh, p. 39.
76 'Business may never see such a deal again', *Weekend Australian*, 5–6 December 2009, Finance pp. 22–23.
77 Ross Garnaut, Interview with Agnes Koh, 19 April 2010, cited in Koh, p. 38.
78 Garnaut, 'Australia counts itself out'.
79 Crowley, 'Climate policy failure', p. 12.

80 Manning, 'One climate policy'.
81 Grattan and Arup, 'Rudd in ETS backflip'.
82 'Gillard caves in to big end of town', *GLW*, 7 July 2010, p, 3.
83 *Weekend Australian*, 10–11 July 2010, p. 12.
84 Tony Maher, Speech to Australia Institute forum, 6 August 2010, reported in *Search News*, vol. 17, no. 4, August 2010, pp. 2–3.
85 Thomas, 'Carbon price plan'.
86 Courtice, 'Labor Party'; *Age*, 16 September 2010, p. 1.
87 AIGN/BCA, Carbon Pricing Forum Program.
88 BCA, Submission on the proposed architecture and implementation arrangements for a carbon pricing mechanism, letter from Jennifer Westacott to Greg Combet, 26 May 2011.
89 S Maher, 'Gillard's clean energy crusade'.
90 Alcorn, 'Coal industry job scare baseless'.
91 S Maher, 'Gillard's clean energy crusade'.
92 *Australian*, 11 July 2011, p. 8.
93 *Age*, 11 July 2011, p. 6.
94 Arup, 'Plant owners'.
95 Quoted in Kirsch, p. 91.
96 Bunting, p. 239.
97 Jowit and Wintour.
98 Quoted in Bradford.
99 Bunting, pp. 95, 98.
100 Guy Pearse, Interview with Agnes Koh, 16 February 2010, quoted in Koh, pp. 50–51
101 Koh, p. 64.
102 http://www.cleanenergycouncil.org.au (viewed 1 June 2009); Hickson, p. 30; Bunting, p. 165.
103 Bunting, p. 58.
104 ibid., p. 246.
105 *Wall Street Journal*, 7 January 2011, Marketplace, p. B1.
106 http://www.betterplace.com/australia (viewed 1 August 2009).
107 Baer and Singer, p. 180.
108 Rosewarne, pp. 46–7.
109 Greg Hunt, Interview with Agnes Koh, 18 March 2010, quoted in Koh, p. 43.
110 Koh, pp. 43–4.
111 Guy Pearse, Interview with Agnes Koh, 16 February 2010, quoted in Koh, p. 44.
112 Ross Garnaut, Interview with Agnes Koh, 19 April 2010, quoted in Koh, p. 44.
113 Howes.
114 Bunting, p. 105.
115 Howes.
116 McNeil, *The Clean Industrial Revolution*, p. 204.
117 ibid., pp. 77, 80, 168.
118 ibid., pp. 124, 138.

119 http://www.theclimategroup.org (viewed 22 February 2009).
120 http://www.theclimategroup.org/programs/australia/ (viewed 5 October 2011).
121 http://www.theclimategroup.org/our-news/news/2011/9/22/the-critical-role-of-ccs-in-the-clean-revolution/ (viewed 5 October 2011).
122 Dilworth, p. 398.
123 Salleh, 'Climate strategy', pp. 138–9.
124 Clark and York, p. 394.
125 Eckersley, 'Greening the Modern State', pp. 82, 102.
126 Eckersley, *The Green State*, esp. p. 11.
127 Baer and Singer, p. 180.
128 http://www/theaustralian.com.au/business/news/European-emission-trading-rocked-by-scandal-over-recycled-carbon-permits/story-e6frg90o-1225842148852 (viewed 15 April 2010).

Part II
Climate Politics in the Middle

6

The Australian Greens

The Australian Greens (Greens) have become an increasingly significant presence in parliamentary politics over the past decade. The major players in the middle layer of climate politics, they also operate to some extent from above, since achieving a 'balance of power' position at the 2010 election. Currently, the Greens have: nine Senators and one House of Representatives member in federal parliament; twenty-four representatives in state and territory parliaments; and more than a hundred local councillors. This chapter examines the Greens' policies; how concern about both social justice issues and climate change have increased electoral support for the Greens; and how they have argued and advanced their position on climate change in the public domain.

Social Movement, Social Justice, Social Democracy: 1992–2007

According to Narelle Miragliotta, the issues which catalysed activists to form the Greens in 1992 were similar to those in Europe, notably the peace movement and 'the broader social justice movement'.[1] Stewart Jackson agrees the Greens started 'as a grassroots party with its roots in new social movements'.[2] The social movement origins of the Greens were certainly evident in the early days of party formation when they argued that parliamentary politics should never be regarded as the main game but simply a necessary addition to other forms of activism.[3] A virtue might have been made of necessity. It was 1996 before Bob Brown was elected to the Senate. He has been a Tasmanian Senator and Greens' leader ever since, only gradually being joined by other Greens.

The Greens have long been on the case of climate change, its top priority.[4] When Brown spoke of the seriousness of climate change in the Senate in 1996, Labor and Liberal members laughed at him.[5] Also in 1996, Brown and Peter Singer argued:

> To get Australia back on track with the Toronto target, the Greens would follow the lead of those European nations that have introduced carbon taxes. This would give a boost to energy efficiency and the use of renewable energy like wind and solar power. We would compensate low-income earners for the disproportionate burden that the carbon tax may cause, if they drive older, less fuel-efficient vehicles, or are unable to afford to insulate their houses or switch to more efficient methods of heating. On the whole, though, carbon taxes are fair because they are a way of making the price of energy include the costs it poses on everyone else.[6]

However, the Greens' initial electoral advance was caused by public concern about social justice issues the Labor Party was no longer championing in the early twenty-first century. At the 10 November 2001 federal election, the Greens stood out for opposition to the Afghanistan war and mandatory detention of asylum seekers—and more than doubled their vote.[7] Brown commented: 'Had the Labor Party taken on Mr Howard over his inhumane policy on asylum seekers, it would be in office today.'[8] According to Mike Steketee: 'Labor MPs and officials concede privately what the Greens are claiming: that left-wing Labor voters told them they were protesting against Labor's support for the Government's policy on asylum seekers and the war in Afghanistan.'[9]

The Greens' longstanding commitment to peace and humanitarianism, economic equality and social justice, enabled them to capture support from Labor Party voters who felt the Greens' program more closely resembled that of the Labor Party they had once known but no longer loved. Tad Tietze argues the rise of the Greens has to be understood in the context of a crisis in social democracy, because the Greens have attracted supporters by promulgating social democratic policies abandoned by Labor and have therefore been the main electoral beneficiary of discontent with Labor.[10] Ben Spies–Butcher's analysis of data from the 2004 and 2007 federal elections found that Greens voters' political attitudes were on average the most left of any party's voters and they were most likely to identify as left-of-centre. Damien Cahill and Stephen Brown likewise found that Greens voters are motivated by 'traditional social democratic agenda': they favour redistribution of wealth and increased social spending rather than tax cuts.[11]

A disaffected Labor voter examining Greens' principles, policies and parliamentary speeches would conclude they were green social democrats. Their guiding principles are: 'Peace & Non Violence'; 'Grassroots Democracy'; 'Social & Economic Justice'; and 'Ecological Sustainability'. The Greens argue that excluding broader social, economic and political issues prevents the opportunity to deal with the cause, rather than just the symptoms, of environmental problems. Since their inception, the Greens have been involved in campaigns and policy initiatives promoting public transport, public health, public education and the rights of welfare recipients and trade unionists; and opposing privatisation and regressive taxation like the GST; and demanding much greater taxation on corporations, especially mining companies. Brown states: 'Our aim must be to accommodate both social and environmental wellbeing.'[12] The two are inextricably linked. For example, the Greens oppose privatisation of utilities, as a matter of social democratic principle but also because privatisation encourages energy consumption. Likewise, the Greens' sustainable transport policy calls for 'a comprehensive public transport system, with critical components publicly owned and controlled'—subsidised to discourage motor vehicle usage.[13]

The traction gained at the 2001 federal election was confirmed in a spectacular by-election in October 2002. Aided by preferences from an independent trade union candidate, the Greens' Michael Organ won the 'safe' Labor House of Representatives seat of

Cunningham in the Illawarra. Although this seat was lost at the 2004 federal election, the Greens continued their electoral advance. They received 7.67% of the primary vote: almost double that received in 2001. Their primary vote increased again in 2007, despite the relative popularity of Labor. The growth at each election from 2001 was due primarily to continuing defections from Labor, younger people entering the electoral rolls and the disintegration of the Australian Democrats.

Which type of former Labor voter turned Green? And who else assisted the Greens' initial electoral surge? Numerous studies pondered this question.[14] The strongest support for the Greens has come from people born after 1965 and people with a tertiary education. According to Jeff Sparrow: 'The massive bias in Green support towards youth suggests that, while some Green voters come from the new middle class, many are in fact on the lower rungs of their profession and as such part of the working class.'[15] Ben Spies-Butcher's analysis of data from the 2004 and 2007 federal elections did indeed find that Greens voters were more likely to be well-educated professionals but that, once education is controlled for, they were less likely to be of a higher-income bracket. Greens voters also tended to be younger, female and have no religion.[16] Yet the average age of a Greens voter in the 2004 election was forty-four.[17]

Brown, in particular, appeals strongly to left-wing voters. Described by his biographer as a 'gentle revolutionary',[18] he appears uncorrupted by the parliamentary club, remaining 'a genuinely decent politician'.[19] He argues that strong protests outside parliament, including law-breaking, strengthen the bargaining position of those negotiating on behalf of the environment inside parliament, 'where the vested interests of the wealthy often usurp public interest and commonsense'.[20] In September 2000, Brown joined with global justice protesters at the S11 protests against the World Economic Forum meeting at Melbourne's Crown Casino. Tietze recalls how Brown reinspired the demonstrators after picket lines were breached by violent police attacks.[21] Brown has also firmly supported the possibility of industrial action in the cause of climate change mitigation. In the Senate on 28 June 2007, he announced:

> Green Bans are going to become increasingly important as we head into an era of climate change over the next 10 years. The Greens policy is to allow workers to make climate change not just a household issue, but a workplace issue. The Greens have a very clear policy on this, that is to allow workers to have the internationally recognised right to strike for whatever matter they choose, if that's an environment matter, so be it.

Brown, unlike Democrats' founder Don Chipp, leads a party that has flourished from the bottom up with a strong grassroots base and a distinctive ideology and vision. Far from disintegrating as the Democrats did, it is much more likely to continue to grow. 'They wanted to keep the bastards honest', observes Brown. 'We want to replace them.'[22]

The 'Continue Polluting Regardless Scheme': 2007–2010

From an electoral advance based primarily on the defection of Labor voters concerned about social justice issues, the more recent increase in Greens' support is based on growing anxiety about climate change. It has not been difficult for the Greens to position themselves electorally as the only party genuinely committed to effective policies to combat global warming. When the Garnaut Review was released in 2008, there was an immediate leap in support for the Greens. Gary Morgan's headline announcing the poll result was 'ALP maintain strong lead over Coalition as Garnaut Report gives boost to Greens'. Morgan reported that 'Among the minor parties, support for the Greens (up 3.5% to 11.5%) is at a new record high after the release of the Garnaut Report' and it concluded 'The debate has caused an increase in support for the Greens.'

The Greens have benefitted from the major parties' shared perception that too much action on climate change is bad for political business. Deputy Leader Senator Christine Milne commented in 2008:

> Over recent decades, a political orthodoxy has been built up that environmental action will be economically damaging, and climate action most of all, and that protection of

> short-term economic interests sits at the very top of all priorities lists. There is a very strong associated assumption that any government that dares to take serious action to reduce greenhouse emissions, and suffers the assumed economic consequences, will be voted out of office quick smart.[23]

Membership of the Greens has grown along with their increasing electoral support. Growing from about 1500 members in 1999, they now have 10 000 members.[24] A survey of activists members in 2008 found the average age was fifty-two, they were 52% male and 48% female, and 77% of them had a university degree.[25] Growth and success have subjected the Greens to significant pressures to change, as Jackson observes, and they are now 'moving from a social movement party to one more resembling an electoral professional party'.[26] The Greens nonetheless retain strong connections with the grassroots climate movement, which serves as a counterweight to pressures exerted by parliamentary processes to become more pragmatic.

The Greens made it clear from the outset of deliberations on the CPRS that they would only support a serious and well-designed ETS. In the lead-up to the release of the legislation, the Greens' website stated:

> Emissions trading is one of the many measures that the Greens have long been campaigning for to put Australia on track to build a new, post-carbon economy. But, in order to be effective, it must be carefully designed to avoid rorting of the system by polluters, and must be one of a package of complementary measures rather than being seen as a 'silver bullet'.
>
> Emissions trading can be an effective tool to achieve two important aspects of the climate policy framework: putting a *price* and a *cap* on carbon emissions. Where a carbon tax is technically simpler (it sets the price and lets the individual players decide how much to reduce emissions), emissions trading has the benefit of providing environmental certainty (it sets a cap on total allowed

pollution, establishes the rules, and lets the market determine the price of polluting).

> An emissions trading scheme (ETS) must be designed to reduce emissions at least cost and in the most equitable way. While the second and third points are vital to good design, emissions trading is about reducing emissions—if it does not achieve that goal, then it is worthless. The ETS must be environmentally credible first and foremost, and its design and its complementary measures must ensure equity and least cost emissions reductions.
>
> The key to ensuring an ETS is environmentally credible is to make the scheme's cap in line with a credible national target to reduce emissions. Once the cap is set, the design of the scheme must back up emissions reduction efforts rather than undermine them.
>
> The Greens believe that a bad emissions trading scheme can easily be counterproductive, reducing our ability to cut greenhouse emissions fast and effectively. The Greens will not support such an emissions trading scheme.[27]

When the detail of the CPRS legislation was revealed, the Greens asserted that it overlooked the risks associated with catastrophic climate change, handed billions of dollars to the 'big polluters' and undermined international efforts to avert global climatic and environmental disaster.[28]

As part of the debate on the CPRS, the Greens submitted a Safe Climate Bill in September 2009, which called for a more stringent ETS without free permits, compensation, and a higher carbon price:

> Where the Rudd Government's CPRS—the *Continue Polluting Regardless Scheme*—locks in failure on the climate crisis by sidelining the science and sandbagging old polluters at the expense of the sunrise industries, the Greens' *Safe Climate Bill* gives us a real chance of success by aiming for the goal we need to achieve and then setting out how to get there.
>
> Our goal is not simply to reduce carbon emissions. The true goal we must aim for is to pass on to our children,

> and our children's children, the safe climate that has nurtured us and made human civilisation possible.

The bill proposed that, 'in the context of a global agreement, Australia's emissions will be reduced to 40% below 1990 levels by 2020, and that the 350 ppm goal is enshrined in law'. It was based on the following principles: 'Making polluters pay, not paying polluters'; 'repowering Australia with renewable energy'; 'save energy, save money, save the planet'; 'travelling with a light footprint'; and 'protecting our green carbon'. It called for 'a massive transformation in the Australian economy ... that will require the creation of hundreds of thousands of new jobs', including a 'renewable energy revolution', 'infrastructure spending on public transport, not more roads', stopping the expansion of coal mines and coal exports and halting the destruction of native forests. Significantly, it stipulated that the bill 'puts equity at the heart of climate action, investing in upgrading homes of low-income Australians for energy efficiency, rolling out transport alternatives in disadvantaged areas, retraining workers in polluting industries for the clean jobs of the future, and funding climate adaptation and emissions cuts in developing countries'.[29]

The Greens Climate Change and Energy policy, finalised in November 2009, provided three pages of detailed proposals about how to achieve: net zero greenhouse gas emissions no later than 2050 with a minimum of 40% reduction on 1990 levels by 2020; the reversal of Australia's growing demand for energy through demand management and increased efficiency of supply and end-use; future energy needs to be met using sustainable, renewable energy sources; that all countries develop and meet emissions reduction targets; and actions to adapt Australia to the impacts of climate change. The principles that informed the policy included: 'the equity principle must be at the core of climate change negotiations and measures'; and 'the cost of reducing greenhouse gas emissions must be distributed fairly, both domestically and between nations'.[30]

The Greens continued to negotiate with the government in the hope that the final form of the CPRS legislation would be strong enough to support. The Morgan Poll released 15 January 2010 found 55% of Greens supporters approved of the CPRS.[31] They introduced twenty-two amendments to improve the legislation, for example, including an unconditional 25% reduction in emissions, the auctioning of all permits, greatly reduced compensation to EITE industries and the inclusion of transport.[32] Instead, the government preferred to treat with the Coalition parties, so the CPRS became further weakened. The Greens decided it was worse than no scheme at all, because, as they explained on their website, bad policy design could make the problem worse, by locking in failure in terms of emissions reduction. 'That is why the Greens could not support the Rudd government's CPRS when the government refused to accept any of our sensible amendments that would have turned the bill from a barrier to action into at least a small step in the right direction.'[33]

The Greens sought to persuade the government to break the deadlock with the Coalition by implementing an interim carbon tax or 'carbon price', a position supported by Garnaut. On 21 January 2010 the Greens announced this 'interim carbon price proposal', starting in July 2010 at $23 per tonne and rising each year 'to get Australia moving while international mechanisms are determined'.[34] Milne stated its advantage was that 'it sends a signal to industry that there will be a carbon price whether as a levy or as part of an emissions trading system—but it does not lock in a weak target' and it also required the cuts to be made domestically and not by way of 'imported permits'.[35] After this interim scheme, 'We can then discuss the longer term solutions Australia will need over the coming two years, secure in the knowledge that a carbon price is already in place, helping to unleash innovative and job creating climate solutions'.[36] The Greens' proposal eliminated compensation to coal-fired power plants, but retained compensation to EITE industries but at a lower rate than under the original CPRS. The Greens' scheme also called for financing climate change adaptation and mitigation in developing countries to the tune of $1.27 billion per annum, indicative of their concern with climate justice.

After Rudd's announcement on 27 April 2010 that the CPRS would be shelved indefinitely, Brown said the government had caved in to right-wing criticisms.[37] Again, the Greens proposed an interim carbon tax of $23 per tonne, rising each year by 4% plus the consumer price index, until the next round of international negotiations, but this was ruled out by the government.[38] *Sydney Morning Herald* reporter Paddy Manning regretted it was not considered: 'The Greens proposal would have generated a $4.4 billion surplus over 2010–12, compared with an estimated $1.5 billion loss for the government's scheme, because it capped assistance to heavy polluters. In fact the proposal was much better than nothing: it was a plausible, market-based response to climate change.'[39]

Labor blamed the Greens for not backing the watered-down CPRS.[40] Yet the Greens' support in the polls rose to record heights. In May 2010, primary support for the Greens, identified publicly with proposing a better, stronger scheme, rose to 13% according to Nielsen and 16% according to Newspoll.[41] A Newspoll on 1 June 2010 found the 16% Greens' primary vote still held—the highest for the Greens ever recorded; and Nielsen the same week recorded 15% for the Greens. Morgan attributed this spike to the government's 'dumping' of its ETS policy.[42] Compared with the 2007 election, the Greens primary vote had risen 8.2% and Labor's primary vote had fallen 8.3% in the same period.[43] The Greens continued to explain their reasons for opposing the ill-fated legislation: 'a badly designed emissions trading scheme, such as the government's CPRS with its woefully weak targets, multibillion dollar polluter handouts and unlimited overseas offsets would actively block progress by locking in the current polluting economy for years to come'.[44]

Balance of Power and Carbon Pricing: 2010–2011

On 21 August 2010, Adam Bandt became the first Green to be elected to the House of Representatives at a general election. Bandt won the previously safe Labor seat of Melbourne with a primary vote of over 36%, a massive swing of more than 13% to the Greens. The previous incumbent, outgoing Finance Minister Lindsay Tanner, warned electors that the Greens were 'activists in politicians' clothing'.[45] Significantly for the future, the Greens received over 20% of the vote in eight lower house seats, over 15% in thirty-six seats and over 10% in ninety-two seats. Nationwide, for the House of Representatives, over 1.45 million people, 11.76%, gave their primary vote to Greens, a 4% swing. If Australia had proportional representation for this house, seventeen Greens would have been elected. For the Senate, more than 1.66 million Australians voted 1 for the Greens, 13.11% of primary votes, a 4% swing and the highest ever third-party vote in

Australian federal politics, three times that of the National Party. A Greens Senator was elected in every state for the first time and the first ever in Queensland and Victoria. With three Greens Senators not up for re-election in 2010, and six elected in 2010, these nine Greens Senators hold the balance of power in the Senate from 1 July 2011: Bob Brown and Christine Milne from Tasmania; Sarah Hanson-Young and Penny Wright from South Australia; Rachel Siewert and Scott Ludlam from Western Australia; Lee Rhiannon from New South Wales; Richard di Natale from Victoria; and Larissa Waters from Queensland.[46]

Why had so many people voted Green? For a start, the policies they took to the election, as in all previous ones they had contested, were more social democratic than Labor's. The Greens wanted Australians 'to have universal access to primary dental care', a policy recently abandoned by Labor. On education, they announced they would abolish all TAFE and HECS fees, and provide two years free kindergarten to all Australian children.[47] The Greens offered, as always, green social democracy: 'The Greens stand for much more than just cutting carbon pollution, securing our water supplies and protecting our environment. Think better public schools, more funding for hospitals and fixing our broken mental health system.'[48] And, unlike the other parties, the Greens opposed the war in Iraq, advocated humane treatment for refugees and asylum seekers, and proposed an end to all forms of legal discrimination based on sexuality.[49]

However, the Greens' greatest electoral advantage was now the climate change issue. From a party that had grown initially by offering a social justice agenda critical of both neo-liberalism and discrimination, it was now increasing its vote further in the wake of the Rudd government's backflip on its CPRS. The Greens' climate policies brought to the 2010 election were: a two-year carbon tax starting at $23 a tonne of carbon, increasing each year, leading into an ETS that would 'punish' not 'pay' polluters; $5 billion in government-backed loan guarantees for base-load renewable energy projects; and Infrastructure Australia to map Australia's renewable energy resources and prepare a scoping study to generate 100% renewable energy by 2030.[50] Bandt's campaign for Melbourne, and that of other Greens candidates with a possibility of election, attracted media attention and public interest, and served as an important force in keeping climate politics alive during an election in which both major parties preferred to downplay or deny the issue.

According to Nielsen polling in 2010, people aged 18–24 were twice as likely to vote Greens as those 25 and over, but the Greens vote was increasing in all age groups.[51] Clearly, assumptions about youthfulness of Greens' voters need to be qualified. So too should stereotypes about inner-city latte-sipping Greens voters, based on the fact that inner-city electorates close to universities, socially mixed, but in the process of gentrification are becoming Greens' electoral heartlands.[52] (At the March 2011 NSW state election, the Greens won Balmain, their first Legislative Assembly seat.) Bandt describes his newly acquired Green seat:

> Melbourne has been a Labor electorate for its 106 year history, and is a broad mix of progressive professionals, young students, public housing residents and working class and linguistically diverse communities. All these groups in various ways have felt abandoned by the major parties' neoliberal policies and failure of leadership on key environmental and social questions.[53]

Although inner-city suburbs are indeed home to 40% of Greens voters, other areas of strength are rural areas or villages and small country towns, such as those on the strongly Greens north coast of NSW.[54] They have also attracted working-class support because of their opposition to neo-liberalism.[55] Some unions donate funds to the Greens rather than Labor at election times. Bandt explains: 'The backing of unions like the Electrical Trades Union reflected this shift away from Labor because of its failure to represent labour values.' Their financial and moral support was an important factor in his successful campaign, he adds.[56] It is a more complex picture than stereotypes suggest.

Given the evenness of the numbers in the House, Bandt's election turned out to be historic in a second way. Along with Independents Andrew Wilkie, Tony Windsor and Rob Oakeshott, his support for Labor over the Coalition was crucial in assisting the formation of the Gillard government. Leading climate movement campaigner David Spratt argues the Greens' refusal to endorse the CPRS contributed to their increased electoral support and therefore a 'hung parliament in which Labor had to elevate The Greens to the centre of climate-policy making in order to maintain power'.[57]

The Greens negotiated an agreement to support the Gillard government on crucial votes of supply and confidence. This signed agreement committed the government to establish a cross-parliamentary committee, also involving experts from outside parliament, to work out a carbon price mechanism. This was described by Milne as 'a vital part of the climate solution, helping to drive transformation of the economy from a resource base to a flourishing, clean, renewable base'. The agreement also committed the government to: deliver improved transparency and integrity to parliament; pursue policies such as universal dental care; debate Australia's involvement in the war in Afghanistan; and referenda to recognise indigenous Australians and local government in the Constitution.[58] Bandt wrote in *Green Party News* in November 2010 that 'it's been hectic', but underlying the Greens' approach were the values he outlined in his first speech to parliament: 'sustainability, equality and compassion'.[59]

The Greens entered 2011 expressing 'real optimism about climate action both at home and globally'. The Cancun climate meeting in December 2010 was notable for the fact that governments, including the new Australian government, acknowledged that the emissions reduction targets so far put on the table were completely inadequate. On the domestic front, the positive development of the Greens-initiated multi-party climate committee was important, 'but we still need to see more ambition from governments, lifting our sights to achieving a safe climate, not just dealing with the "political problem" of climate change'.[60]

Come Carbon Sunday, Brown hailed the final announcement of the carbon price details as 'a world-leading outcome' that would lead to better outcomes at Durban, the next international conference on climate change. Perhaps bearing in mind concerns emanating from sectors of the grassroots climate movement, Brown, with Milne and Bandt at his side, dismissed criticism that the new carbon tax was similar to the CPRS. Despite having made compromises in formulating the carbon pricing mechanism, Milne, who had been the party's lead negotiator on the multi-party climate committee, said the

Greens were 'delighted' with the deal they had struck, because it would not close the door to even stronger, more ambitious action in the future. Milne stated: 'This historic agreement delivers the biggest investments in renewable energy and biodiversity Australia has ever seen.' This *Age* report noted that, while some commentators had warned that compromising with government risked weakening the Greens vote, on the evening of Carbon Sunday Bandt had a warm reception from Greens supporters at an event in Melbourne's CBD.[61] The Morgan Poll result issued on 9 September 2011 indicated the Greens polling 12.5% of primary votes, up 1.5% on the previous poll.

Realos or Fundis?

Tietze points to 'ideological confusion at the heart of the Greens project: specifically, the rejection of class as a central organising feature of society and the appeal to universal values, which avoids a serious assessment of how to deal with the concentration of economic and political power in society'. This lack of a critique of neo-liberalism, let alone capitalism, has led to the incorporation of pro-market ideology in the Greens' approach, notably their in-principle support for emissions trading as a key driver of the transition to a low-carbon economy. It was only pressure from 'the small but significant climate action movement' that led the Greens to break with Rudd's CPRS and sharpen its critique of cap-and-trade mechanisms.[62]

A different perspective is provided by Peter McMahon, who argued in 2006 for the Greens to 'make their move' as the only party able to confront the global warming problem, because it was the only parliamentary party critical of neo-liberalism. Both major parties are now based on neo-liberal ideology, which assumes conditions enabling limitless growth, so neither can respond meaningfully to the problem. For the Greens to succeed in providing the solution, they must 'shift away from their softly, softly approach to make the problems plain and the solutions apparent' and 'situate themselves as the party with the most relevant but radical ideas to deal with what is undoubtedly the most critical challenge facing modern civilization'. This was feasible, because of the Greens' critical distance from neo-liberalism: 'If the Greens can shape their core concerns to construct a coherent alternative ideological position, then they can seize the initiative and present a new way for Australia.'[63]

There is no shortage of Greens critical not just of neo-liberalism but also of capitalism and economic growth. For example: 'Overcoming our obsession with perpetual economic growth is one of the first things that should change.'[64] The Greens are not a homogeneous political entity. According to Jenny Stock, by 2006 the Greens exhibited 'deep divisions (variously described as purists versus pragmatists, and resource ecologists versus radical ecologists), and are further divided on how to best make the compromises necessary to operate in the formal political system'.[65] Jackson's survey of activist members reveals the wide range of political beliefs.[66] Miragliotta's 2010 study of radicalism and pragmatism in the Greens describes it as 'One Party, Two Traditions'.[67] In 2012, Sally Neighbour drew attention to the long-running tension over organisational matters between the Australian Greens and the NSW Greens, who have supported what Brown describes as an 'old-fashioned' states' rights agenda that prevents the Australian Greens operating effectively as a nation-wide party.[68]

Diversity is a characteristic shared with the German Greens, divided as is well known between 'Fundis' (fundamentalists) and 'Realos' (realists). Within the Australian Greens, the radical ecologists and eco-socialists tend to function at the periphery, for most Greens are primarily interested in regulating capitalism and making it more socially just and environmentally sustainable. However, even mainstream, 'Realo' Greens are responsive to issues relating to climate justice, a concern that arises naturally out of their longstanding commitment to social justice. While the divisions within the Greens are real and significant, they are not played out as dramatically as in the German Greens.

A crucial factor in the internal debates within the Greens is pressure brought to bear by the grassroots climate movement in assisting more radical elements within the Greens to prevail policy-wise over those preferring a softer approach. The structure of the Greens allows it to be responsive to the broader climate movement. Greens' members participate directly in the organisation, in line with the party's principle of 'grassroots democracy'. The organisational manifestations of this principle include: formal commitments to affirmative action; decision-making by consensus processes; member-initiated plebiscites; and explicit guarantees of the right of members to participate in the formulation of policy and strategy.[69] Former Democrats Senator and Greens candidate for Brisbane in 2010 Andrew Bartlett points out that the Greens 'have a grassroots nature that the other parties now completely lack', which gives them greater diversity and dynamism.[70] However, within the NSW Greens there is evidence that the grassroots democracy principle is observed more in rhetoric than in reality, with the faction around Rhiannon operating, according to longtime former Greens MP Ian Cohen, like 'a small cadre of tightly bound, ideologically attuned people controlling the committees and decision-making structures of the party, and it's actually the antithesis of grassroots democracy'.[71]

With the possible exception of NSW, grassroots activity within the party facilitates connections with grassroots activism outside of it, because the experience of Greens members in the environmental movement, and now the climate movement, brings perspectives from these movements directly into the heart of the Greens. Ian Rintoul insists the Greens have little 'organised presence' within the climate movement,[72] but Greens members are usually conspicuous at climate rallies and protests, carrying Greens' placards, scattered as individuals in the larger assemblage rather than forming a bloc in the style of the far-left parties. The Greens might operate at the middle level of climate politics, but their connections with the movement from below are significant and reciprocal. Greens MPs often speak at climate conferences and rallies. Climate activists frequently work in support of Green candidates; and the vast majority of them would give their primary vote to the Greens.

Grassroots climate movement pressure is also vital in strengthening the bargaining power of Greens in negotiating for stronger climate change policies with other political forces. In April 2011, Bandt emphasised that progressive social reforms require a social movement: the government had to be 'the midwife of the zero carbon society' but it would be much easier for the Greens to extract adequate carbon pricing legislation from the government if there were thousands on the streets regularly and vociferously protesting for an even higher price of $70 per tonne.[73] As the major player in the middle level of climate politics, the prospects for the Greens to force the pace from above is dependent to a large extent on the strength of the climate movement from below.

Notes

1 Miragliotta, 'Similar but different', pp. 6–8.
2 Jackson, p. 15.
3 Burgmann, *Power, Profit and Protest*, p. 203.
4 Jackson, p. 9.
5 Bennett, 'Top 10 reasons'.
6 Brown and Singer, p. 14.
7 *Australian*, 12 November 2001, p. 7; 13 November 2001, p. 4.
8 *Australian*, 12 November 2001, p. 7.
9 Steketee.
10 Tietze, pp. 27–28.
11 Cahill and Brown, p. 262.
12 Brown, *Memo for a Saner World*, p. 70.
13 Quoted in Matthews, 'What is "sustainable" transport?'.
14 Cahill and Brown, pp. 261–263.
15 Sparrow, pp. 206–7.
16 Cited in Tietze, p. 33.
17 Research by Vromen, cited in Cahill and Brown, p. 261.
18 Norman.
19 Bennett, 'Top 10 reasons'.
20 Environmental Defender's Office, p. 3.
21 Tietze, p. 32.
22 Neighbour, pp. 21–2.
23 Milne, 'Even the Eskimos agree', p. 6.
24 Tietze, p. 32; Neighbour, p. 21.
25 Jackson, pp. 3–4.
26 ibid., pp. 15, 2.
27 http://www.greens.org.au (viewed 1 June 2008).
28 Australian Greens, 'Rudd's white flag of surrender'.
29 The Greens, *Safe Climate Bill.*
30 Australian Greens, 'Policy: Climate Change and Energy'.
31 http://www.roymorgan.com/news/polls/2010/4456 (viewed 1 February 2010).
32 http://greensmps.org.au/content/media-release/greens-release-amendments-make-cprs-work (viewed 15 February 2010).
33 'Carbon levy information'. http://greens.org.au/content/Greens (viewed 15 February 2010).
34 *Green Party News*, June 2010, p. 1.
35 Quoted in *Green Party News*, June 2010, p. 1.
36 Quoted in Alcorn, 'Greens put carbon tax on the agenda'.
37 Grattan and Arup, 'Rudd in ETS backflip'.
38 *Sydney Morning Herald*, 1 May 2010. http://www.smh.com.au/business/

one-climate-policy-and-it-only-comes-in-green-20100430-tzjz.html (viewed 2 May 2010).
39 Manning, 'One climate policy'.
40 'Rudd goes from keen to mean with Greens', 2 May 2010. http://www.smh.com.au/opinion/politics/rudd-goes-from-keen-to-mean-with-greens-20100501-u03v.html (viewed 20 July 2010); Johnson, 'The CPRS'.
41 Hartcher, 'Brown leads Greens to the promised land'.
42 Davis, 'A "pox on both their houses"'. http://www.roymorgan.com/news/poll/2010/4510 (viewed 15 June 2010).
43 *Weekend Australian*, 5–6 June 2010, p. 2.
44 'The Greens and emissions trading—your questions answered'. http://greensmps.org.au (viewed 30 May 2010).
45 'Rudd goes from keen to mean with Greens', 2 May 2010. http://www.smh.com.au/opinion/politics/rudd-goes-from-keen-to-mean-with-greens-20100501-u03v.html (viewed 20 July 2010).
46 Bennett, 'The election that was', pp. 4–5; *Green Party News*, November 2010, pp. 1–8.
47 'Vote 1 The Greens', Australian Greens election leaflet, 21 August 2010.
48 Bennett, 'Top 10 reasons'.
49 ibid.
50 *Age*, 21 August, p. 3.
51 *Age*, 12 June 2010, p. 9.
52 For a classic example, see Black.
53 Bandt, p. 15.
54 Research by Turnbull and Vromen, cited in Cahill and Brown, p. 261.
55 Lavelle, p. 10.
56 Bandt, p. 15.
57 Spratt, 'A carbon price, and then … ?', p. 17.
58 Milne, 'Balance of power comes to Canberra', pp. 6–7.
59 *Green Party News*, November 2010, pp. 3–4.
60 *Green Party News*, January 2011, p. 1.
61 Veness, p. 7.
62 Tietze, pp. 34–5.
63 McMahon, pp. 24, 26, 27.
64 Kelso, p. 13.
65 Stock, p. 266.
66 Jackson, pp. 3–4.
67 Miragliotta, 'One party, two traditions', p. 585.
68 Neighbour, pp. 27–8.
69 Miragliotta, 'Similar but different', pp. 8–9.
70 Bartlett, p. 13.
71 Quoted in Neighbour, p. 24.
72 Rintoul, 'Greens balance'.
73 NTEU, Climate Change Forum, Melbourne, 28 April 2011, Burgmann, participant-observer notes.

7
ENGOs and Think-Tanks

The issue of climate change is so pressing that non-environmental NGOs, such as Oxfam and World Vision, have developed policies on it.[1] However, it is naturally ENGOs that have made the running on the issue, to which they now devote considerable energy and resources. Most of the ENGOs discussed are established, well-known organisations; some are new players on the scene, formed specifically in response to climate change. Their stances and policies on climate change reflect their underlying political orientations and particular interests. Not all significant ENGOs are discussed. An omission is the Wilderness Society, which has campaigned hard for forest and bushland protection to increase carbon absorption, as part of the solution to climate change.[2] Friends of the Earth is discussed in Part III, as a major force in the grassroots climate movement from below.

Australian Conservation Foundation (ACF)

The ACF, established in 1965, describes itself as 'Australia's longest running and most respected national environment organisation'. It lobbies governments, government departments and industry; educates the public through its many publications; maintains a high media profile; endeavours to expand its membership and support base; and sells products such as t-shirts and posters with an environmental theme.[3] Many of the publications downloadable from its website are concerned with global warming. Since 2008 the ACF has been a member of the Southern Cross Climate Coalition, which also includes the ACTU, Climate Institute and the Australian Council of Social Service (ACOSS).

Even if market forces cannot solve the problem of climate change, the ACF does not believe that capitalism must be abolished to save the planet. According to Elim Papadakis and Anya Moore, 'the ACF, like business groups and government agencies, has questioned the idea that the environmental paradigm is essentially incompatible with the dominant paradigm based on economic growth'.[4] Essentially, the ACF looks to government regulation of business to enforce ecological standards. However, during Howard's prime ministership, the ACF despaired somewhat of government, so liaised more directly with business interests, attempting to persuade companies to embrace environmental issues of their own accord, that environmental sensitivity leading to technological innovation could be good for business.[5] It accepts contributions from corporate supporters, though not ones clearly implicated in pollution.[6]

The ACF supports in principle an ETS that would purportedly drive down emissions, avoid loopholes and giveaways, and generate revenues to invest in a transition to a 'clean economy'.[7] In September 2008, it called on the Rudd government to commit Australia to a 2020 target to reduce emissions by at least 30% on 2000 levels, and by 40% if other developed countries did the same. It also urged commitment to funding carbon reduction schemes in developing countries. The two key tests to be set for the CPRS were whether it was 'environmentally effective' (guided by the most recent science so as to avoid costly impacts) and 'internationally credible' (given that Australia and the other developed countries have created more than 75% of global carbon pollution and continue to create much more per person than in developing countries).[8] ACF Executive Director Don Henry stated that Garnaut's conservative targets would 'condemn the Great Barrier Reef and the Murray Darling to destruction'.[9]

When the CPRS details were unveiled, Henry stated the scheme was 'a significant step forward on climate change' but argued its target to cut emissions by just 5–15% by 2020 against 2000 levels was 'far too weak to deal with the problem and will not position Australia to be a leader in this year's crucial international negotiations on climate change'.[10] Nonetheless the ACF backed the CPRS as formulated at that stage on the grounds that the reduction targets could subsequently be improved.[11] The ACF argued the Australian parliament should: commit to an unconditional emissions reduction of at least 25% by 2020; cut the compensation for the big emitters and allocate at least 10% of permit revenue to help developing countries with climate change; allocate substantial funding to building resilience to climate change in Australian ecosystems through a Biodiversity and Climate Change Fund; set world's best practice national energy and fuel-efficiency standards; and strengthen and pass the Renewable Energy Target legislation, ASAP.[12]

The ACF withdrew its support for the CPRS on 25 November 2009 after the government agreed to compromise with the Liberal Party by weakening the CPRS. The ACF criticised the opposition proposals for directing general budget revenues 'towards the balance sheets of Australia's largest polluting industries'.[13] During the 2010 election, the ACF invidiously compared the policies put forward by Abbott and the Coalition parties with the more progressive stance on climate change mitigation assumed by the Conservative Cameron government in the UK.[14] In 2011 it welcomed the Gillard government's carbon pricing legislation, but warned: 'A fixed price on carbon, while an important step that will help business adapt, does not, on its own, ensure carbon pollution levels will fall, so it's vital that we move from a fixed price to a legislated cap on pollution levels.'[15] More confidently, it anticipated that the Clean Energy Finance Corporation would jump start a 'clean energy boom in Australia'.[16]

The ACF has indicated concern with climate justice in two main ways. First, it urges increased assistance to Australia's Asia Pacific neighbours to cope with climate change:

> The world's poorest people—many of whom are already experiencing the effect of climate change—will be the worst hit under climate change with impacts for the developing world predicted to be far more devastating than in industrialised nations.

> Australia is certainly vulnerable to the impacts of climate change, but not as vulnerable as many of our neighbours in the Asia Pacific.

Amongst other measures, it calls for an increase in Australia's overseas development assistance and for climate change considerations to be integrated into all relevant parts of the assistance program; and for Australia's immigration program to be reviewed to consider mechanisms to support people displaced by climate change within the region.[17]

Second, in conjunction with the ACTU, it developed a 'road map for a cleaner, stronger economy'. Acknowledging the anxiety felt by many workers about potential job losses with any move to a low-carbon economy, it commissioned economic modelling to assess how best to protect jobs across all regions of Australia. Its report concluded: 'A targeted investment program to reduce Australia's pollution and build our competitiveness in a global clean energy economy delivers more jobs, lower debt, savings for households from lower energy costs, a higher standard of living and a healthier and more resilient environment.'[18]

Also focused on equity issues at home, the ACF produced a document with ACOSS and Choice, to help prepare households for climate change, based on the principles of 'efficiency, equity, immediacy'. It urged special consideration for low-income and disadvantaged households on the grounds that: low-income earners tend to live in areas more likely to be adversely affected by climate change; on average, low-income earners spend a greater proportion of total weekly household budget on energy and water; and lower-income households are currently less able to introduce measures to improve energy efficiency. In designing an ETS, it was therefore critical the costs of such a scheme not be borne disproportionately by low-income households.[19]

Greenpeace Australia Pacific (Greenpeace)

Greenpeace is part of Greenpeace International founded in Vancouver in 1971. Established in Australia in 1977, it joined with the Pacific region in 1998 to form Greenpeace Australia Pacific, which had more than 100 000 supporters by 2008.[20] Greenpeace describes itself on its website as: 'an independent campaigning organisation that uses non-violent direct action confrontation to expose global environmental problems and to force solutions which are essential to a green and peaceful future'. Its core values are Independence ('We do not accept money from governments, corporations or political parties because it would compromise our core values'); Bearing witness ('Philosophically and tactically, our peaceful protests work to raise awareness and bring public opinion to bear on decision-makers'); and Non-violent direct action ('We strongly believe that violence in any form is morally wrong and accomplishes nothing').[21] A small number of highly active individuals, often employed as full-time campaign staff, carry out the guerilla tactics for which Greenpeace is famous, while its vast membership base provides mainly money, without being involved in decision-making.[22]

Its actions are typically dramatic and eye-catching. For example, on 20 October 1997, Greenpeace's 'Climate Rescue Team' climbed onto the roof of the Prime Minister's Sydney residence and erected solar panels to advocate the use of solar power rather than fossil fuels. They were arrested and charged, but the magistrate released them on a good behaviour bond, commenting: 'I accept that you were acting truthfully in terms of your own convictions ... I guess Australian society would be a lot duller if there weren't some people who were prepared to say what they think.'[23] In August 2002, three Greenpeace climbers scaled the flagpole at Parliament House in Canberra and unfurled a banner reading 'Ratify Kyoto Now', resulting in their arrest along with four other Greenpeace demonstrators.[24] Greenpeace's spectacular actions against coal are described in Chapter 12.

Non-violent direct action is not its only tactic. Carefully researched briefs are presented to courts, the press and governments to support Greenpeace campaigns for changes in public and private policies. It participates in meetings from UN conferences to town hall gatherings; it produces documentary films and delivers lectures in schools. It sells badges, fridge magnets, stickers, tea towels, t-shirts, sweat-shirts, posters and car bumper stickers. Greenpeace aims to educate people and to identify causes and expose culprits. As part of a worldwide organisation, Greenpeace gives priority to campaigns that can be addressed on a global scale. One of its long-running international campaigns has been to: 'Stop climate change by phasing out fossil fuels (oil, coal and gas) and replacing them with clean renewable energy, such as wind and solar power.'[25]

Greenpeace made these recommendations to the Rudd government in 2008:

> Legislate a greenhouse gas reduction target of greater than 40% below 1990 levels by 2020.
>
> Establish an emissions trading scheme that delivers a decrease of our emissions in line with legislated interim targets.
>
> Legislate a national target for 40% of electricity to be generated by renewable energy sources by 2020.
>
> Massively invest in the deployment of renewable energy and strongly regulate for energy efficiency measures.
>
> Establish an immediate moratorium on new coal-fired power stations and extensions to existing coal-fired power stations, and phase out existing coal-fired power stations in Australia by 2020.
>
> Set a target of 2% per year to reduce Australia's primary energy demand.
>
> Ensure transitional arrangements for coal dependent communities that might be affected by the transition to a clean energy economy.
>
> Redirect all public subsidies that encourage the use and production of fossil fuels towards implementing energy efficiency programs, deploying renewable energy and supporting the upgrading of public transport infrastructure.[26]

Greenpeace was critical of the CPRS. John Hepburn of Greenpeace asserted: 'It's clear that Rudd has been listening to the big polluters and this is another shift towards the interests of the polluters rather than climate action.'[27]

Greenpeace maintains Australia is well positioned to implement renewable energy resources, such as solar photovoltaic, concentrating solar thermal, wind, wave and geothermal energy. It strongly opposes CCS and drew attention to the problems associated with Rudd's proposed Clean Coal Initiative:

> First, it dangerously distracts us from efforts we should be making now to convert to renewable energy-based power generation ... Second, CCS, like nuclear power, is too expensive, too risky, safe storage is not certain, and it can't do the job to cut emissions in the timeframe we need. Third, there are big questions in any case over liability in the case of failure to work. Any money, public or private, that goes into CCS is wasted, a missed opportunity to roll out the renewable solutions that are ready to go right now.[28]

The growth of renewable energy solutions, Greenpeace stresses, means that strong action to tackle climate change is not a jobs *versus* environment issue but a jobs *and* environment issue.[29]

Greenpeace commissioned the Centre of Full Employment and Equity at the University of Newcastle to model the effect on employment of a transition from coal-fired power to renewables in the Hunter and Wyong region, an area where coal is integral to the economy and community. The report in July 2008 found such a shift would create between 7500 and 14 300 jobs—a net gain in jobs of between 3900 and 10 650—creating a 'renewable "Silicon Valley" '. It criticised the federal government for having no policies to assist those working in coal-fired power generation to gain employment in new, lucrative renewables, despite the fact that: 'The transition to a renewable energy economy will bring huge economic benefits as well as being far cheaper to future generations than the cost of dealing with climate change impacts.'[30]

Greenpeace argued that unions and environmentalists were 'critical participants in a just transition process' and a just transition process targeting global warming offers scope for transforming the traditional agenda of unions, bringing them into collaborations with environmental organisations. 'Green-labour alliances can inspire the broad-based community awareness and actions needed to make a just transition to renewable energy and new green jobs—collective social action on issues of ecological sustainability and related social development.'[31] Likewise, Greenpeace International described the GFC as 'a perfect opportunity to develop stimulus packages that cut greenhouse gas emissions and create green jobs'.[32]

Having firmly rejected the CPRS, Greenpeace was relatively pleased with the Gillard government's Clean Energy Future legislation. Its response was enthusiastic—and eco-modernist. It would keep Australia 'in line with climate science while growing our economy'. Smart businesses would 'quickly find ways to pollute less so they pay less' and it would create 'good clean energy jobs'. All in all, it was 'a big step forward'.[33]

WWF-Australia

WWF-Australia is part of the WWF International Network, the world's leading, independent conservation organisation. Founded in 1961, it is active in over 100 countries, supporting around 1300 conservation and environmental project, and has five million supporters. Initially called the World Wildlife Fund, in 1986 it changed its name to World Wide Fund for Nature, though World Wildlife Fund remains its official name in Canada and the United States. The Australian office, established on 29 June 1978, simply uses the initials WWF-Australia on its website and publications, adorned by its familiar panda logo. It is Australia's second largest ENGO, surpassed only by Greenpeace; its 50 000 supporters provide nearly 70% of its annual income.[34]

Its earliest projects include research to save endangered species; and, as its name suggests, its approach to climate change emphasises its impacts on species and places, such as the Great Barrier Reef. In 2011, the International Year of Biodiversity, WWF-Australia announced it will 'lead efforts to conserve our most outstanding natural environments, recover threatened species and build resilience within our ecosystems to the impacts of climate change' and urge the government to strengthen laws that protect biodiversity and plan for climate adaptation.[35]

Its website explains that, in Australia and throughout the oceanic region, 'we work with governments, businesses and communities so that people and nature can thrive within their fair share of the planet's natural resources'. WWF-Australia 'prides itself on engaging successfully with governments, communities, businesses and individuals'.[36] The extent of WWF-Australia's engagement has brought criticism from other sections of the climate movement, most notably in 'Taming the Panda', produced by Clive Hamilton and Andrew Macintosh of the Australia Institute. They describe how WWF-Australia, in contrast to the ACF and Greenpeace, maintained a close relationship with the Howard government, supporting many of its environmental policies, even bestowing on it three 'Gift to the Earth Awards', and refraining from criticism of its failure to ratify the Kyoto Protocol.[37]

WWF-Australia's support for the CPRS was also criticised; so too its support for a national CCS strategy, in alliance with the Climate Institute, the ACA and the Construction, Forestry, Mining and Energy Union.[38] WWF-Australia even conflates CCS with renewables. For example, its 2010 annual report noted it 'continued to advocate increased support for renewable energy sources, including low-emission solar thermal and Carbon Capture and Storage (CCS) power plants'.[39]

Even when voicing shared opinions, its style can offend other activists. In an introduction to its 2010 annual report, modestly entitled 'WWF leadership. A beacon in the dark', it expressed its 'sense of disappointment ... with the Australian Government and Opposition failing to show leadership on climate change' and added: 'In the lead-up to the Federal Election, WWF-Australia played a crucial role in trying to convince the major parties, and voters, of the urgent need to act now and put a price on carbon.'[40]

WWF-Australia approved Gillard's Carbon Sunday announcement of the carbon price details, but called for a target of 90% reduction by 2050.[41] When this Clean Energy Future legislation was introduced into parliament in September 2011, WWF-Australia welcomed it as an important step towards protecting Australia's environment from the negative effects of carbon pollution: 'We urge the public to get behind this important reform to protect our reefs, our turtles, our economy and our kids' future.'[42] Its priorities are clear.

WWF-Australia's best-known contribution to climate politics is Earth Hour. This is now a global event held on the last Saturday of March, which requests households and businesses to turn off their non-essential lights and other electrical appliances from 8.30 to 9.30 pm local time, to raise awareness about the need to take action on climate change. It was conceived in 2007 by WWF-Australia and the

Sydney Morning Herald; and 2.2 million Sydney residents and 2100 businesses participated by turning off all non-essential lights. Following Sydney's lead, other cities around the word adopted the event from 2008 onwards. In Australia that year, 58% of people in capital cities joined in by switching off their lights.[43] By 2010 more than 4300 cities and towns in 126 countries participated. WWF-Australia CEO Greg Bourne stated: 'People all over the planet have again shown that the spirit of Earth Hour comes from millions of small actions that we can all take as one.'[44]

The tokenism and individualism of Earth Hour, and its dubious connections with polluting industries, has drawn attention from climate activists. For example, *Green Left Weekly* noted that the *Sydney Morning Herald* on 17 March 2008 inserted a forty-page, full-colour Earth Hour Magazine in its 17 March 2008 issue, 'festooned with full-page ads by electricity suppliers such as Energy Australia, Integral Energy and Country Energy—the ones that hawk all that coal power—car companies such as Toyota, Fiat and Hyundai, and even Cascade (100% carbon offset beer!)'. Although 59% of Sydneysiders reportedly participated in Earth Hour 2008, electricity in Sydney and the Illawarra dropped by a mere 2–3% and 3.6% in Australia as a whole.[45]

WWF-Australia proudly promotes Earth Hour, describing it as 'The world's biggest climate change campaign', achieving 'an amazing "lights out" in all major cities around the globe'. In 2010 'Australians joined millions of people in 4300 cities around the world to curb their energy consumption for one hour but also took into their homes WWF's more enduring message—to reduce their impact on the environment every hour of every day'. Through Earth Hour, it explains in management-speak, 'we will empower action, celebrate what can be done, redouble our efforts around what hasn't been done, and ultimately mobilise people through a sense of shared community and overarching values'.[46]

Southern Cross Climate Coalition (SCCC)

In 2008, the ACTU, along with the Climate Institute, ACF and ACOSS, established the SCCC as a climate change coalition dedicated to leading debate on 'a fair response' to climate change policies. Its founding statement, published 6 July 2008, explained its rationale:

> We share the view that Australia faces major climate change risks unless the nation makes an urgent and orderly shift towards a low-carbon economy. We acknowledge that the transition poses challenges for some of our communities. But we believe that it will also create great potential for long-term prosperity and jobs growth. We aim to build on the Australian tradition of working together through challenging times. We seek to advance constructive and long-term proposals for Australia and to work with government, industry, our constituents and the community to unlock the substantial economic, social welfare and environmental opportunities that will emerge from Australia's response to climate change.[47]

As a coalition involving social, union and environmental organisations, it was well placed to stress that collaboration across all sectors of Australian society would be critical to achieving an effective, fair and lasting response to climate change. 'We understand that effective action on climate change means much more than reducing greenhouse emissions. It means grabbing hold of new economic and job creation opportunities at the same time as improving the living standards of all Australians fairly and sustainably.'[48] In a statement released on 7 July 2008, the SCCC urged Rudd to 'stand his ground against special pleading from certain sections of industry, take the call for urgent action from Garnaut to the G8 next week, and do everything possible to get a broadly based emissions trading scheme fully implemented'.[49] The SCCC was criticised by others in the climate movement for not opposing the CPRS outright after its details were unveiled.[50]

Since its inception, the SCCC has argued the benefits of green jobs and lobbied the government to do more to create them, in the interests of both the environment and the economy.[51] For example, it urged the Australian government to 'build on the promise of the $3.8 billion Energy Efficient Homes program by ensuring the federally funded school buildings and community and defence houses are energy and water smart, and by introducing further green economic stimulus measures'.[52] Shortly after the GFC struck, the SCCC called for a 'green new deal'—'an urgent stimulus package of investment into a program of retrofitting housing and commercial and public buildings to make them more water and energy efficient, and into infrastructure assets that will reduce our carbon footprint'.[53]

In early November 2009, it organised a forum in Brisbane on climate change and jobs to pressure politicians and businesses to take action. The forum maintained that up to one million green jobs could be created in the next 20 years if Australia's leaders got serious about tackling climate change; and Australia could become a leader in six key industries within the growing global $6 trillion green products and technology market: renewable energy, energy efficiency, production of bio-materials (including plastic from plant starch), green buildings, sustainable water systems and waste and recycling. It maintained an effective ETS was crucial to unlock billions of investment dollars needed to drive the transformation to a low-carbon economy and create hundreds of thousands of jobs. Climate Institute chief executive John Connor told the forum: 'It's time for the dinosaurs in politics and business to evolve and stop blocking the industries and jobs which will cut the greenhouse gases harming our environment.'[54]

In 2009 the SCCC, along with the Property Council, the Australian Green Infrastructure Council and the Institute of Superannuation Trustees produced a *Joint Statement: Towards a Green New Deal.* It called for three main objectives: *retrofitted buildings* to enhance energy and water efficiency carried out nationwide in residential, commercial, and public sectors, with assistance for low-income people as the first to undertake household efficiency audits; *sustainable infrastructure* like public transport, freight rail and small renewable-energy installations—solar, wind, geothermal—to reduce the carbon footprint; special attention to the construction industry and materials sector; and *green industries* for the manufacture of internationally competitive new products and services, projecting 500 000 green jobs, with an 'immediate effort invested in green skills for Australia's trades men and women'. Salleh described this as a 'job stimulus package' that objectifies nature as 'out there', in line with the humanity versus nature contradiction, and pays too little attention to including youth, indigenous people and women in the new green jobs.[55]

In November 2010, the SCCC outlined 'four foundations to drive a fair and inclusive transition to a low pollution economy': 1. reducing our economy's dependence on pollution; 2. ensuring fair and inclusive action; 3. unlocking clean energy jobs; and 4. strengthening international action. Elaborating upon the second foundation, it observed:

> Low income earners tend to live in situations more likely to be adversely affected by climate change and are less able to move or make other necessary adjustments to their living circumstances. Low income earners are likely to be affected first and hardest by climate change. Low income households should be assisted with affordable measures that contribute to mitigation and adaptation.

For example, low-income earners are more exposed to energy price increases, so policy responses need to be well designed.[56]

The SCCC welcomed Gillard's Clean Energy Future package 'as a critical first step forward' on climate change mitigation. 'The price on pollution and its complementary measures lay the foundation for reducing emissions and remaining competitive in the emerging global low pollution economy.'[57] In August 2011 it assessed this package against the criteria of its 'four foundations'. Considering 'fair and inclusive action', it welcomed the measures announced to provide assistance to households via changes to the transfer and taxation systems, but noted that low-income households typically spend a much higher proportion of their income on essential goods and services, particularly energy and food, that will be disproportionately affected by a price on carbon pollution. 'The assistance offered to some recipients of income support, in dollar terms, may be insufficient, initially and over time. The SCCC will closely monitor development of the package in legislation and as implemented to ensure that those least able to afford price increases are not burdened unfairly.'[58]

Safe Climate Australia (SCA)

Al Gore launched SCA at a Business Breakfast for a Safe Climate on 13 July 2009 in Melbourne that brought together a thousand 'community leaders', including those representing 'some of the largest and most influential corporations in Australia'.[59] On a previous visit to promote his film, Gore had recruited eighty-four Australians to his 'army' of climate change ambassadors, including farmers, business leaders and celebrities such as the Cat Empire's lead singer, Felix Riebl; by mid-2007, this group had already given 693 presentations to more than 40 000 people.[60]

A small group of climate deniers protested the launch of SCA.[61] Following the launch, Gore was asked by reporters if the CPRS constituted an adequate ETS, Gore replied:

> It's not what I would have written. I would have written it as a stronger bill, but I'm realistic about what can be accomplished within the political system as it is ... I am sincerely convinced that the right way forward is to get to the maximum that the political system will allow us to accomplish and begin the change, and then, as we gain experience with it, toughen it, strengthen it, make it better based on experience as business and industry learn how to adjust.[62]

The Chairman of SCA is Ian Dunlop, a former Shell executive, chair of the ACA 1987–88 and the Australian Greenhouse Office Experts Group on Emissions Trading 1998–2000, and a Fellow of the Australasian Institute of Mining and Metallurgy and the Energy Institute (UK), Deputy Convenor of the Australian Association for the Study of Peak Oil, and a Member of the Club of Rome, amongst other honours. Dunlop identified the new group as an 'apolitical organisation that wanted to produce a practical plan'.[63] SCA's other directors include: Dr Joe Herbertson AM (former head of BHP's corporate research laboratories, now Executive Chair of Crucible Carbon Pty Ltd, which conducts consulting, research and business development activities in support of sustainable, carbon neutral futures); and prominent grassroots climate activist Philip Sutton.[64]

The *Greenlivingpedia* describes SCA as an NGO 'formed and steered by a foundation group of concerned scientists, community and business leaders with a shared understanding of the scientific imperatives for emergency action to restore a safe climate'.[65]

According to the SCA website, it aims 'to mobilise Australia's extensive technological, economic and political expertise and resources in planning the transition of the Australian economy to zero net carbon, the sequestering of dangerous levels of existing carbon from the atmosphere, and in assisting the building of a global consensus for restoring a safe climate'.[66] As an awareness and fund raiser for the organisation, on 2 November 2009 a relay team of thirty-five runners consisting of people working in emergency services, including fire fighters, a policeman and a nurse, set out on a 7000-kilometre run, from Cooktown in far north Queensland, through Brisbane, Sydney, Canberra and Adelaide, terminating at St Kilda Beach in Melbourne on 29 November. This national Run for a Safe Climate linked 'Australia's unique, world-renowned natural icons, most of which will be devastated if atmospheric greenhouse gas levels are not brought back to a safe climate threshold'.[67]

In 2009 SCA launched two key projects: (1) the Climate Science Imperative to review 'the most recent evidence on climate change trends, risks and scenarios'; and (2) the Australian Safe Climate Transition Plan, 'a "whole-of-society" plan for transitioning to a net zero-carbon economy at emergency speed and implementing large-scale sequestration of existing atmospheric carbon pollution'. This Safe Climate Transition Plan falls within the parameters of ecological modernisation—it claims to 'encompass an all-sector approach including stationary energy, energy efficiency, housing and commercial buildings, transport, industrial and extraction processes, land use and agriculture', which is intended to maintain a 'high standard of living, security of energy and food supply, access to mobility and comfort and well-being for all'.[68]

The SCA supports a cut in CO_2 levels to 280–325 ppm to create a 'safe climate'.[69] Despite Dunlop's previous connections with the oil industry, he was critical of the CPRS for not going far enough:

> The Government's stated objective is for strong action, but the targets, compensation and escape clauses in the CPRS imply exactly the opposite ...The target for stabilisation of atmospheric carbon to avoid catastrophic consequences and maintain a safe climate is now probably a concentration of less than 300 ppm carbon dioxide, not the outdated

> 450–550 carbon dioxide in current proposals. This means emissions reductions by Australia must be in the range of 45 to 50% by 2020 and almost complete decarbonisation by 2050, rather than the 5 to 15% by 2020 and 60% by 2050 currently being proposed.[70]

In 2010, SCA launched the How Fast? project, organised by Herbertson and Sutton. Responding to the recent incidence of extreme weather events globally, such as the 2010 heat wave in Russia and northern China and flooding in Pakistan, this project aims to provide a systematic assessment of how fast society needs to act to restore a safe climate.[71] Sutton's involvement in SCA provides a link between this middle-level player and the many grassroots networks in which Sutton is involved.

Think-Tanks

Think-tanks, as they are popularly called, are often constituted as NGOs, but differ from normal NGOs in their focus on research and development of ideas, on thinking rather than doing. With climate change a subject relatively new to human cognition and governments still pondering policy responses, think-tanks have an important role. Corporations have poured huge funds into the denialist think-tanks discussed in Chapter 5. On much more limited donations, progressive think-tanks attempt to promote climate change awareness and constructive responses.

These think-tanks are not necessarily left wing. For example, the Wentworth Group of Concerned Scientists is a deliberately small, elite and politically moderate think-tank, consisting of 'leading Australian scientists, economists and business people with conservation interests'.[72] Its director is Peter Cosier, who was environmental policy advisor to former Coalition Minister for the Environment, Senator Robert Hill. Tim Flannery and David Karoly are amongst the eleven members of the Group (who are all male). The Group publishes reports, such as 'Australia's climate is changing Australia: the state of Australia's water' (November 2006) and 'Optimising carbon in the Australian landscape' (October 2009). Cosier is an ecological modernist: 'Climate change is not a conflict between economic growth and the environment. We can fix this problem, not by turning our backs on the industrial revolution, but by building on what it has created for us. We just need to change the way we power our machines.'[73] He advocates decarbonising energy production and restoring a positive carbon balance in forests and agricultural land. Humanity has 'all the technologies in place today to fix the problem'; but CCS technologies, he warns, remain unproven.[74]

Another respectable and politically moderate think-tank is the Grattan Institute, formed in 2008 and based in Melbourne. As an independent public policy think-tank, it is not primarily concerned with climate change; but it has a particular interest in policy relating to energy and has supported the extraordinary initiative of Beyond Zero Emissions, discussed in Part III.[75] It has also produced extensive research on the best forms of carbon pricing.[76] It supports carbon pricing but was highly critical of the CPRS, stating that 'much of the protection proposed for the major emissions intensive industries is unnecessary or poorly targeted' and the bill would delay the structural adjustment required to move to a low carbon economy.[77] It also described the assistance to polluting industries outlined in the Gillard government legislation as 'excessively generous—bordering on protectionism', but concluded that, overall, the package 'manages the tension inherent in climate change mitigation policies: to set out a clear goal and the policy's likely consequences, but also leave room for innovation and creativity. The proposed mechanism provides much of the direction and confidence that the economy needs.'[78]

The two best-known think-tanks concerned with climate change are the Climate Institute and the Australia Institute, though the latter also works on a broader range of issues.

The Climate Institute

Established in late 2005 in Sydney, the Climate Institute describes itself as 'a non-partisan, independent climate change focused research organisation', concerned to 'drive innovative and effective climate change solutions'.[79] It aims to make a leading contribution to climate change policy debate in Australia through evidence-based research, working with a range of partners, 'including government agencies, academic institutions, businesses, industry groups and non-government organisations'. Major areas of work include: climate change impacts in Australia and abroad; emissions trading policy design; the economics of climate change policies; managing the social costs of climate change policies; energy efficiency policy design; and international climate change policy.[80] Its vision is for 'a resilient Australia prospering in a low carbon global economy, participating fully and fairly in international climate change solutions'. It is funded primarily by the Poola Foundation (Tom Kantor Fund).[81] It belongs to an international consortium called the Global Climate Network, which has member organisations in the UK, USA, China, India, Germany, Brazil, Nigeria and South Africa.[82] Its CEO, John Connor, is a lawyer who has worked as an environmental consultant to NGOs, governments and business. Other board members include: Andrew Demetriou, CEO of the Australian Football League since 2003; Clare Martin, former Labor Chief Minister of the Northern Territory 2001–2007; Professor Tony McMichael, former Director of the National Centre for Epidemiology and Population Health at the ANU; and Dr Graeme Pearman, former head of the CSIRO Division of Atmospheric Research.[83]

At the 2007 election, the Climate Institute announced its research indicated 'a mismatch between public hunger for leadership and the parties' responses to climate change', but Labor was doing better than the Coalition, because of its commitment to ratify the Kyoto Protocol 'and a stronger commitment to clean and renewable energy'.[84] Climate Institute enthusiasm for clean technology is expressed within an eco-modernist framework: 'taking decisive action now will open up new clean energy opportunities, creating up to 34 000 new jobs in the electricity sector alone by 2030, and ensure our industries remain globally competitive as the world switches to cleaner energy and cleaner economies'.[85]

The Institute supported the CPRS, while commenting that it required greater support for renewable energy. In a policy brief 'Helping unleash International Clean Technology Investment', it argued that climate change mitigation 'will require a multi-trillion dollar investment in clean technology in developing countries over the coming decades'. It aims to facilitate investment in such clean technology. It has formed partnerships with various corporate bodies, including AGL, one of Australia's largest fossil fuel companies. Its support for the CPRS and connections with the likes of AGL, have provoked criticism within the wider climate movement.[86]

At the 2010 election the Institute rated the parties on a points basis. Out of 170, the Greens scored 143, Labor 48 and the Coalition 22.5.[87] However, its response to the Gillard government's carbon price was enthusiasm for its effects on both climate change and the economy. Connor stated it would begin to address climate change and also ensure Australia remained 'economically competitive in the face of other countries ramping up their own low-carbon policies'.[88] Its assessment of the details announced on Carbon Sunday concluded that, overall, 'the carbon pollution agreement is a vital and long overdue step forward with great potential for Australia to cut pollution, clean up the economy, reduce energy waste and join the global effort to tackle climate change'. With 'an appropriate investment mandate and governance', it had the potential to remove barriers which prevent the full potential of Australia's world class renewable energy resources from being realised. 'The investment mandate must be focused on near-zero pollution technologies such as large scale solar rather than support the further expansion of technologies like gas generation and production.'[89]

However, the reduction targets were deemed inadequate. The Climate Institute approved Garnaut's recommendation of at least 25% reductions on 2000 levels by 2020 and 90% by 2050. It was concerned that the household assistance might not reduce incentives to pollute but acknowledged that, as the assistance was not subsidising energy bills directly, consumers would still experience price signals and many will reduce their energy consumption accordingly. Targeted transitional assistance to EITE industries was warranted. 'However, this assistance should be balanced against the need to support workers, households and communities, as well as investing in clean energy and stronger global action. Assistance to industry should be targeted at trade exposure and maintain investment signals for low-pollution investments that meet world's best practice.' As it currently stands, the Climate Institute felt 'certain companies will be over-compensated, risking tax-payer funded production subsidies that will increase pollution'.[90]

Australia Institute

The Australia Institute, based in Canberra, describes itself as 'the country's most influential progressive think-tank', which 'conducts research on a broad range of economic, social and environmental issues in order to inform public debate and bring greater accountability to the democratic process'. If 'genuine progress' is to be achieved, 'conscience, equity and concern for the future must be the guiding principles of our democracy'; insightful questions combined with excellent research make possible socially just, environmentally responsible and economically viable solutions. 'The Institute is determined to push public debate beyond the simplistic question of whether markets or governments have all the answers to more important questions: When does government need to intervene in the market? When should it stand back? And when regulation is needed, what form should it take?'[91]

The Australia Institute was formed in 1994 by Clive Hamilton, whose ideas are discussed in the next chapter. Hamilton served as its director until February 2008, when he was succeeded by Richard Denniss. The Australia Institute regularly produces publications, both short and long, on climate policy matters. As far back as 1997 it published a detailed policy paper on ecological tax reform in which it recommended a carbon tax of AUD$23 per tonne.[92] It has continued to argue the need for either a strong ETS or carbon tax, but was deeply critical of the CPRS. Denniss argued the CPRS 'diverges significantly from the textbook model of an ETS', because its targets do not conform to the climate science, it does not cover various sources of emissions (such as agriculture), its structure destroys incentives for voluntary action, it grants free permits to the biggest emitters, it would cap permits at $40 and the 'administrative costs of the scheme would be disproportionate to the small benefits associated with a 5% emissions-reduction target'.[93] On 16 December 2008, Denniss wrote a scathing report on the CPRS for Crikey.com, entitled 'By 2020 no Australian polluter will live in poverty', also noting that 'Under the CPRS, the harder we try to save energy, the cheaper the permits bought by the big polluters will be.'[94]

The Australia Institute succeeded in drawing public attention to this fact that, once the CPRS set a national reduction target, any efforts by individuals to reduce 'their' emissions would simply free up spare pollution permits for other polluters. Having identified the problem, the Institute proposed a solution in the form of a 'cap and slice' amendment, which would ensure that voluntary efforts by individuals or communities to reduce their emissions (for example, installing solar panels, investing in a community wind farm or setting a state-wide target such as the ACT's 40% reduction target) actually resulted in slicing the amount of pollution permits that would be issued rather than simply traded from one polluter to the next. It alleged that Senator Wong was determined to deflect attention away from this fault in the CPRS.[95] Denniss wrote on 2 February 2009 on Inside.org.au:

> Let's be clear about this. When the CPRS comes in, it won't matter how many solar hot water systems are installed, how many air conditioners sit unused on a summer's day or how small our cars are. Unless the CPRS is either amended or defeated in the Senate, it will be impossible for individuals to reduce Australia's emissions below the 5% target decreed by the government.[96]

After the 2010 election, the Australia Institute observed the ALP had promised inaction on climate change until at least 2013 and the Coalition's inaction promise was open-ended, whereas the Greens had campaigned for a carbon tax. 'The election result saw a big increase in the Green vote and the public's failure to endorse the "agenda" of either major party.' This August 2010 policy brief strongly recommended a carbon tax, rather than an ETS. The principles underlying carbon pricing should include: removal of subsidies that encourage the use of greenhouse-gas-emitting fuels; a price on emissions; investment in public transport and other public goods or natural monopoly infrastructure to ensure consumers can easily respond to price signals; regulation to enhance energy efficiency when existing market failures reduce the ability of higher energy prices to reduce energy consumption; and certainty for investors about the direction of legislative change.[97]

It argued a carbon tax of $25 per tonne would not be a 'great big tax on everything' but could be 'good for the hip pocket' if the enormous revenue raised was used to improve the financial position of an average family by more than $1000 per year. 'The $13 billion in revenue should not be seen as a cost that is taken out of the economy but as a redistribution of $13 billion from polluters and towards other groups in society that are deemed to be most deserving of assistance.'[98] One of the major problems with the CPRS had been that the polluters themselves were the major recipients of the proposed compensation package. In addition to emissions reduction, the social justice benefits of a carbon tax could be immense. 'Introducing a carbon tax is both economically efficient and, if the compensation package is well designed, highly equitable.'[99]

When the carbon price details were announced, the Australia Institute deemed the 'modest carbon price' good news because it 'would neither impoverish Australians nor bankrupt our economy', but bad news because it won't save the planet either'. It was, however, a decided improvement on the CPRS, because it would do more to transform the energy sector than that was designed to achieve. One big difference is the commitment to buy out up to 2000 megawatts of coal-fired electricity generation capacity, likely to lead to the closure of Hazelwood power station, one of the biggest single emitters in the world. Under the CPRS, no coal-fired power stations were expected to close until 2034. Another major difference is the significant increase in funding for all stages of renewable energy, including research, development and commercialisation, and a new fund to help leverage private sector funding for commercialisation. While the big polluters will still receive 94.5% of their permits free at the commencement of the scheme, there are new review arrangements which mean compensation payments may fall in five years time, whereas the CPRS locked in compensation until 2020.[100]

This assessment of the Clean Energy Future package approved the household assistance and its design, insisting there was no validity to the argument that if households receive enough compensation to cover the cost of any energy price rises then the scheme will make no difference: 'history tells us that many people will save a bit more electricity and pocket the compensation to spend on something more fun'.[101] It was especially jubilant that its 'long, and at times lonely, prosecution of the need to ensure that the voluntary efforts of individuals and communities results in reductions in Australia's emissions rather than just freeing up permits for big polluters might have finally paid off'. All purchases of GreenPower will be additional to the national target, as will other forms of voluntary action. Unlike the CPRS, the deal brokered between government, Greens and

Independents 'acknowledged the need to empower individuals and communities to do their bit extra'.[102] And, especially significant, there was 'upward flexibility' on emissions reduction targets, because the 2020 targets will be recommended by an independent commission, whereas the CPRS 'gave polluters certainty' and 'locked in failure'.[103]

Labor's Cheer Squad or Independent Climate Activists?

The CPRS debate exposed significant divisions and weaknesses on the part of ENGOs. The Climate Institute and WWF-Australia supported it as 'a step in the right direction'. The ACF only withdrew support late in 2009. Greenpeace and FOE opposed it, because the proposed cuts did not go far enough.[104] Most activists in the grassroots climate movement were dismayed by the complicity of ENGOs with such a flawed scheme. In the 'Climate Reader' compiled in January 2009 by the Victorian Climate Action Centre and Carbon Equity, FOE Climate Justice Director Damien Lawson criticised ENGOs, which urged people to elect an ALP government and 'then talk softly to the new government behind closed doors rather than continue the mobilisation'.[105]

Agnes Koh argues the empirical data suggests there is a puzzling relationship between Labor governments and ENGOs that came to bear in the way the CPRS resulted in excessive compromises towards the carbon lobby. A pro-Labor partisan-streak among some ENGOs not only clouded their objectivity on the CPRS but also ensured they were unable to lobby the Labor government effectively, because the government was confident of their support so was unlikely to yield to their demands. As a 'fragmented coalition', ENGOs as a whole were then at a disadvantage in comparison with the carbon lobby.[106] Spratt agrees, referring contemptuously to 'three large, rusted-onto-Labor groups providing cover for an appalling policy that won't guarantee a reduction in Australia's emissions for decades'.[107]

Koh's interviews with senior climate change advisers in early 2010 reveal the extent of the problem. Garnaut stated to her: 'This government feels it has the support of environmental groups just anyway! So it does not have to worry about them. Whereas it worries about criticisms from business so it gives in to them.'[108] Hamilton was appalled that the ACF, WWF-Australia and the Climate Institute offered their support to a CPRS slanted towards the interests of the big emitters.[109] Pearse observed: 'I think there's a partisan streak in some of those organisations particularly the ACF and Climate Institute, a pro-Labor mentality and they'll deny it till they're blue in the face but it's there.' This caused what he described as a 'disappointing split' among ENGOs: 'the fact that three prominent brands made a decision to back the CPRS legislation—prior to the amendments—was extremely unfortunate and gave unwarranted credit to the government at that time and managed to camouflage all the failings in the scheme's design'.[110] Pearse commented of the ACF and WWF-Australia that 'you won't hear "boo" from them about the emissions we're exporting through our coal which are bigger than Australia's total emissions'. The climate movement would remain a futile movement until the ENGOs 'stop ignoring' the more complex issues and flaws within the CPRS and RET, such as 'permits to polluters and outsourcing Australia's emissions abatement efforts'.[111]

As the Clean Energy Future legislation was presented to parliament, Pearse conceded it was better than the CPRS but still contained many of its faults, so he wonders why all the 'cheering' from the ENGOs? One reason, he concludes, is the increasing tendency of environmental groups to focus on incremental wins. 'Rapid transitions *to* renewables and *away* from fossil-fuel exports are considered unthinkable ... Settling for much less ambitious goals and overstating their significance is easier.' And: 'Most "suit-wearing" greenies also sport a neo-liberal faith in markets, with many building careers promoting the idea that emissions trading is the solution to climate change.'[112]

Notes

1 *GLW*, 17 February 2010, p. 3.
2 For example, http://www.wilderness.org.au/articles/embracing-forest-protection-as-a-climate-solution-not-just-an-offset-mechanism (viewed 2 June 2010).
3 Burgmann, *Power, Profit and Protest*, pp. 189–90.
4 Papadakis and Moore, p. 350.
5 Burgmann, *Power, Profit and Protest*, p. 191.
6 Fitzgerald, pp. 159–160.
7 ACF, 'What is an emissions trading scheme?'.
8 ACF, 'Cut carbon pollution'.
9 Quoted in Clarke, 'How to crash a planet'.
10 Spratt, 'Kevin Rudd'; Henry, 'Risking a new era'.

11 *Age*, 5 May 2009, p. 1.
12 http://www.acf.org.au (viewed 5 August 2009).
13 ACF, 'Projected financial impact', p. 6.
14 Denise Boyd, Campaigns Director, ACF, 'Conservatives step up to serious action on pollution', Email, 9 August 2010.
15 http://acfonline.org.au/articles/news.asp?news_id=3428 (viewed 5 October 2011).
16 Quoted in Ryan.
17 ACF and other organisations, *Australia Responds*, p. 2.
18 ACF/ACTU, *Creating Jobs*, p. 5.
19 ACF/ACOSS/Choice, *Energy & Equity*, pp. 2–3.
20 Greenpeace, 'Fact Sheet'.
21 http://www.greenpeace.org/Australia/en/about (viewed 4 October 2011).
22 Doyle, p. 81.
23 J Johnson, 'Non violent action and the law', pp. 50–51.
24 Vromen et al, p. 257.
25 http://www.greenpeace.org.au (viewed 1 February 2002).
26 Greenpeace, 'Energy revolution', p. 8.
27 Quoted in Spratt, 'Kevin Rudd'.
28 Quoted in Kevin, p. 150.
29 Greenpeace, 'From a coal-fired economy', p. 7.
30 ibid., p. 1.
31 ibid., p. 7.
32 Greenpeace International, p. 3.
33 Greenpeace, 'The Carbon Price—a two-minute explanation', Video, 12 July 2011. http://www.greenpeace.org/australia/en/photosandvideos/videos/Carbon-Price---a-two-minute-explanation/ (viewed 4 October 2011).
34 http://www.wwf.org.au/about_us/ (viewed 5 October 2011); http://en.wikipedia.org/wiki/World_Wide_Fund_for_Nature (viewed 5 October 2011).
35 WWF-Australia, p. 13.
36 http://www.wwf.org.au/about_us/ (viewed 5 October 2011).
37 Hamilton and Macintosh, p. 17.
38 Diesendorf, *Climate Action*, p. 142.
39 WWF-Australia, p. 7.
40 ibid., p. 2.
41 *Age*, 11 July 2011, p. 9.
42 'Carbon price bill marks important milestone', 13 September 2011. http://www.wwf.org.au/news_resources/media_centre/?2980/Carbon-price-bills-mark-important-milestone (viewed 5 October 2011).
43 http://en.wikipedia.org/wiki/Earth_Hour (viewed 5 October 2011); 'Earth Hour History'. http://www.earthhour.org.au/earth-hour-wwf/earth-hour-history (viewed 1 June 2010).
44 http://www.earthhour.org.au/latest-news/earth-hour-rolls-round-the-world (viewed1 June 2010).
45 Townsend.
46 WWF-Australia, pp. 3, 6.

47 *Sydney Morning Herald*, 6 July 2008. http://news.smh.com.au/national/range-of-groups-form-climate-coalition-20080706-32gi.html (viewed 1 June 2010).
48 ibid.
49 www.climateinstitute.org.au (viewed 29 August 2008). A four-page joint statement is available at: www.actu.asn.au/Images/Dynamic/attachments/6023/SCCCstatement-final.pdf (viewed 1 June 2010).
50 Spratt, 'Kevin Rudd'.
51 S Maher, 'Green jobs'.
52 http://www.actu.org.au (viewed 29 August 2008).
53 Burrow, p. 8.
54 AAP, 'Australia "could lead way in green jobs" '.
55 Salleh, 'Green New Deal', p. 17.
56 SCCC, *Stronger, Fairer, Healthier*.
57 http://www.climateinstitute.org.au/our-publications/reports/870-building-the-foundations-for-a-low-pollution-clean-energy-economy (viewed 5 October 2011).
58 SCCC, *Building the Foundations*.
59 http://www.safeclimateaustralia.org (viewed 4 August 2009).
60 Minchin.
61 Morton, 'Climate change most dangerous event'.
62 'Gore forms Safe Climate Australia Group'. http:// www.upi.com (viewed 4 August 2009).
63 Fyfe.
64 http://www.safeclimateaustralia.org/about-safe-climate-australia/ (viewed 6 October 2011).
65 http://www.greenlivingpedia.org/Safe_Climate_Australia (viewed 6 October 2011).
66 http://www.safeclimateaustralia.org/about-safe-climate-australia/ (viewed 6 October 2011).
67 http://www.climatepositive.org/run-for-a-safe-climate/ (viewed 6 October 2011).
68 http://www.safeclimateaustralia.org (viewed 4 August 2009).
69 ibid.
70 Dunlop.
71 SCA, 'Safe climate restoration'.
72 http://www.wentworthgroup.org (viewed 24 February 2010).
73 Cosier, p. 4.
74 ibid., p. 6.
75 http://www.grattan.edu.au/about_us.html (viewed 6 October 2011).
76 http://www.grattan.edu.au/programs/energy.php (viewed 6 October 2011).
77 'Grattan Institute Report on Emissions Trading Scheme and Carbon Reduction.' http://www.grattan.edu.au/publications/026_energy_report_22_april_2010.pdf.
78 Morrow.
79 http://www.climateinstitute.org.au/ (viewed 7 October 2011).

80 http://www.climateinstitute.org.au/policy-a-research (viewed 7 October 2011).
81 http://www.climateinstitute.org.au/about-us/about-us (viewed 7 October 2011).
82 www.globalclimatenetwork (viewed 13 June 2009).
83 http://www.climateinstitute.org.au/about-us/about-us/303-climate-institute-board (viewed 7 October 2011).
84 Climate Institute, 'World's first climate changed election?'
85 'Myth Busters', 11 September 2011. http://www.climateinstitute.org.au/our-publications/reports (viewed 7 October 2011).
86 Butler, 'Green groups take dirty money'.
87 http://www.climateinstitute.org.au/images/starratingdetails16aug2010.pdf (viewed 17 August 2010).
88 http://acfonline.org.au/articles/news.asp?news_id=3428 (viewed 5 October 2011).
89 Climate Institute, 'Passing the pollution policy test', pp. 1, 4.
90 ibid., p. 7.
91 http://www.tai.org.au/?q=node/1 (viewed 9 October 2011).
92 Frankel, p. 197.
93 Denniss, 'Can emissions trading save the planet?', pp. 82–3.
94 http://www.crikey.com.au/2008/12/16/denniss-by-2020-no-australian-polluter-will-live-in-poverty/ (viewed 5 August 2010).
95 Australia Institute, 'A win for voluntary action'.
96 Quoted in Butler, 'Emissions trading'.
97 Denniss and Richardson, pp. 1–2.
98 ibid., p. 7.
99 ibid., p. 9.
100 Australia Institute, 'A clean energy future?'
101 ibid.
102 Australia Institute, 'A win for voluntary action'.
103 Australia Institute, 'A clean energy future?'
104 *Age*, 5 May 2009, p. 1; Spratt, 'Time for a Plan B'.
105 Lawson, 'Looking back', p. 3.
106 Koh, pp. 45, 47.
107 Spratt, 'Suicidal sweethearts'.
108 Ross Garnaut, Interview with Agnes Koh, 19 April 2010, quoted in Koh, p. 45.
109 Quoted in Koh, p. 45.
110 Guy Pearse, Interview with Agnes Koh, 16 February 2010, quoted in Koh, pp. 45–7.
111 ibid., p. 60.
112 Pearse, 'The Climate Movement'.

8
Academics

This chapter discusses the domain of credentialed intellectual authority: the primary battleground where climate scientists and other academically qualified people battle to present the facts about climate change and argue for effective action, against the misinformation generated by climate change deniers.

University Research Centres and Activities

The greenhouse mafia has cultivated connections with universities; some, in keeping with the spirit of the corporatised university, have accepted linkages with the fossil fuel, mining and steel industries. An Australia Institute discussion paper in 2007 observed that: Gordon Martin (chairperson of Coogee Resources) was Chancellor of Curtin University of Technology; Michael Chaney (a Woodside director) Chancellor of the University of Western Australia; Jerry Ellis (a former chairperson of BHP) Chancellor of Monash University; and John Phillips (a former director of Western Mining Corporation) Chancellor of the University of Western Sydney. Fossil fuel companies, it noted, sponsor various university chairs and academic positions. The University of Adelaide has the Santos Chair of Petroleum Engineering, the University of Queensland the Xstrata Chair of Metallurgical Engineering, the University of Newcastle the

BHP Billiton Chair of History and Technology, the University of Wollongong the BHP Professor of Environmental Science, the University of Sydney the BlueScope Steel Professor of Steel Structures. Various fossil fuel companies, such as BHP Billiton and Exxon Mobil, fund research activities in Australian universities.[1]

This greenhouse mafia influence has attracted the attention of climate movement activists. An example is the furore around the Centre for Coal in Sustainable Development (CCSD), a consortium based in several universities, funded by large mining companies, including Rio Tinto, BHP Billiton and Xstrata Coal. It purports to 'optimise the contribution of coal to a sustainable future'. Barney Glover, Vice-Chancellor of the University of Newcastle, is a board member of the Centre and the University of Newcastle is a partner in the Centre. Other universities involved are the University of Queensland, UNSW, Macquarie University and Curtin University of Technology. Newcastle University students protested the university's partnership in this enterprise in April 2008, their letter of complaint contending that the university 'cannot provide independent research into climate change solutions while it is a CCSD partner alongside some of the world's largest mining corporations (Rio Tinto, BHP Billiton, Xstrata Coal)'.[2]

At the same time, many individual academics and academic research centres examine climate change and climate change adaptation and mitigation in a disinterested manner, free of direct funding from interested parties. They have played a distinctly positive role as players in the middle level of climate politics. Numerous academics in Australian universities are also engaged in teaching students about climate change and lecturing about it in various venues, including workshops and conferences both in the academy and outside of it. Most of these academics are physical scientists of one sort or another. Many others are political scientists, lawyers, sociologists, anthropologists, human geographers, historians and philosophers, and so on. Some are creative artists and art critics. For example, in 2008 RMIT University hosted an exhibition called 'HEAT: Art & Climate Change', which displayed twenty-two art works that examined the issue of global warming.[3] On 10 October 2009, the La Trobe University Museum of Art sponsored a symposium on 'Three Degrees of Change', which sought to extend the discourse on climate change to voices beyond the 'political and scientific communities' to others, such as artists, architects and urban designers.

The climate change issue reaches beyond the obvious domains of the sciences and social sciences and explodes disciplinary boundaries. For example, at a conference on 'Imaging the real life on a greenhouse earth' at the ANU on 11–12 June 2008, a group of climate scientists, earth and prehistory scientists, political scientists, an environmental lawyer, health and population experts, and a poet released a joint statement in which they described the urgent need for an effective response to global warming.[4] Another example of cross-disciplinary academic activity is the University of Melbourne's Inaugural Festival of Ideas on 'Climate Change/Cultural Change' on 15–20 June 2009, which acknowledged 'that Australia is on the sharp end of climate change' and debated strategies for mitigation and adaption. Patrick McCaughey, former Director of the National Galley, served as Festival Director. Speakers included Nobel Laureate immunologist Peter Doherty, economist Ross Garnaut, climate scientist David Karoly, public health expert Rob Moodie, novelist Kate Grenville, reproductive biologist Roger Short, political scientist and public broadcaster Waleed Aly, indigenous political philosopher Sana Nakata, poet and children's novelist Lisa Gorton, cultural critic and novelist Sophie Cunningham, and lawyer and novelist Randa Abdel-Fattah.[5]

Because most universities have researchers and research centres engaged in activities that inform the myriad debates around climate change, two examples must suffice: the Centre for Climate Law and Policy at the ANU; and the Melbourne Sustainable Society Institute at the University of Melbourne. There are also academics directly engaged with the grassroots climate movement from below, whose contributions we discuss in Part III.

The Centre for Climate Law and Policy (CCLP), part of the College of Law at the ANU, was created in 2007 to provide a focal point for law and policy research related to climate change, examining issues such as: international climate law and climate litigation; adaptation law and policy; the regulatory framework for emissions trading schemes; renewable energy laws; regulation of transport, forestry and land use emissions; environmental assessment and climate change; the politics of climate law; the relationship between federal, state and local climate measures; and human rights and climate change. It also runs courses in climate law and provides consultancy services. Professor Tim Bonyhady is director, Andrew Macintosh associate director. Its founding sponsor is law firm Baker & McKenzie. It was officially launched on 6 June 2008 by Peter Garrett.[6] CCLP has a working paper series.[7] It argued the abatement targets recommended in the Garnaut Review were inadequate.[8] So too was the structure of the CPRS. Macintosh commented on Crikey.com.au on 24 February 2009: 'Either Wong doesn't understand her own scheme or she is deliberately lying.' Impatience with the inadequate targets accepted by political leaders around the world was expressed in a CCLP paper about the impossibility of keeping warming within 2° C after the Copenhagen conference.[9]

The Melbourne Sustainable Society Institute (MSSI) is committed to research that fosters sustainability in Australia and the Asia Pacific region. In terms of vulnerability to climate change, MSSI states:

> Climate change poses risks to the things people value—such as our health, homes, environment and our way of life. Vulnerability refers to the potential loss or damage to these things we value. Vulnerability to climate change is not equally distributed. Factors such as class, gender, the nature of livelihoods, and location all determine who is at risk and the potential losses they may experience. Research into vulnerability involves understanding changes in both social and ecological systems, although more often than not it is social processes that are the most powerful causes of vulnerability.[10]

MSSI focuses on four themes: agricultural futures, risk and resilience, sustainable cities, and water. Of these, the one on risk and resilience particularly addresses climate change issues, such as climate change vulnerability and bushfire risk. The research leader of the risk and resilience theme is political geographer Jon Barnett, professor in the School of Land and Environment, who studies the social and institutional dimensions of vulnerability and adaptation to climate change.[11] MSSI arranges conferences and public lectures. For instance, on 11 May 2009, it sponsored a public lecture by Bill McKibben from the United States, renowned climate campaigner and founder of the 350.org movement, who spoke on ' "Because the world needs to know": building an international climate movement'.[12] Its website provides a link to a lecture by renowned ethicist Peter Singer, Laureate Professor at the university, on the ethics of climate change mitigation.[13]

Climate Scientists

There are very few climate change deniers with climate science credentials. Only two have received much attention in Australia: William Kininmonth and Ian Plimer. However, the statements of the tiny handful of deniers fuel the outbursts of more typical deniers such as conservative Sydney shock-jock Alan Jones, who asserted on 15 March 2011 that humans produce only 0.001% of carbon dioxide emissions.[14] (Also untruthfully, at the Convoy of No Confidence rally of truck drivers and other opponents of the carbon price outside Parliament House on 28 August, Jones blamed the poor turnout on police having prevented a thousand trucks from crossing the ACT border.)[15]

William Kininmonth, a retired meteorologist, is a vociferous opponent of anthropogenic global warming theory. He had a career in meteorological science and policy spanning more than 40 years. From 1986 to1998 he headed Australia's National Climate Centre at the Bureau of Meteorology, with responsibilities for monitoring Australia's changing climate and advising the Australian government on the extent and severity of climate extremes, including the recurring drought episodes of the 1990s. He served as Australia's delegate to the World Meteorological Organization's Commission for Climatology and was involved in the intergovernmental negotiations for the UNFCCC. Following the release of the IPCC Third Assessment Report in 2001, Kininmonth decided the IPCC conclusion about the largely anthropogenic basis to climate change was fallacious. He published *Climate Change: A Natural Hazard*, in which he finds systematic fault with the IPCC climate system model. The book launch was organised by the denialist Lavoisier Group, of which Kininmonth is a member, and was chaired by mining magnate Hugh Morgan.[16] In the final chapter of his book, he states:

> The assertion that recent warming is unprecedented relies largely on a reconstruction of surface air temperatures covering the last millennium that is fraught with problems of methodology and contentious data. The assertion that the climate system has only limited internal variability is an outcome of constraints on computer models that are necessary to ensure their stability. These assertions ignore a large body of paleoclimate evidence that points to climate having changed abruptly and relatively frequently in the past.[17]

A review of Kininmonth's book in the *Australian Meteorological Magazine* argued he failed to present the case for natural warming and his statements contradicted well-accepted work.[18]

Another academic denier is Ian Plimer, geology professor at the University of Adelaide and, like Kininmonth, a member of the Lavoisier Group. He published a book in 2009 in which he argues that climate changes are not due to anthropogenic activities that create emissions, but are driven by the Earth's position in the galaxy, the Sun, wobbles in the Earth's orbit, ocean currents and plate tectonics.[19] The book's initial print run of 5000 increased to 15 000 copies, boosted by approving reviews in the conservative press and favourable mentions by some federal MPs. David Karoly commented: 'If the book was handed to me as an assignment in my undergraduate earth science class, it would have failed—not because I don't agree with the answers, but it doesn't support the answers with sources.'[20] Michael Ashley, astrophysics professor at UNSW, reviewed Plimer's book:

> Plimer has done an enormous disservice to science, and the dedicated scientists who are trying to understand climate and the influence of humans, by publishing this book. It is not 'merely' atmospheric scientists that would have to be wrong for Plimer to be right. It would require a rewriting of biology, geology, physics, oceanography, astronomy and statistics.[21]

Plimer hosted Viscount Christopher Monckton, a renowned British denier and ultraconservative Catholic, on his tour of various

Australian cities between 27 January and 8 February 2010. Monckton described the Copenhagen Climate Conference as a 'sort of Nuremburg rally' and refers to the UN as the 'New World Order'. While here, Monckton wrote a report with Family First Senator Steve Fielding, addressed to IPCC Chair Pajendra Pachauri, complaining about the 'fradulent' nature of IPCC reports.[22] Gina Rinehart helped fund another tour by Monckton in 2011. Rinehart claims she has never met a leading scientist who believes in the connection between climate change and carbon emissions.[23]

This is strange because the realists definitely have the numbers, overwhelmingly, amongst scientists, especially climate scientists. Key individuals have intervened in public debates about climate change policy. They have also issued joint statements of concern directed at politicians and policy makers, particularly during the Rudd government. For example, in September 2008, sixteen climate scientists collectively stated that, if the government did not legislate to cut emissions by 25% of 1990 levels by 2020, then it 'will leave Australians with a legacy of economic, environmental, social and health costs that will dwarf the scale of national investment required to address this fundamental problem'.[24] On 1 August 2009, fifteen leading Australian climate scientists released a public statement in which they discuss anthropogenic contributors to climate change and the need for Australia and the world to 'act now'. The signatories included David Karoly and Will Steffen.[25]

David Karoly is Professor of Meteorology and Australian Research Council Federation Fellow in the School of Earth Sciences at the University of Melbourne since 2007 and a member of the Australian Academy of Science's National Committee on Earth System Science. He held the Williams Chair in Meteorology at the University of Oklahoma 2003–2007 and was involved, through several different roles, in the preparation of the IPCC Fourth Assessment Report. He admits to having been a climate change sceptic in 1987, but has become a leading public voice calling for a strong climate policy in Australia and is a member of the Wentworth Group of Concerned Scientists. In addition to his speaking engagements in academic settings and public forums, he has addressed climate rallies, including a Shutdown Hazelwood rally in Melbourne in November 2010, where he stated:

> Replacing all of Hazelwood would be an important symbol to Australia and the world that we are prepared to take action to clean up our energy supply. But it would be more than a symbol. It would also reduce future climate change and reduce the numbers of people impacted by flooding resulting from sea level rise.[26]

He also speaks regularly to community groups about climate change, for example, the Macedon Ranges Sustainability Group at the Woodend Neighbourhood House on 31 August 2011.[27]

Professor Will Steffen is the Executive Director of the Climate Change Institute at the ANU. He served on the panel of experts supporting the Multi-Party Climate Change Committee, and was previously Science Adviser to the Australian Department of Climate Change and Energy Efficiency. From 1998 to 2004, he was Executive Director of the International Geosphere-Biosphere Programme, a Stockholm-based network of scientists studying global environmental change.[28] He stresses that, of the developed countries in the world, Australia is the most vulnerable to climate change, with population largely on the coast (as sea levels rise faster than originally thought) and much of the country susceptible to drought. As Deputy Chair of the organising committee of the Copenhagen conference, he returned saying the conference was 'bad news' from a science side, the major messages being that the climate system is moving faster than anticipated, but 'good news' on 'technologies that we can use to reduce and mitigate the impact of climate change'.[29] In February 2011 he was appointed to the new Climate Commission to advise the government, which prompted the denialist Australian Climate Madness website to refer to him on 23 May 2011 as 'the Labor government's chief scaremonger'.[30]

Professor Barry Brook is another climate scientist with a public voice. He is director of the Research Institute for Climate Change and Sustainability in the School of Earth and Environmental Sciences at the University of Adelaide, which embraces a large number of very significant climate change research projects.[31] He has received a number of distinguished awards in recognition of his research excellence, including climate change studies. Brook often does public presentations on climate science, especially in Adelaide and environs. Controversially within climate movement politics, he is a proponent of fourth-generation nuclear reactors as a climate change mitigation technology. He maintains a blog, on which he posted this notice on 4 October 2011:

> I've long argued on this blog that fossil fuel replacement this century could, on technical grounds, be achieved via a mix of nuclear fission, renewables and perhaps also fossil fuels with carbon sequestration, with a high degree of electrification; nuclear would probably end up supplying over half of final energy.[32]

Very hostile to the nuclear option is Ian Lowe. President of the ACF since 2004, Lowe is Emeritus Professor of Science, Technology and Society at Griffith University, and adjunct professor at Sunshine Coast University and Flinders University. He is a Fellow of the Australian Academy of Technological Sciences and Engineering, was named Humanist of the Year in 1988 and an Officer of the Order of Australia in 2001 for services to science and technology, especially in the area of environmental studies. He has been a referee for the IPCC and attended the Geneva and Kyoto UNFCCC conferences. In 2009, Lowe won the International Academy of Science's Konrad Lorenz Gold Medal, carrying the title: 'Hope for the Future for a Sustainable World'.

Lowe has made countless contributions to newspapers, radio, television and periodicals; and is a frequent speaker at climate change gatherings. Lowe has also published a dozen books, many touching on climate change. In *Living in the Hothouse*, he delineates three broad approaches to reducing CO_2 production: expand use of fuels that are not carbon-based; reduce energy use by conservation measures of various kinds; and make preferential use of those carbon-based fuels that produce least CO_2 per unit of useful energy.[33] In the second edition of *A Big Fix: Radical Solutions for Australia's Environmental Crisis*, he called for radical changes that require 'not just new technology, though that is certainly needed, but also a new set of community values and attitudes that will inform our interaction with Earth's natural systems and each other'. He asserts that Australian society has been given 'the illusion' of success by liquidating our natural capital: coal, copper, lead, gas, beef.[34] He laments that most profits from the mining industry benefit overseas financial interests or are 'frittered away in consumptions that provides no enduring benefit'.[35]

Lowe argued that the Howard government's energy policies 'might have been drafted by the coal industry' and he was scathing about Howard's promotion of nuclear power: 'Promoting nuclear power as the solution to climate change is like advocating smoking as a cure for obesity. That is, taking up the nuclear option will make it much more difficult to move to the sort of sustainable, ecologically healthy future that should be our goal.'[36] Although he supports an emissions trading scheme for Australia, he also criticised the Rudd government for having made excessive concessions to carbon-intensive industries.[37]

Dr Graeme Pearman is a Fellow of the Australian Academy of Science and the Australian Academy of Technological Sciences and Engineering, and has been a member and chair of many Australian and international meteorological/climate change committees. He was awarded the CSIRO Medal (1988), a United Nation's Environment Program Global 500 Award (1989), Australian Medal of the Order of Australia (1999) for his services to atmospheric science and promotion of the science of climate change to the public, and a Federation Medal (2003). He was science adviser to Al Gore during his visits to Australia in 2006 and 2007.[38]

Most significantly, Pearman was a CSIRO scientist from 1971 and Chief of the CSIRO Division of Atmospheric Research from 1992 to 2002. Whilst he was chief, he joined the Australian Climate Group. He told ABC Radio on 13 February 2006 that CSIRO management pressured him about this: 'I was initially told that I shouldn't take part in the group, in an email and in phone calls, from my immediate superior at CSIRO.' He referred to the difficulties CSIRO scientists are facing:

> I think that the organisation itself is very frightened to be caught out in a sense saying things that are different to policy. And I think in the particular case of CSIRO, that's exacerbated by the fact that they're so short of resources, the funding, that they are reluctant, very reluctant to do

> anything that might feed back in terms of an effect on their budget ... There needs to be an acceptance, I think, not just by any particular government, but by the community at large that when you have groups of experts such as groups of scientists within the community, their greatest value is if they actually can be pretty fearless in what they say.

Pearman was made redundant by CSIRO in 2003.[39]

Nowadays, Pearman runs his own company, consulting for both the public and private sectors, including briefings to businesses on climate-change science and sustainability to aid risk assessments around the climate change issue. He has become particularly interested in the psychological barriers to climate change mitigation and adaptation, working on such human behaviour issues as an Adjunct Senior Research Fellow at Monash University.[40] He talks to climate change groups, for instance, he gave an address on 'A Psychological Take on Climate Change. The Real Issue is Us!' to Melbourne's Bayside Climate Change Action Group on 31 August 2011.[41]

Scientists and Social Scientists

In the following sections, we take a more detailed look at the ideas of some individual academics who are not specialist climate scientists, but have a presence in the public domain where climate change is debated. They present a wide range of views, indicative of the heterogeneity of the climate movement.

Mark Diesendorf

Dr Mark Diesendorf is deputy director of the Institute of Environmental Studies at UNSW, and formerly a principal research scientist at CSIRO, professor of Environmental Science at UTS, president of the Australasian Wind Energy Association, coordinator of the ACF's Climate Change Program and vice-president of the Australia New Zealand Society for Ecological Economics.[42] He is also a climate activist who generously shares his academic expertise on platforms at climate movement conferences, rallies and demonstrations, building connections between the middle layer of climate politics and the grassroots movement. Broadly speaking, his approach to climate change mitigation is green social democratic: 'Since technological change appears to be easier, faster and already under way, there is a case for focusing technological change while working at a more measured rate to stop growth in population and consumption per person and to redistribute wealth.'[43] Two of his books are valued contributions to both policy debates and climate activism.

In his 2007 *Greenhouse Solutions with Sustainable Energy*, Diesendorf argues that, with political will, Australia could halve its emissions in a few decades, utilising a combination of efficient energy use, solar hot water, gas, bioenergy, wind power, improved public transport and fuel-efficient vehicles.[44] He maintains that three drivers of energy demand are population, affluence, and technology choice. He recommends therefore that Australia implement a population policy that would not encourage parents to have additional children and limit professional and business immigrants (but without limiting refugee intake). Also necessary would be significant increase in the reuse, recycling, and reprocessing of materials; and the development of cleaner domestic industries to replace energy and emissions-intensive industries, such as smelting.[45]

Diesendorf strongly opposes nuclear power, because the alleged safeguards are inadequate, it would be unable to make a contribution to reducing emissions within 20 years and it is a diversion from renewable energy sources, which are the real solution. He argues that 100% renewables could be achieved by 2040 and Australia could eventually export solar power.[46] He deems hydroelectric, tidal and wave power as limited in potential.[47] He views wind energy as one of the least expensive and more environmentally sound of renewable sources of electricity.[48] In addition, biomass from existing agricultural crop residues could generate 39% of the amount of electricity generated in 2003–2004.[49] Solar hot water could provide for the vast majority of hot-water systems in Australia; and solar energy be used more intelligently for passive heating of buildings. Photovoltaic (PV) systems are suitable for residential use and solar thermal electricity (STE) for providing low-cost energy storage.[50] Geodynamics has received research and development grants from the federal government since 2002 and within a decade 'could prove itself to be a potentially large-scale source of base-load power at a price that is competitive with wind power'.[51] Higher urban density is one of the keys to greater use of public transport, a crucial component in the transition to a sustainable society. Cities should be gradually reshaped into a hierarchy of clusters, comprising transit cities, town centres, and local centres. The facilities for cycling and walking should be upgraded and governments should expend considerable funds upgrading railway tracks.[52]

Diesendorf delineates various barriers to 'sustainable development'. Firstly, there are cultural barriers, stemming from the widely held notion that humanity can and should continue with endless growth in energy consumption. Secondly, there are economic barriers, because existing economic structures emphasise energy-intensive mineral exploitation and processing, and the production and export of fossil fuels to the rest of the world. Thirdly, there is inadequate research, education, training and information to encourage energy efficiency and renewable energy technologies, whereas fossil fuel industries are 'well represented and well funded'. Fourthly, there are institutional barriers: the institutions that sell electricity, gas and petroleum products 'exclude most cost-effective opportunities for implementing efficient energy use'. Fifthly, there are political barriers: 'the political power of the big greenhouse-gas-emitting industries', exemplified in the huge government subsidies to them.[53]

The CPRS, according to Diesendorf, had 'so many failings' it was 'beyond redemption'. He contended it would: grant billions of dollars of free emissions permits to big polluters; encourage growth in EITE industries; permit unlimited overseas offsets; ensure that emissions reductions by households would reduce the reduction task of big polluters; not allow renewable energy to compete with coal, because its upper carbon price limit was initially $40 per tonne.[54] Yet he supports carbon pricing in principle as 'a vital policy ... encouraging shifts in the economic structure towards systems that are less greenhouse intensive' but, for a carbon tax to achieve its aims, its revenue would have to be invested in development of greener, renewable forms of energy. 'A carbon tax must be accompanied by government policies to support the construction of the necessary infrastructure.'[55]

His 2009 *Climate Action: A Campaign Manual for Greenhouse Solutions* stresses climate justice imperatives at both international and national levels. Because less developed countries have the least responsibility for causing climate change, the least capacity to alleviate impacts and will likely suffer the most damaging impacts, developed countries have 'a moral responsibility to take the lead in reducing greenhouse gas emissions and to compensate the less developed countries'. In practice this must involve a large transfer of wealth from rich to poor countries.[56] Domestically, he urges a 'socially just transition' that: compensates low-income earners for the impact of a carbon price and mandatory energy efficiency standards, especially those in private rental housing; provides government assistance to relocate and retrain workers displaced from polluting industries or pensions for those who cannot be assisted in this way; and offers incentives for the manufacture of components of new sustainable energy technologies in coal-mining regions.[57]

'The proponents of business-as-usual can see the writing on the wall for greenhouse polluting industries', he writes, so 'some are pushing with desperation to build new dirty coal-fired power stations, oil from coal and shale projects, motorways, smelters, etc., before these projects are widely recognised as monstrosities and banned'. The climate action movement must shut down these leftovers from the smokestack age. 'At the same time it must work constructively to create an ecologically, economically and socially sustainable society. This is the struggle for the future of the Earth and humanity.'[58]

Recently, he has admitted the fragile alliance of minority Labor government, Greens and three Independents 'offers the possibility of small improvements in climate/energy policies, such as the introduction of a low carbon price', but maintains that the 'only pathway to radical change is through substantial growth in the social movement for climate action'.[59] Like many green social democrats, he senses that greater legislative momentum from above depends on the strength of the movement from below.

Tim Flannery

Tim Flannery is an internationally renowned field zoologist, palaeontologist, explorer and conservationist, who regularly speaks on ABC Radio, BBC Radio and national public radio in the USA. Formerly director of the South Australian Museum, he became a professor at Macquarie University in 2007. He was granted the Australian

Humanist of the Year Award in April 2005 and Australian of the Year in 2007. He is a member of the Wentworth Group of Concerned Scientists and serves as Vice-Chair of the Australian Climate Group. He was Chair of the Copenhagen Climate Council in 2009. In early 2011 Flannery was appointed Australia's Climate Commissioner, to lead a panel of science, business and economics specialists to advise the community on climate change and 'build a consensus on how to reduce Australia's carbon emissions'.[60]

Flannery achieved much attention for his 2005 book *The Weather Makers: The History and Future Impact of Climate Change*, which forcefully describes the findings of climate science on the anthropogenic basis of climate change. His 2010 book *Here on Earth: An Argument for Hope* is also inviting much attention. Apart from producing more than 130 peer-reviewed scientific articles, many of them on climate change, he also publishes widely in the public domain, outlining his views on climate change and how to handle it.

Flannery calls for an Earth Commission for Thermostatic Control that would regulate emissions around the world; and insists a decline in global population is crucial in solving the climate crisis. Given record-breaking temperatures on every continent and tens of thousands of people marching in the streets to protest government inaction on climate change, Flannery considers the fossil fuel industry is on global notice: business cannot continue as usual.[61] He argues climate change will destroy the civilised world if an effective policy of mitigation is not implemented within the next two decades.[62] He advocates an ETS, but with significant government intervention: 'to make deep cuts in our emissions, carbon trading must be accompanied by far-sighted government regulation.'[63]

Flannery maintains three fallacious premises underpinned Howard government climate policies: renewable sources of energy could not provide base-load electricity so had only a limited role to play; funding and developing new technologies without imposing carbon taxes or trading schemes could avert catastrophic climate change; and funding 'clean coal' initiatives was sufficient to avert climate chaos.[64] Australia is in a 'unique position to develop and rapidly deploy clean coal technology', and Flannery supports this, but insists it is necessary also to develop renewable sources of energy, particularly solar and geothermal.[65] He also favours use of nuclear power.[66]

Flannery's most extraordinary contribution to climate politics is his proposal for the creation of 'Geothermia', a new city on the NSW/South Australia/Queensland border that would harness the 'fabulous riches' of the Cooper Basin: geothermal energy, natural gas and one of the world's richest solar energy resources.[67] Flannery argues that such a city could be completely energy-efficient and energy self-sufficient—and provide a model for future urban development around the world. He outlined this ecotopian vision in a major article in the *Sydney Morning Herald* on 21 December 2006 and again in his *Quarterly Essay* in 2008, 'Now or Never. A Sustainable Future for Australia?'

Noting that Australia was seen by many as 'the rogues who threaten the planet'—because of refusal to ratify the Kyoto Protocol, coal exports and the highest per capita emissions in the world—Flannery declared we need a new vision 'as grand and purposeful as the Snowy Mountains scheme ... that will polevault Australia from denying laggard to the forefront of innovation and best environmental practice'.[68] Australia could move swiftly away from its polluting energy base through development of the Cooper Basin. The overall energy embedded in this 'remarkable reserve' is three times larger than that embedded in Australia's total known reserves of natural gas and could power the entire Australian economy for around 75 years. The hostile surface presents another resource: 240 watts of sunlight falling on every square metre of gibber and sand almost every day of the year. Yet this is dwarfed by the potential of the region's geothermal resource, which could yield enough electricity to power all of Australia at the price we now pay for electricity generated by burning brown coal.[69]

He asked his readers to imagine: an Australia that decides to build on this natural wealth in clean energy, creating a centre for minerals processing and natural gas conversion by using emissions-free power; linking the north–south railway with Queensland and Western Australia to bring bauxite and other minerals to the processors, and then market them through the Port of Darwin; exporting gas and processed minerals from the mammoth Olympic Dam mine nearby; and providing Australia with a fully linked freight rail and electricity grid powered from zero-emissions sources. 'If you can do this, then you can imagine a nation transformed from climate-change pariah to leader in the fight for the survival of our planet.'[70]

All of this would require a new city in the desert. Let's call it Geothermia, he writes. What might it look like?

> I imagine solar-collector arrays towering over a low-rise city, providing shade and conserving soil moisture ... Perhaps the infrastructure would be underground, and the region planted with desert trees and shrubs to soften the local climate. Geothermia would be a city not of thousands, but of hundreds of thousands: a place with its own critical mass ... it would need to be a fully sustainable city—Australia's very first. Building it could teach us how to live in this harsh land of ours. And the building of Geothermia could itself become an industry, generating mountains of intellectual property applicable in desert cities around the world.[71]

If we are to maintain our prosperity and our environment 'then we laggard Australians must grasp such possibilities and make them real'.'[72] Should we succeed, the benefits would be enormous, for what Denmark is to wind so we would become to these most promising of new energy sources. 'In future, hot rocks and the sun may employ more Australians than coal ever did.' In building Geothermia we would become, for the first time, real Australians, because by learning how to live sustainably we would finally have a long-term future on this continent.'[73]

Flannery was lampooned by right-wing columnists, who ridiculed Geothermia without explaining its basis in geothermal energy. Andrew Bolt in the *Herald Sun* gave vent to a denialist spiel against 'global warming catastrophist Tim Flannery' who, 'in a panic urges us to embrace a new Utopia' that might be 'uneconomic'.[74] Tim Blair referred to 'celebrity bone-doctor Flannery' and said of Geothermia that a better name might be Fairybreadville, a rival city to Chocolatecrackletopia or Muskstickland. Blair presented a diary of a Geothermian into the future, where energy sources are practically non-existent, the major industry is 'arts grants' and the official unemployment level is 'everybody'. The entry for 3 May 2052 read: 'Major traffic snarl on the way home. Two recumbent bicycles got run down by a vintage Prius on the Hugo Chavez motorway, blocking traffic all the way back to that windchime factory near the Al-Quaeda Memorial. They should ban those spark-guzzling monsters.' He concludes with a Note: 'Geothermia was declared closed in 2070, shortly after the ambitious city's primary power supply collapsed due to unauthorized use of a Breville sandwich maker.'[75]

Clive Hamilton

Dr Clive Hamilton, formerly a development studies academic then public servant in the Bureau of Industry Economics and the Resource Assessment Commission, was director of the Australia Institute before becoming professor of public ethics at the Centre for Applied Philosophy and Public Ethics at the ANU in 2008. He has been an outspoken critic of the stances on climate politics taken by governments, Coalition and Labor; the pernicious influence of the greenhouse mafia; and the slowness of Australian society to respond to the looming crisis. Major works on climate politics include: *Running from the Storm: The Development of Climate Change Policy in Australia* (2001); *Scorcher: The Dirty Politics of Climate Change* (2007); *Requiem for a Species: Why We Resist the Truth About Climate Change* (2010). There are many articles, essays and media interviews too.

Hamilton's influential *Growth Fetish* (2003) also cuts to the heart of both cause and inability to solve the global warming problem. He deplores how political parties the world over believe the first objective of government is economic growth. The growth fetish notion appears in subsequent writing about climate politics. The CPRS, for example, indicated growth fetish. Both major parties believe it has 'magical powers that provide the answer to every problem' and that to question it is to question progress and prosperity. 'The fetish with economic growth is the principal obstacle to coming to grips with the threat of global warming.'[76] He fears too that the forces in favour of climate change mitigation, recognising that growth fetishism stands as an immovable obstacle to action on climate change, have also capitulated and begun to argue that 'we can have the best of both worlds, a healthy atmosphere and robust economic growth, and indeed that promoting renewable energy in place of fossil energy would accelerate economic growth'.[77]

Above all, Hamilton points the finger at greenhouse mafia influence. In February 2009, he described their relationship with the Rudd government:

> It was apparent early in 2008 that behind the scenes the fossil fuel lobby was organising. They martialled their troops and rearmed themselves with arguments, fighting funds, lobbyists and dodgy economic studies. They rebuilt their networks in government and the public service, insinuated themselves into policy processes, schmoozed back-benchers and dined privately with ministers and their staff. They whispered about how important the old energy industries are to the economy, how Labor voters value their jobs, and how they will take their business offshore.

Hamilton admits he was optimistic initially about Rudd government intentions on climate change mitigation, but about a year later recognised that the greenhouse mafia had conquered the new government.

> The climate emergency has turned into a crisis of democracy. The government is meant to protect the interests of the people, but it has protected the interests of the big polluters. The Government is in the thrall of a powerful group of energy companies and it is apparent even to the most dim-witted observer that these corporations are, as Thoreau wrote, 'more interested in commerce than humanity'.[78]

In mid-2010 when the Labor government was defending its inaction by reference to decreasing concern about climate change, Hamilton made the salient point that the softening of public attitudes did not erode the case for protecting ourselves from climate disruption. 'Public opinion has never stopped a drought.'[79]

On 5 December 2009, Hamilton ran as the Greens candidate in the Higgins by-election, the safe Liberal seat in inner-eastern Melbourne vacated by Peter Costello. Victorian climate groups formed a coalition to run a 'Vote Climate in Higgins Campaign'.[80] Labor did not field a candidate, helping Hamilton achieve a swing to the Greens of 21.7% in the primary vote and 39.8% in the two-party preferred vote. However, despite hopes Hamilton might clinch the seat, the Liberals won the seat with 54.6% to Hamilton's 32.4 per cent.[81] The Greens described it as 'a massive effort' showing 'extraordinary commitment' that contributed to the 'outstanding result'.[82]

While disparaging about the CPRS, Hamilton does favour a strong, well-designed ETS. In a recent interview in *Green Left Weekly*, he stated such a scheme would reduce emissions effectively 'because it puts an explicit, legally enforceable limit on emissions and then allows the market to set the price for carbon emissions'. In the case of a carbon tax on the other hand, Hamilton maintains that 'the market has the certainty of a fixed price and the environment has to absorb the uncertainty of how much carbon will be put into the atmosphere'.[83] He is deeply critical of carbon capture and storage as a pointless diversion from serious emissions reduction strategies. He regrets that many corporate leaders and politicians, the latter probably in thrall to the former, argue that the 'future of the world has come to depend on one factor above all others, the successful development and deployment of carbon capture and storage'.[84]

However, even an ideal ETS is not enough. Hamilton urges a broad range of supplementary activities, such as: 'public support for the development and commercialisation of low-emissions technologies'; the 'abolition of the system of fringe-benefits tax on company cars, which provides incentives to drive further'; 'congestion charging in cities, along with the abolition of toll rebates'; 'the introduction of vehicle fuel-efficiency standards'; 'investment in rail' and the 'construction of a very fast train from Sydney to Melbourne, through Canberra, and then extended to Brisbane'; a 'massive public investment in urban public transport'; and 'limiting the expansion of airports'.[85] Since Australia is experiencing 'substantial environmental damage' from its growing population and most immigration takes the form of business migration, the country should reduce its standard migrant intake from about 300 000 to 50 000 a year, but increase its humanitarian migrant intake to 50 000 'to expand the number of asylum seekers coming into Australia'.[86]

Hamilton has also paid careful attention to the social psychology of climate change. The deniers find a response amongst many people, because the message of climate science is, literally, frightening and the denialist story comforting. He refers to the phenomenon of 'climate numbing', which occurs because 'truly opening oneself to the science demands a distressing transformation of our unspoken assumptions about the future, ones conditioned by two centuries of technological advance and higher living standards. It is too difficult to contemplate the possibility of the end of progress …' However, amongst scientists who cannot avoid the implications of the science, 'there is a palpable mood of panic'.[87] In *Requiem for a Species* he observes: 'Without some unforeseeable stroke of luck, a warming of four degrees and more appears very likely. The best estimate is that we will reach that level in the 2070s or 2080, although if things go badly it could be as soon as the 2060s.'[88] Grim though the future of humanity seems further into the twenty-first century, Hamilton urges us not to despair but to move forward, ending his book with the counsel that 'only by acting, and acting ethically, can we redeem our humanity'.[89]

Ted Trainer

Ted Trainer is a social scientist and research fellow at UNSW. Since his influential 1985 book *Abandon Affluence*, he has argued that free-market capitalism and the wasteful consumer culture it spawns, is unsustainable; and the growing concern more recently with the issue of climate change has sharpened Trainer's critique.[90] An important premise in Trainer's thought is that the 'very high levels of production and consumption and therefore of energy use that we have in today's consumer-capitalist society cannot be sustained by renewable sources of energy'.[91] He acknowledges the superiority of renewable sources of energy over fossil fuels, and examines in great detail the contribution wind, solar thermal, photovoltaic solar, tidal and geothermal could make.[92] However, he concludes that the climate movement makes faulty assumptions about the degree to which renewable sources of energy can meet present levels of demand.[93]

This circumspection regarding renewable sources of energy makes Trainer especially concerned to find ways to restrain capitalism's tendency to overproduction and the associated encouragement of consumerism, which has been so destructive of the planetary environment. Trainer argues that 'the fundamental cause of the accelerating destruction of the global ecosystem is overproduction and overconsumption'.[94] Developed societies such as Australia exceed sustainable levels of production and consumption. Australians, he points out, 'can only live as affluently as we do because we are taking and using up most of the scarce resources and preventing most of the world's people from acquiring anything like a fair share'.[95]

Trainer calls for 'appropriate development' for both 'rich' and 'poor' countries. For poor countries, this would include a focus on local economic self-sufficiency; the utilisation of 'low, intermediate, and alternative technologies processing locally available resources'; and a commitment to environmental sustainability.[96] For rich countries, it would entail an enormous reduction in consumerism,[97] ideally a 'zero-growth economy' in which societies 'will work hard at reducing the amount of producing and consuming going on'.[98] Such a 'conserver society', committed to greater economic equality and with 'much lower than present rates of per capita income and resource use', would contribute to the creation of a 'world order that is peaceful, just, ecologically sustainable and in which inequality and poverty have ceased to exist'.[99]

Starting from the assumption that 'consumer-capitalist society' cannot be repaired, Trainer proposes a blend of social planning, cooperative economic endeavours and only a small 'enterprise sector'.[100] He elaborates upon this notion of the 'Simpler Way', which incorporates the following principles: far simpler material living standards; high levels of self-sufficiency within households, neighbourhoods, towns and nations; relatively little trade or transport; small-scale cooperative economies in which most of the things we need are produced by local labour from local resources; much greater use of renewable sources of energy; and a commitment to human rights and social justice, particularly with regard to developing countries.[101]

Describing himself as more anarchist than Marxist,[102] Trainer imagines a high degree of self-government in the society of the

'Simpler Way', but also government regulation to restrain market forces:

> The present desire for affluent-consumer living standards must be largely replaced by a willingness to live very simply, cooperatively and self-sufficiently. People must be conscientious, caring responsible citizens, eager to come to working bees, to think about social issues, and participate in self government. They must be sociologically sophisticated, aware of the crucial importance of cohesion, cooperation, conflict resolution, etc. They must have a strong collectivist outlook. They must understand and care about the global situation. Above all they must willingly choose and find satisfaction in materially simpler lifestyles.[103]

Neo-liberal globalisation has exacerbated the problems caused by consumer-capitalism. 'Inequality and polarisation are accelerating. The state has ceased to be concerned with redistribution of wealth.' Economic historian Karl Polanyi, he writes, stressed how misguided it is for a society to allow the market to be dominant and insisted that unless market forces are under 'tight social control' they will destroy society and its ecosystems; everything will be open to sale for maximum profit. All this is 'sociologically appalling', he writes: 'Great damage is being done to social cohesion, public spirit, trust, concern for the underdog, good will and concern for the public interest.' He outlines the preconditions for a better sustainable society.

> You cannot have a satisfactory society made up of competitive, self-interested individuals all trying to get as rich as possible! In a satisfactory society there must be considerable concern for the public good and the welfare of all, and there must be considerable collective social control and regulation and service provision, to make sure all are looked after, to maintain public institutions and standards, and to reinforce the sense of social solidarity whereby all are willing to contribute to the good of all.

It is not possible, Trainer insists, to have 'a good society' unless we make sure that considerations of morality and justice and the good of society are the primary determinants of what happens. 'In other words there must be much social control and regulation of the economy.'[104]

However, Trainer is also clear that there are limits to what regulation can achieve and those who believe 'technical fixes' are possible are expecting a green state to achieve the impossible. Climate change cannot be solved by attempting to regulate an economy 'that is still anarchic and still profit-motivated'. Nor can it be solved without jeopardising affluence and growth. He regrets that 'most people on the left ... believe that if we just get rid of capitalism and re-regulate and so on we can all live affluently'. This, he maintains is the 'crunch point' that the left has to think about, because 'the most fundamental issue on the global scene is the impossibility of growth and affluence'.[105]

Since the global economy is massively unjust, rich countries must 'move down to living on their fair share of global wealth'.[106] Trainer's critique of unsustainable consumption is based in a climate justice perspective, which he regrets is 'too small' within the international climate movement and its implications about how we live not fully appreciated by people on the left:

> They have a very good understanding of global justice but they think if we get rid of the domination of the capitalist class then we can liberate the capacity of new technologies so we can all live affluently. But if you go by the available global resources you realise that is not possible and we have to go for a simpler way kind of solution.[107]

In June 2011, Trainer argued that the clear and urgent task for the Eco-village and Transition Towns movements is to establish examples of the way out of the global predicament: development of local systems designed to meet our needs 'in conditions of frugal and cooperative self-sufficiency'.[108]

Notes

1 Hamilton and Downie, pp. 10–12.
2 Cunich.
3 Linda Williams, 'Reshaping the Human Self-Image'.
4 Glickson, *Imagining the Real.*
5 Baer, participant-observer notes; http://ideas.unimelb.edu.au/2009/about/welcome.html (viewed 30 June 2009).
6 http://law.anu.edu.au/cclp/index.asp (viewed 10 March 2010); http://law.anu.edu.au (viewed 23 September 2011).
7 For example, Macintosh and Wallace; Macintosh and Woldring; Macintosh, 'Keeping warming within the 2°C limit'.
8 Wallace, 'Garnaut's Targets'.
9 Macintosh, 'Keeping warming within the 2°C limit'.
10 http://www.sustainable.unimelb.edu.au (viewed 26 October 2011).
11 ibid.
12 Baer, participant-observer notes.
13 http://mpegmedia.abc.net.au/rn/podcast/2010/03/bia_20100307.mp3 (viewed 12 October 2011).
14 Quoted in K Williams.
15 ABC News, 29 August 2011.
16 'The global warming sceptics', *Age*, 26 November 2004. http://www.theage.com.au/articles.2004/11/26/1101219743320.html (viewed 18 July 2009).
17 Kininmonth, p. 207.
18 Walsh.
19 Plimer.
20 Morton, 'The sceptic's shadow of doubt'.
21 Ashley.
22 Hamilton, 'Viscount Monckton'.
23 Benedek.
24 *Australian,* 29 September 2008, p. 1; http://www.theaustralian.com.au/news/nation/climate-fight-only-worth-10-to-aussies/story-e6frg6nf-1111117611839 (viewed 2 June 2010).
25 Wilkinson, 'Scientists hit back'.
26 Karoly.
27 NTEU, Community Announcements, Email, August 2011.
28 http://climatecommission.gov.au/about/will-steffen (viewed 12 October 2011).
29 Interview in *ANU Reporter*, Winter 2009, p. 14.
30 http://www.australianclimatemadness.com/2011/05/will-steffen-labors-alarmist-in-chief/ (viewed 12 October 2011).
31 http://www.adelaide.edu.au/climatechange/research/ (viewed 15 June 2010).
32 Brook.
33 Lowe, *Living in the Hothouse*, p. 122.
34 Lowe, *A Big Fix*, pp. 3, 31.
35 ibid., p. 32.

36 Lowe, 'Reaction time', p. 19.
37 Lowe, *A Big Fix*, p. 49.
38 http://www.climateinstitute.org.au/about-us/about-us/303-climate-institute-board (viewed 7 October 2011); http://theconversation.edu.au/profiles/graeme-pearman-4465 (viewed 12 October 2011).
39 'Scientists make CSIRO censorship allegations', *AM*—Monday, 13 February 2006, 08:15, Reporter: David Mark. http://www.abc.net.au/am/content/2006/s1568322.htm (viewed 12 October).
40 http://theconversation.edu.au/profiles/graeme-pearman-4465 (viewed 12 October 2011).
41 NTEU, Community Announcements, Email, August 2011.
42 Diesendorf, *Climate Action*, frontispiece; http://www.ies.unsw.edu.au/staff/mark.html (viewed 11 October 2011).
43 Diesendorf, 'Strategies for radical climate mitigation', p. 104.
44 Diesendorf, *Greenhouse Solutions*, p. 5.
45 ibid, pp. 85–6, 268.
46 Mark Diesendorf, St James Ethic Centre and Wheeler Centre, public debate on 'Australia should embrace nuclear power', Melbourne Town Hall, 4 March 2010 (Baer, participant-observer notes); Mark Diesendorf, plenary address, Climate Action Summit, Canberra, 31 January–2 February 2009, quoted in Butler, 'Climate Action Summit'.
47 Diesendorf, *Greenhouse Solutions*, pp. 171–2.
48 ibid., p. 127.
49 ibid., p. 151.
50 ibid., p. 169.
51 ibid., p. 175.
52 ibid, pp. 204–209.
53 ibid., pp. 287–92.
54 Diesendorf, *Climate Action*, p. 105.
55 ibid., pp. 97–8.
56 ibid., pp. 16–17.
57 ibid., pp. 113–14.
58 ibid., pp. 198–99.
59 Diesendorf, 'Strategies for radical climate mitigation', p. 112.
60 http://www.timflannery.com.au/about-tim (viewed 11 October 2011); ABC News, 10 February 2011.
61 Flannery, 'Burnt and hostile'.
62 Flannery, 'Now or never', p. 9.
63 ibid., p. 35.
64 Flannery, *An Explorer's Notebook*, pp. 243-44.
65 Flannery, 'Now or never', p. 30, 33.
66 ibid., p. 35.
67 ibid.. p. 37.
68 Flannery, 'Burnt and hostile'.
69 Flannery, 'Now or never', p. 35; Flannery, 'Burnt and hostile'.
70 Flannery, 'Burnt and hostile'.
71 Flannery, 'Now or never', pp. 37-8.

72 ibid., p. 38.
73 Flannery, 'Burnt and hostile'.
74 Bolt.
75 Blair.
76 Hamilton, *Requiem for a Species*, p. 35.
77 ibid., p. 41.
78 Hamilton, 'Climate emergency', p. 2.
79 Quoted in *Weekend Australian*, 5–6 June 2010, Inquirer p. 5.
80 Vote Climate in Higgins Campaign, leaflet, n.d., 1p.
81 http://www.abc.net.au/elections/federal/2009/byelections/higgins_result.htm (viewed 28 April 2010).
82 http://greens.org.au/node/5416 (viewed 28 April 2010).
83 'Listen to the climate scientist', *Green Left Weekly*, 29 September 2010, p. 11.
84 Hamilton, *Requiem*, p. 41.
85 Hamilton, 'Climate change', pp. 153–154 .
86 'Listen to the climate scientist', *Green Left Weekly*, 29 September 2010, p. 11.
87 Hamilton, 'Climate change and its implications', p. 9.
88 Hamilton, *Requiem*, p. 196.
89 ibid., p. 226.
90 Butler, 'Abandon affluence'.
91 Trainer, *Renewable Energy*, p. 2.
92 ibid., pp. 41, 57, 72, 91, 107, 109.
93 ibid., p. 7.
94 Trainer, *Saving the Environment*, p. 8.
95 ibid., p. 10.
96 Trainer, *Developed to Death*, pp. 199–201.
97 ibid., p. 196.
98 Trainer, *The Conserver Society*, p. 108.
99 Trainer, 'A green perspective on inequality', p. 124.
100 Trainer, *Towards a Sustainable Economy*, p. 143.
101 Trainer, *We Must Move to the Simpler Way*.
102 Butler, 'Abandon affluence'.
103 Trainer, *We Must Move to the Simpler Way*.
104 ibid.
105 Quoted in Butler, 'Abandon affluence'.
106 Trainer, *The Transition*.
107 Quoted in Butler, 'Abandon affluence'.
108 Trainer, The window of opportunity, pp. 1, 3.

9
The Union Movement

Capitalist economies are characterised by underuse of labour resources and overuse of environmental resources. Corporations tend both to reduce labour costs and use the cheapest production methods possible, regardless of ecological consequence. Thus employment options are restricted at the same time as the planetary environment is degraded.[1] James O'Connor expanded upon the 'second contradiction' of capitalism in 1998. The first contradiction of capitalism is overproduction: the relentless profit-driven expansion of markets. The resulting oversupply of products causes profit rates to decline and economic downturns, as demand cannot equal supply. The second contradiction within capitalism is its tendency to destroy the environment on which it depends for its profit-making, thus hastening its own destruction. It is caused by 'underproduction' of the conditions of production: capitalist enterprises externalise costs and thus fail to renew or repair them. These conditions therefore become 'underproduced', whether they are ecosystems from which materials are extracted or contaminated by industrial pollution, the health of workers or state-supplied infrastructure. A crisis of undersupply of these crucial conditions occurs because capitalist firms are impelled by the profit-motive to avoid and resist environmental protection and public spending: 'There is no profit in maintenance or preservation, or actions taken, and resources expended, to prevent bad things happening that would otherwise occur. The profit is in expansion, accumulation, and marketing something old or new at lower costs.'[2]

The two contradictions within capitalism generate social movements: labour movements by its propensity to create economic inequality; and ecological movements by its predisposition to degrade the environment. O'Connor therefore urges the essential compatibility of red and green social movement aims. To cement such a natural alliance, he suggests a focus on the principle of 'Preservation First' against accumulation-driven capitalism, so business would not be subject to the contradictions of capitalism. In place of production for profit, he advocates that business should be engaged 'to preserve, defend, and enhance the conditions of life and life itself'.[3]

Green Jobs

This 'natural' red–green alliance is starting to emerge, judging by the practice of union and green groups cooperating together to produce a new discourse around green employment, which has started to replace the formerly hegemonic rhetoric of 'jobs versus environment'. This alliance is obviously encouraged to the extent that business is engaged to 'to preserve, defend and enhance the conditions of life and life itself'. Renewable energy targets, accompanied by carbon pricing, are part of such engagement. A lengthy CSIRO report in 2008 concluded that a rapid transition to 'a low-carbon, environmentally sound society' would have little or no impact on national employment and there would be much to gain: 'greater wellbeing, cost-saving and greater competitiveness and reduced environmental impact'.[4] In July 2009 Rudd criticised those who 'constantly scaremonger about the possible loss of jobs through the transition to a lower carbon economy' for failing to talk about 'the new clean energy jobs of the future'—and he launched Labor's national conference with a pledge to create 50 000 green jobs or training opportunities, at a cost of $100 million.[5]

Though politicians talk much and do little, the newfound desirability of government policy discouraging business-as-usual overuse of resources and underuse of labour—due to both climate change and economic recession—nonetheless promotes the increasingly popular phenomenon of green-collar workers or green jobs. The term 'green collar worker' was first used in hearings before the US Congress in 1976.[6] But who are green-collar workers? An extremely complicated taxonomy was developed in 2009 by the Environment Institute of Australia and New Zealand, which decided there were three factors that described a green-collar worker: the skills and responsibilities of the individual; the industry and nature of the organisation for which they work; and whether the job and the organisation tend towards the environmental or sustainable end of the green spectrum.[7] The ACF definition emphasises outcomes: 'Green jobs or green-collar jobs contribute to better environmental outcomes or increased sustainability. Green-collar jobs range from low-skill, entry-level positions to high-skill, higher-paid jobs, and include opportunities for advancement in both skills and wages.'[8]

The fact that employees, unlike employers, have an interest in increased use of labour and decreased use of resources was stressed as long ago as the late 1970s by an organisation established by unionists in partnership with the ACF: Environmentalists for Full Employment. Its National Convenor was Jack Mundey, secretary of the NSW Builders Labourers' Federation during its famous 'green bans' period 1971–1975, when building workers refused to construct environmentally damaging development.[9] Mundey argued:

> There is a growing awareness that technology used primarily for profit, erodes jobs, and even worse, denies job opportunities for future workers. Furthermore most high technology consumes enormous capital and resources. Overall the need for humankind to reject blind adherence to all economic growth, to examine in detail which economic growth is socially desirable, beneficial and necessary, is now greater than ever before. The need to harmonise with nature is essential if civilisation is to live on in the 21st century.[10]

Mundey outlined the urgent task for Environmentalists for Full Employment: 'The carefully orchestrated myth that the fight for a decent environment increases unemployment must be exploded.' His organisation argued that capital ear-marked for environmentally harmful projects such as increasing electricity generation or uranium mining could provide much more social benefit and many more jobs if invested in alternative environmentally sound activities.[11]

The tradition of manual workers campaigning on environmental issues was continued around this time in the Movement Against Uranium Mining, with transport and maritime workers' refusal to handle uranium restricting uranium mining and export. The frontal challenge of the green bans movement and the union campaign against uranium mining prompted business organisations to commence a propaganda war designed to alienate unionists and environmentalists from each other. The theme was, predictably, that environmental protection costs jobs.[12] Dave Kerin, a Builders Labourers' Federation member and environmental activist, argues that from the mid-1970s 'employers saw an opportunity to divide workers from environmentalists, and so began an internal conflict within the workers' movement which presented the problem as "jobs versus environment" '.[13] The union movement, Mundey noted, was 'caught in a false dichotomy, presented with choosing between jobs or the environment ... a contradiction that has been fostered by opportunist politicians, corporations or government bureaucracies'.[14] To counter that impression and provide positive solutions to ecological crisis, he argued unions had to engage creatively in designing socially responsible job-creation schemes. 'So much damage has been done to the planet that just to repair the damage will require millions of workers.'[15]

Companies engaged in unsustainable development continued to engage in 'job blackmail' to bend governments to their will. It was difficult for unions to contest this dominant 'jobs versus environment' notion. Nonetheless Australian unions were amongst the first in the world to push the issue of green jobs and were instrumental in encouraging renewable energy and green business initiatives in the 1990s, in advance of green capitalists.[16] In 1991 the Australian Council of Trade Unions (ACTU) identified green employment as a key issue for the labour movement.[17] In 1993, the ACTU and the ACF joined forces to set up a 'green jobs' program, announcing they 'recognised that they had a common interest in the creation of jobs which are environmentally beneficial'. The ACF/ACTU Green

Jobs Unit, funded by the federal Departments of Employment, Education & Training and Environment, Sport & Territories, was established in Melbourne to conduct research and training programs, organise job placements, and promote green jobs throughout the country, especially for unemployed workers, young people, Aboriginal, and disabled people.[18]

As these green jobs initiatives were wound down by the Howard government from 1996, a union/green group called Earthworker was formed in 1997, sponsored by the Victorian Trades Hall Council. By 2000, Earthworker had three environmental organisations affiliated, and fourteen unions or divisions of unions representing workers in construction, manufacturing, services, tertiary education, nursing, journalism and entertainment.[19] Earthworker argued that issues of sustainability and sustainable work had to become as basic a part of the union movement's agenda as wages and conditions. It was envisaged that this material project would encourage a discursive transformation, creating what Earthworker called 'a jobs *and* environment dialogue'.[20] One of its success stories was its 'Solar Wind and Water Alternative Energy Plan', which brought together unions, alternative energy activists, academics, corporations, and local and state governments to create a wind industry in Victoria, cheaper access for householders to solar hot water units and thousands of jobs in manufacture, installation and administration. It was also involved in ACF and construction industry union initiatives to create a greener construction industry, in design and products used.[21]

Despite continuing efforts on the part of corporations and media to pit 'greenies' and workers against each other, green jobs' initiatives and links between unionists and environmentalists persisted; and now green employment initiatives are flourishing. Some unions are even going it alone. Backed by the Gippsland Trades and Labour Council and inspired by the 200 000 cooperatives in Venezuela run under workers' control, the Earthworker Social Enterprise suggests organised labour itself could create green jobs and push Australia towards a sustainable future. While others in the Latrobe Valley heard climate change sounding the death knell for the local economy, the Earthworker Social Enterprise created an alternative vision of 'Workers leading the way to a green economy'.[22] This Latrobe Valley union initiative drew the attention of a newspaper on the other side of the world. The *Guardian* reported on 7 September 2010:

> A co-operative of union activists is working with local government, farmers and workers to establish a factory that makes solar water heaters, which would be funded with pension funds and state incentives for renewable power. Their aim is to provide jobs for former mine workers, reduce energy costs and cut emissions. Dave Kerin said it was necessary to find an alternative way to address the climate crisis, because political methods were not working.[23]

When government support failed to eventuate, in 2011, Kerin established the Earthworker Cooperative, supported by the Australian Manufacturing Workers' Union (AMWU), to enlist 100 000 Australians to contribute $20 to establish the 'Eureka's Future Workers Cooperative' in Morwell, to manufacture and then sell solar hot water systems collectively. More generally, the Earthworker Cooperative is 'a social enterprise with a mission to create realistic solutions for transitioning Australia's workforce into a low carbon economy'. It aims to establish more cooperative factories in regional areas where large proportions of the workforce are dependent on carbon intensive industries/energy generation.[24] Kerin stated at the Earthworker Cooperative's launch on 11 August 2011 on the steps of Victoria's Trades Hall: 'We are going to grow Australian manufacturing jobs which produce the renewable energy goods and *work* our way out of the climate emergency.'[25]

The debate has shifted in ways favourable to green employment for two obvious reasons. Climate change has focused attention on the destructive nature of certain industries. The GFC has encouraged scepticism about the capacity of free-market capitalism to bring prosperity and security to all classes of society. In this context there is growing space for argument that challenges capitalism's tendency to underuse labour resources and overuse environmental resources. Class and nature as issues that used to pit labour movement activists and greenies against each other have become more reconcilable. This becomes even more evident if global warming and global economic crisis are not seen as unrelated events, each separately providing 'political opportunity' for environmentalists and unionists. Both have been brought to the planet by unrestrained capitalism.

Ariel Salleh has criticised the new green Keynesianism for reinstating an eco-modernist strategy, directed at saving capitalism.[26] This is to a large extent true, but its potential for transcendence of the red/green binary is nonetheless a positive outcome that could overcome barriers between labour and environmental movements that have prevented more effective critique of capitalism's second contradiction. This chapter examines the ideas enunciated by Australian unions as the best means to save both the planet and peoples' livelihoods, in the period since 2006, by looking at the ACTU; the Construction, Forestry, Mining and Energy Union Mining and Energy Division (CFMEU ME), the union most directly involved in greenhouse gas emitting industries; and efforts to build the 'natural' red–green alliance depicted by O'Connor.

The Australian Council of Trade Unions (ACTU)

The ACTU was active and agitated about climate change more than twenty years ago. In 1991 it adopted a 'Policy on Greenhouse' that called for the immediate implementation of all available low-cost measures to reduce greenhouse emissions; supported federal government initiatives in the international negotiations on a climate change convention; called on governments to fund research into energy efficiency and renewable energy technologies; acknowledged that reduction of emissions would require a mixture of pricing, taxation and regulatory measures, and argued price measures alone would have an adverse effect on low-income earners and that regulatory and planning measures would be the most efficient and effective means; and called on federal and state governments to develop a national approach to encourage improved performance in all energy production, distribution and use, for example, in power stations, vehicles, buildings, plant and equipment and domestic appliances.[27]

In July 1992 the ACTU published *The Greenhouse Effect: employment & development issues for Australians*. In the introduction 'Greenhouse—An Incentive for Social and Economic Change', ACTU assistant-secretary Jennie George announced:

> The enhanced greenhouse effect is with us now. Its impacts, and measures to counteract it, will increasingly affect Australian society and the everyday lives of working people ... it means moving away, over time, from heavy reliance on our fossil fuel industries to a more diverse and less resource-intensive economy.
>
> Australian working people and their trade unions must be active in this process. It is working people who will bear many of the costs of measures to reduce greenhouse gas emissions, and who will be responsible for implementing the necessary changes in the workplace.

George strongly urged trade unions, individual unionists and the wider community to take up the challenge presented by the greenhouse effect, because: 'It is likely to be one of the greatest tasks facing Australian society and industry as we enter the twenty-first century.'[28] The concluding chapter on 'Trade Unions' Vital Role' urged high-level union collaboration with governments, business, and academic, scientific, consumer, environment and welfare groups, to determine 'how measures can be implemented in a way which maximises benefits, minimises costs and which at least maintain social equity'. It called on union members to make greenhouse issues part of regular union activities, to work with environment groups on it and aim to develop 'an awareness in environment groups of the importance of employment, industry development and social equity'.[29]

Had this 1992 vision been implemented, it would have positioned Australia as the global leader in climate change mitigation and greatly reduced the cost of transition to a low-carbon economy. The ACTU remained concerned about the issue and has recently become highly vocal. On 16 February 2007 the ACTU convened a national Union Forum on Global Warming and Sustainable Energy to develop solutions to climate change. ACTU President Sharan Burrow announced the main aim was for the union movement to take a stronger role in providing leadership on the issue: 'we must act now—not in 10 years time'. She reiterated the connection between ecological sustainability and employment protection: 'the development of renewable energy sources and cleaner technologies offer the prospect of new growth areas for jobs and the Australian economy that need to be urgently explored.'[30]

The following day the ACTU published its 'Principles and Policy on Global Warming', which described global warming as 'the pre-eminent policy challenge of our time' and stated: 'Unless decisive action is taken now to reduce greenhouse gas emissions, the planet we bequeath to future generations will be harsher and more hostile to the human condition than that which we have inherited.' Concerted national and international effort to reduce emissions carried potential for better jobs: 'Industry must face up to global warming and be accountable for investing in sustainable jobs rather than raising the fear of job losses and expecting government handouts. Australian unions are committed to playing a constructive role in meeting the challenge.' This publication emphasised climate justice issues: the 'environmental consequences of our production and consumption must be reckoned alongside the efficiency of its generation and the fairness of its distribution'. It reiterated the ACTU's commitment to 'reduce poverty, raise living standards, create decent jobs, provide opportunity to all'.[31]

In July 2008, the ACTU established the SCCC, discussed in Chapter 7. A few months later, as the GFC struck, the ACTU also collaborated with the ACF to release *Green Gold Rush: How An Ambitious Environmental Policy Can Make Australia a Leader in the Global Race for Green Jobs*. Linking job-creation with climate change mitigation, this booklet identified six industries where strong climate action and industry policy could create an additional 500 000 green jobs by 2030: renewable energy, energy efficiency, sustainable water systems, biomaterials, green buildings, and waste and recycling. 'In this time of economic uncertainty, one of the few good news stories is the continued prospects for the growth of green industries.' With the right policies, 'we have an opportunity to turn action to combat our environmental challenges into growth of the green economy, securing jobs and industry well into the future'.[32]

The ACTU submission on the CPRS maintained a 30% emissions reduction below 2000 level by 2020 was possible 'without major technological breakthroughs or lifestyle changes and at minimal cost to working people'. It also argued any compensation to companies must be conditional and tied to investments in renewable and low-carbon technologies.[33] Burrow took issue with industry voices 'dominated by those peddling fear in order to maximise compensation'. She asked: 'surely they must now be prepared to face up to the reality of global warming and be accountable for investing in more efficient practices ... how about some maturity from the business community and some shared responsibility for creating a smarter economy that builds a sustainable future for all?' She claimed that by 2030 there could be 850 000 jobs in green industries; and policies should ensure the investment necessary to win a big share of what would be a $US3 trillion global green market by 2020. 'The global financial crisis and the potential impact on employment show another way that we can make green investment work for the economy, the environment and jobs.'[34]

The ACTU Congress in June 2009 endorsed a seven-page policy document on 'A Fair Society: Environment and Climate Change Policy', reiterating the ACTU's commitment to 'urgent and decisive action to transition Australia to a low carbon economy' and rejecting the notion that such action would damage the economy, because 'as the global economy switches to a low carbon future, the result will create multi-trillion dollar markets for low carbon and clean energy technologies, infrastructure and consequently production processes', in which Australia should compete.[35] However, in addition to the usual commitment to 'promote growth in green jobs in associated/new industries and services', this document also inserted a commitment to 'ensuring that the transition process protects jobs in carbon intensive industries' and to 'support those industries to achieve best of sector standards to remain competitive'.

Previous ACTU policy statements had emphasised the need to 'protect jobs', with the strong implication that job protection would be ensured overall by increased green employment opportunities. This 2009 document supported the International Trade Union Confederation's 'just transition' policy statement, yet its own position of protecting jobs in carbon-intensive industries was now out of line with the Confederation's emphasis on transition away from such forms of employment. It urged 'a strong emissions reduction cap' in the projected ETS, and consequent auctioning of permits 'must be just, support vulnerable Emissions-Intensive Trade-Exposed industries during transition but be designed to ensure all industries are required to make the changes necessary to get to best practice standards to ensure global competitiveness and survival of Australia's traditional industries'. Also, where the ACTU had previously urged a steeper emissions reduction target than the government's, and regardless of other countries' actions, this document supported the government's position of a reduction target of 25% by 2020 'in the event of a comprehensive global agreement to stabilise emissions at 450 ppm or better'.[36] On the other hand, this document did state that market measures alone would be insufficient to achieve the necessary changes, and must be complemented by direct government regulation and incentives for investment in new green jobs and training programs targeting areas from which carbon-intensive jobs are at risk.[37] The ACTU and its affiliates committed to 'co-ordinate industry-wide campaigns for industry policy, investment and/or regulation necessary for making the transition to a low carbon economy'.[38]

Late in 2009 Burrow led representatives of Australian unions at the Copenhagen Climate Summit. Admitting this did not deliver the anticipated binding global treaty to combat global warming, the ACTU stated it would continue to lobby governments worldwide for strong climate action, including 'a just transition' for workers, 'which means embedding decent labour standards and good quality jobs as the new clean energy economy emerges at home and around the world'.[39] Around this time, the ACTU, in conjunction with the ACF, released a report on 'Creating Jobs. Cutting Pollution', which pointed out: 'Action to reduce pollution can go hand-in-hand with job creation and a prosperous and environmentally healthy Australia. Australia could create more than 770 000 extra jobs by 2030 by taking strong action now to reduce pollution.'[40]

It appears that the fiasco of Copenhagen might have strengthened those voices within the ACTU that had caused the 'watering down' of the ACTU's position on climate change evident in the 2009 policy document. The ACTU website for several years until mid-2010 had announced that: 'Climate change is the most pressing social and economic challenge we face, and Australians will be hit hard if we do not respond urgently to climate change.' It called upon the government 'to develop policy and drive investment towards new and cleaner technologies, and new industries—to establish Australia as a global leader and to take advantage of economic opportunities'.[41]

However, currently the ACTU website states that: 'Climate change is one of the most pressing economic, social and environmental challenges we face ...' The significant alteration here is that climate change is now only one of the most pressing challenges. It also incorporates the additions in the 2009 policy document, shying away from acknowledging the need for transition away from coal mining and coal-fired power generation:

> Unions are determined that adjusting to a low pollution economy must focus on support for emission-intensive and trade-exposed industries, measures to protect existing jobs, programs to attract investment in clean energy and production, and assistance to low income households. Some industries and regional economies will be impacted more acutely than others. It is imperative that these communities are assisted to successfully adapt their industries and build new ones to ensure decent living standards, job opportunities and services continue to thrive in these areas.[42]

One might perceive here the hand of the Mining and Energy Division of the CFMEU, which represents workers in the hugely significant coal-mining and coal-power electricity generation sectors. Yet, as we see in the following section, this union, especially under the leadership of Tony Maher, has never denied the science of global warming and the need to move to a low carbon economy. According to Maher, speaking at a climate change conference convened by the National Tertiary Education Union (NTEU) 28–29 April 2011, the commitment to protect existing jobs was made to counter employer attempts to scare workers into opposing mitigation policies.[43] Another possible influence is the Australian Workers Union, which represents workers in aluminium smelters and other carbon-intensive industries. National Secretary Paul Howes, involved in the replacement of Rudd with Gillard when Rudd was proposing the mining super-profits tax, opposes restrictions on uranium exploration.[44] On the other hand, Howes publicly supports the Gillard government's carbon pricing and dismisses aluminium industry claims that its recent problems are caused by this.[45]

Apart from this identifiable backtracking, the ACTU website otherwise reiterates the message that 'It is simply not true that the shift to a low pollution economy will result in job losses. Research commissioned by the ACTU and ACF demonstrates that Australia still has an unparalleled opportunity to create hundreds of thousands of "green collar" jobs.'[46] In February 2011 the ACTU reassured members that workers' interests were being represented as the government negotiated the proposed carbon price, which was 'a vital step forward in taking action on climate change'. It stated: 'Unions representing workers in all sectors and industries are more than willing to make a strong start on the transition to a low pollution economy as long as the requirements for a "just transition" are identified, examined in detail and properly addressed.'[47] In April 2011 the ACTU published a colourful pamphlet entitled *Climate Change is Union Business*, which argued: 'Job creation and action on climate change are closely connected.'[48] ACTU support for the Gillard government's carbon price stressed the connection between climate change mitigation and green jobs with 'just transition'. New ACTU President Ged Kearney stated:

> Australia's future economic and environmental security requires the Federal Government taking action to reduce carbon pollution and grow cleaner industries. Australia can create up to a million clean energy jobs with the right policies, but we need them to be implemented as quickly as possible to drive investment so we can begin the economic transformation and support workers to adjust.[49]

CFMEU Mining and Energy Division (CFMEU ME)

If any union had good reasons to be in the business of climate change denial, it would be the CFMEU ME, concerned as it is with protecting workers in sectors of the economy most responsible for greenhouse emissions. This has not been the case. The CFMEU ME has long acknowledged the seriousness of global warming and the role of fossil fuels in causing emissions; but aims to ensure that the necessary process of reducing these emissions does not disregard the rights of workers in these industries to economic security and alternative employment options. This union's position has often been misunderstood and wilfully misinterpreted, including by mining companies wanting to protect business-as-usual jobs.

In 1992, the union's researcher, Peter Colley, wrote one of the first union publications on climate change anywhere in the world (the ACTU's *The Greenhouse Effect*); and the union represented Australian unions at the UN Earth Summit where the UNFCCC was adopted. In 1997 it was part of an international union contingent in Kyoto when the Protocol to the UNFCCC was adopted.[50] It supported ratification of the Kyoto Protocol and argued for social justice to be a key consideration in development of climate change responses. In 2001 it co-wrote the climate change policy of the International Federation of Chemical, Energy, Mine, and General Workers. Union general secretary Tony Maher claims that from 1990 to 2006 the CFMEU EU 'championed a progressive policy in the international trade union movement' and 'coal miners were consulted and asked to support the policies adopted by the union'.[51]

In November 2006 the union conducted a national plebiscite of all members to respond to its policy position on climate change, which recommended: ratification of the Kyoto Protocol, because the reasons for not ratifying it were 'spurious at best and more likely just plain lies'; an ETS; massive investment by the industry in low-emission coal power; an energy-efficiency program; and action in transport and agriculture, including reduction of land clearing. It took this policy to members at workplace meetings around the country. Maher commented: 'Our members are sick of being demonised and seen only as part of the problem. They are proud of the fact that the union is positioning miners as being part of the solution. It has been an empowering experience.' The policy was carried nearly unanimously. It was welcomed by the ACF, WWF-Australia and the Climate Institute but Howard commented it would destroy jobs. Maher noted: 'John Howard's lies go unchecked by the industry he is pretending to support'.[52]

Late in 2006 the union launched a discussion paper to reinvigorate 'its public role in the climate change debate'.[53] It called for all stakeholders to work together 'to address the threat to humanity and the environment that is posed by global warming'.[54] It continued to consult with its members. In May 2007, the union's National Convention of around 400 delegates from every coal mine and power station in the country unanimously endorsed a million-dollar television advertisement to refute 'this falsehood that climate change inaction protects jobs. Nothing could be further from the truth.'[55] The ad ran on television stations in mining regions in late 2007. Featuring hard-hatted employees in industries highly complicit in emissions, its message was: 'Protect our Jobs and Communities. Choose a Government that's serious about climate change.'[56]

On 16 August 2007 Maher delivered a major speech to the NSW Minerals Council Environment Conference. 'History will harshly judge the Howard Government's decade of denial and inaction. Whether it harshly judges the industry or the union will depend on what we now do.' He observed that in all the discussions he had had with politicians, business leaders, environment groups and workers, 'nobody has ever argued that signing the Kyoto Protocol would cost jobs. Yet John Howard continues to say it ... the industry stands by and lets patent untruths be told.' Finally, he commented that mineworkers were empowered by the climate change debate and were having their say through their union: 'it's a campaign not just for our jobs; it's a campaign for the future of the industry and the environment'.[57]

The union's 'Climate Change Policy' released in September 2007 emphasised that burning fossil fuels such as Australia's coal played a major role in greenhouse emissions. It argued: 'We must contribute responsibly to seeking solutions to the clear threats presented by human-caused climate change' and 'our members and all Australians need to be mobilised in their homes, communities and workplaces to discuss climate change and how to combat it'.[58] Its November 2007 Briefing Note commenced with the observation that the world needed Australian coal but emissions from its use were contributing to climate change. 'They must be substantially reduced.' Its conclusion reiterated the social justice aspect of mitigation strategies: 'We must all learn to live and work differently. The challenge must be met equitably and fairly if it is to win the widespread community support essential for success.'[59]

The union's questionable enthusiasm for CCS was outlined in December 2007. 'Carbon Capture and Storage: making it happen' noted that the Stern Review had calculated that, even with reorientation to renewables, most of the world's energy needs would still be supplied by fossil fuels after 2050, so it stressed the need to develop cleaner ways of obtaining such energy.[60] In conjunction with WWF-Australia, it produced a paper in 2009 on CCS called 'Power to Change' in which it stated that it recognised 'the severity of the global warming problem' and supported the increased deployment of renewable energy and the introduction of emissions trading, but also saw the development of CCS as critical to the future of coal use in a carbon-constrained world. 'Global warming is a huge threat to the future of humanity and the planet. The CFMEU does not resile from the challenge—we have faced many great challenges in advancing the interests of working people over more than 100 years.' Its reasoning was essentially that it was clear that the developing world will rely on coal and gas for decades to come, so 'we must reduce emissions from their use if we are to stop dangerous climate change'.[61]

On 10 September 2008, the union's submission to the Green Paper on the CPRS acknowledged 'the imperative to act quickly' to meet 'the greatest challenge of our generation, and perhaps of the entire twenty-first century'.[62] Global warming, and responses to it would affect every household and every workplace. 'It is therefore "core union business".' In this context the goal for unions is not to support or parrot other stakeholders, governments or political parties:

> The task of unions is to ensure that working families are protected from global warming to the greatest extent possible, and that responses to global warming take into account the imperative to meet the concerns of working families and utilise the capacities they have to positively contribute to solutions. Mitigating global warming is not a technical exercise consisting of tweaking economic instruments and technologies. In many respects it will be a massive exercise in social engineering. It will fail if it does

not integrate the concerns and capacities of working people.[63]

Under the heading 'Social justice', it stressed that, beyond its core constituency of workers, unions recognise the need to provide opportunity and protection to the disadvantaged in our society, including those not in the workforce—the young, the old, the disabled etc. It is also the case that many working people and their families are disadvantaged due to lack of skills and opportunity. 'Global warming responses that do not have a strong social justice component will tend to exacerbate disadvantage, as costs will be imposed that the disadvantaged have less capacity to meet.'[64]

The union's position has been misrepresented in the media, perhaps wilfully. In July 2009, the *Australian* seized on a comment by Tony Maher to run a headline: 'Green jobs dopey—union leader.' The point Maher was making was that white-collar workers were much more comfortable with the notion of 'green jobs', whereas he was concerned with fitting his blue-collar constituency into 'a restructured low-carbon economy' and ensuring that green industries could provide attractive jobs: 'A coalminer or a power station worker isn't going to leave their job on $120 000-plus with well-regulated shift arrangements and decent conditions to install low-wattage light bulbs or insulation.' Moreover, the existing coal and steelmaking industries had an important part to play 'in producing a low carbon future', because renewable energy infrastructure would require massive amounts of steel that should be produced by Australian workers.[65]

At the SA Unions Climate Change Conference in Adelaide on 24 September 2009, Maher elaborated on the peculiar position in which his members find themselves He noted that the CFMEU faces several challenges, including that many of its members are 'killing the planet' and are working for an industry with a poor image. Maher commented that the union constitutes 'fertile ground for scepticism'. A 2007 member survey indicated that about half of them had felt victimised by 'anti-coal groups'. He pointed out that miners have been 'demonised', but unfairly, because the CFMEU favoured ratification of the Kyoto Protocol, the implementation of an ETS and investment in 'low emissions power'. On the energy debate, Maher argued that the 'truth is that we need them all, CCS, geothermal, uranium, renewables'.[66]

Recently, Maher argues that the fate of the Rudd government indicates that mining companies are growing in power. They demanded $10 billion in free permits to compensate for the 30% price fall they experienced during the global crash in early 2009. When the Rudd government agreed to only 10% of this, they moved to kill the CPRS altogether. Malcolm Turnbull was serious about a bipartisan policy on carbon reductions, and so mining companies supported internal Liberal Party moves against him. Rudd's RSPT only amounted to an extra 8% taxation, but, based on their success with the CPRS, mining companies decided to kill the tax and, if possible, the Labor government.[67] On 6 July 2011, Maher called on the ACA to 'end its carbon scare campaign', noting 'the coal industry is making record profits—more than $1 million per worker'.[68]

Meanwhile, in the Latrobe Valley, the CFMEU has been working with community groups to develop a 'transition plan' to phase out coal while ensuring workers in the coal industry are not left behind. Branch secretary John Parker stated: 'The social contract of the coal industry will have waned in the next five years unless they come up with something that doesn't produce the carbon or captures the carbon. They will have missed the boat.'[69]

Beyond the Red and the Green

In 2009, Emergency Services Officers at Hazelwood power station were on strike for almost four months, demanding better pay and conditions. International Power responded by hiring non-union labourers not adequately trained for the work. On Wednesday 8 July 2009 environmental activists joined the picket line at Hazelwood as part of the annual Students for Sustainability conference. Defying all dietary stereotypes, the union cooked a vegetarian BBQ for the students. Not covered by mainstream media, Indymedia reported: 'it was clear that the interests of climate change activists and power workers are aligned. There'll be no jobs on a dead planet, and there'll be better workplace conditions in a renewable sector with workers control!' Again, on 27 July, environmental activists met at FOE headquarters in Collingwood before dawn for a long day-trip to support the striking Hazelwood workers.[70]

The 'jobs versus environment' propaganda constructs a stereotype of inner-city latte-sipping greenies who care nothing for coal miners' livelihoods and a stereotype of manual workers hostile to environmental issues and subsequently to climate change mitigation policies. Yet greenies of all sorts have diligently pursued just transition strategies; and, as Tony Maher, said on ABC radio in 2008: 'to think that people don't care for the environment just because they have heavy industry jobs is a lot of rubbish'.[71] Fred Rose's study of labour and environmental movements in the USA examined the 'politics of division that confronts the public with tragic choices between protecting economic well-being and preserving the environment'. Powerful interests are served by pitting citizens and their legitimate issues against each other: 'Dividing movements and social groups is a well-established strategy for undermining opposition and distracting public attention from issues of power and common interests.'[72]

The current climate and economic crises point to the common interests of labour and environmental movements in challenging capitalism's tendency to underuse labour and overuse environmental resources. Not only are there no jobs on a dead planet, it is now increasingly acknowledged there are more jobs in a sustainable green economy than a 'brown' one. Targetting the 'twin crises of economy and ecology' is both necessary and possible.[73] By prioritising the use of labour rather than resources, a sustainable green economy could guarantee stable and high levels of employment without the need for the harmful overall economic growth and expansion that underpins employment security in a brown economy. The rhetoric of 'green jobs' inspires climate activists and has broad acceptance amongst unionists, because it does not require a shift away from the traditional union emphasis on employment security.

In July 2008, the AMWU conducted a poll of its members, which revealed that 93% of the members working either directly or in subsidiary energy intensive industry supported action on climate change if there was government support for new environmentally sound manufacturing industries. Members also indicated they were much more concerned about climate change impacts and the kind of world future generations would inherit than the effects of climate change mitigation measures on jobs. AMWU National Secretary Dave Oliver said the results of the poll were clear: 'Manufacturing workers want Australia to act to tackle global warming. Ninety per cent of members agree that the Federal Government must invest in creating jobs in new manufacturing industries that won't contribute to climate change.'[74] These results are remarkably similar to those revealed in a survey of 2000 members conducted by the NTEU late in 2009, which discovered more than 97% were concerned to fight climate change and that over 75% felt this was a 'core union issue'.[75]

The Queensland Public Sector Union (QPSU) acted as a trail-blazer for white-collar workers to demonstrate their commitment to climate change mitigation by working with an ENGO on the issue. Early in 2008, the QPSU surveyed its members and found that, for 80% of them, tackling climate change was a high priority. In March 2008 it launched the Climate Heroes campaign, supported by the ACF.[76] The campaign commenced with more than 500 'Climate Heroes', supported by 29 000 QPSU members and supplied with materials by the ACF, committed to 'hero missions' every four weeks around themes like waste reduction, transport, energy efficiency and even climate-friendly food. The program stated: 'The 500 climate heroes will collectively work with thousands of people across their workplaces, each making small but important changes to their daily practices. When large numbers of people work together like this, even small changes can have a big effect.'[77]

The ACTU and ACF decided to build on the success of this QPSU initiative. In August 2009, Union Climate Connectors was launched to provide an accessible forum for all union members to participate in climate change action, with the ACF to assist in training and to provide resources.[78] Burrow's email to Australian unionists inviting them to sign up as a Climate Connector emphasised: 'Unions have a proud history of campaigning on environmental issues. But rarely has the stakes been as high as they are now. That is why unions around the country are joining together to encourage members to be part of the climate solution.'[79]

The Climate Connectors website affirms the centrality of unions in meeting the challenge of climate change: 'Union members have a vision for a clean energy future, with clean, green workplaces; and have a leadership role to play in developing and demanding the action we need to get there.' It emphasises the positive outcomes for workers: 'We know that those solutions have the power to create millions of jobs in a clean energy economy—jobs for us, and jobs for future generations.' It situates Climate Connectors within the traditions of unionism: 'In taking up this challenge, we draw on a proud union history of struggle and achievement—through our collective efforts we have lifted living standards and improved the quality of working life; and we have campaigned for fairness and justice in our communities.'[80]

The campaign's activities include learning, communicating, implementing changes and lobbying. The ACF provides four Action Kits (Energy, Food, Jobs and Water), web tools, phone advice, and training in how to campaign for climate change mitigation in both the workplace and larger society.[81] The Jobs Kit gave details of the potential for green jobs in particular industries and occupations and started with the observation:

> We have to stop polluting Australia and damaging our planet. By being smart we can take action on climate change to cut pollution and create new, and better jobs. It takes 49 tradies and apprentices to build 1000 solar panels—and we need millions! Wind turbines have over 10 000 parts. Who's going to make and install them? ... Clean energy technology ... is an exploding field that will touch every industry. The good news is, it's not 'jobs or the environment.' Cutting pollution creates jobs.[82]

Climate movement organisations have been quick to promote green Keynesianism. For example, the Green Institute organised a Green New Deal conference 23–25 October 2009, which saw about a thousand people crammed into the largest lecture theatre at the University of Melbourne to hear Bob Brown's public lecture on 'The Greening of Politics'. There are also regular conferences organised jointly by unionists and climate activists.

An example is the Trade Union Climate Change Conference on 9 October 2010, sponsored by the Victorian Trades Hall Council and the Climate Emergency Network (CEN) and attended by over seventy people from fourteen different unions. Amongst the climate movement speakers, Matthew Wright from Beyond Zero Emissions presented its plan for 100% renewable stationary energy by 2020.

Chris Breen, climate activist and AMWU Print Regional Councillor, argued that a price on carbon would just raise prices for everyone without necessarily reducing emissions and called for direct regulation of emissions and government investment in renewables and infrastructure in the same way that governments in the past had 'built every bit of coal-fired power stations'. Union speakers included AMWU official Mike Nicolaides and CFMEU ME Latrobe Valley President Luke Van der Meulen.[83]

Nicolaides stated that the AMWU supports carbon pricing, but prefers a carbon tax to an ETS, so long as it involves ambitious targets for both emissions reductions and increased renewables. In addition to a carbon price, it favours direct government intervention, since market forces alone cannot solve the climate change problem that has been caused by market failure. 'Our members have had enough of market forces', he added. The AMWU rejected the 'jobs versus environment' notion, since the development of renewables would utilise more labour; the jobs would be different but there would be 'lots of them', and, with some adaptation and retraining, the skills needed for a green economy were already there in the brown economy. The costs of transition to renewables should not fall exclusively on workers displaced from the brown economy, so the wider society had to demand just transition strategies; and unions had to be in the thick of the debate about how to shape the transition and ensure the new green jobs were also union jobs.[84]

Van der Meulen agreed free markets were the problem not the solution to climate change, that capitalism was literally 'killing us'. Governments were simply puppets of big business, especially the fossil fuels corporations, hence the 'miserable 20% reduction target'. The Labor Party was beyond redemption, so 'we need to build socialism'. In the meantime, 'we can have workers' cooperatives under capitalism' and he spoke enthusiastically about the Earthworker Social Enterprise. His union favoured systematic development of 'climate change jobs', relocating workers from petrol and coal industries to solar and other renewable industries; and just transition strategies to ensure that money was directed to green jobs and compensation to displaced workers not power corporations. It was important, he said, that anti-Hazelwood protesters made their support for just transitions clear to the local community.[85]

The conference ended with an open discussion on union strategy and a resolution commending the BZE plan, demanding government commitment to substantial emissions reductions and policy to develop and direct new industries for renewable energy. It called for 'just transitions', because 'it must not be workers and their communities that are forced to accept lower living standards and be disadvantaged'; and 'One thing is clear—we cannot rely on the free market to support those who are most vulnerable and disadvantaged.' After an intense debate, the final version of the resolution included an addition that read: 'We reject market based solutions like carbon pricing and instead call for regulation, and direct government investment in renewable energy, to really ensure the "polluters pay" for the transition we need.'[86]

A Union Climate Action Network (UCAN) was established following this conference, which has met regularly, usually in the Victorian Trades Hall. It campaigns in support of the BZE plan 'within unions and for unions to fight for real solutions, both for workers and the climate'. It stands for 'A just transition to jobs-rich renewable energy solutions—Union involvement and unionised jobs—A ten year transition to 100% renewables—Government regulation not market failure—No new coal.'[87] On 29 September 2011, it helped organise the Rally for Renewable Energy, Jobs and Regional Communities on the steps of Victoria's parliament, prompted by the Baillieu government's restrictions on wind farms and cuts to the solar feed-in tariff.[88]

Another significant conference bridging the red–green division was the Union and Community Summer School 2010 at Melbourne Trades Hall, 10–11 December. A central focus was 'Unions and the climate emergency', which discussed four important matters: jobs *and* environment (breaking down union suspicion of environmentalism); how to create climate jobs; building links with the grassroots climate movement; and the role of social/cooperative production. Speakers included AMWU Environment Officer Colleen Gibbs, NSW South Coast Labor Council secretary Arthur Rorris, Workers' Solidarity Network member Joe Montero, CFMEU Victorian Secretary Luke Van der Meulen, FOE Climate Campaign coordinator Cam Walker, Eureka's Future campaigner Dave Kerin, NTEU Victorian secretary Colin Long, and Western Australian Maritime Union of Australia delegate Sam Wainwright. A workshop on the green bans of the 1970s, featuring Jack Mundey, showcased the significant historical connections between the labour and green movements.[89]

Union action on climate change, in partnership with climate activists and networks such as CEN, also aids unions as organisations. Unions have suffered significant membership decline in the past quarter century. Marj O'Callaghan, an educator/organiser with the ACTU, argues unions must do more than bargain for better pay and conditions or they will fail to speak to workers who are satisfied with their work. So 'values campaigning' initiatives such as Union Climate Connectors provide unions with a 'positive agenda' that creates opportunities to organise new groups of workers and 'broaden the understanding of what unionism, collective action and solidarity are about'.[90] Climate change mitigation is especially well suited to values campaigning, because climate change is educational in and of itself: it is not only an issue relevant to the immediate interests of unions but one that provides a cogent critique of neo-liberalism, which has discouraged timely solutions to the global warming problem. Climate change mitigation suggests the advantages of social democratic collectivist values. These are the values that underpin unionism and wherever and whenever these values are weakened, unionism suffers, as we have seen in the past few decades.

Ariel Salleh argues the first requirement of a green new deal should be to help people understand how the dominant global capitalist system relies on the abuse of nature and the exploitation of labouring bodies.[91] The 'jobs and environment' initiatives do fall short of offering any significant challenge to capitalism, but they certainly assist to break down the divisions between unionists and greenies that have for so long strengthened those who profit from the exploitation of both labour and nature.

Notes

1 Polanyi; Schnaiberg; Benton; O'Connor, 'Capitalism, Nature, Socialism'.
2 O'Connor, *Natural Causes*, p. 317.
3 ibid., pp. 314, 316, 319.
4 Hatfield-Dodds et al, pp. 29–30.
5 Quoted in Coorey.
6 Connection Research, p. 12.

7 ibid, p. 23.
8 http://www.acfonline.org.au (viewed 1 June 2010).
9 M Burgmann and V Burgmann.
10 Mundey, 'Compatibility of the 3 "E's" Vital to Survival', p. 1.
11 Environmentalists for Full Employment.
12 Burgmann et al, *Unions and the Environment*, pp. 4–6.
13 Kerin, 'Challenging the myth', p. 19.
14 Quoted in *Age*, 19 August 1989, Extra p. 20.
15 Mundey, 'From grey to green', p. 20.
16 Burgmann et al, *Unions and the Environment*, pp. 23–6.
17 ACTU, *The Greenhouse Effect*, inside back cover.
18 Green Jobs Unit; Burgmann et al, *Unions and the Environment*, pp. 23–24.
19 McNaughton, pp. 37–38; de Boehmler.
20 McNaughton, p. 44.
21 Earthworker, 'From Fossil Fuels to Renewables'; *Age*, 15 November 1999; McNaughton, pp. 44–45.
22 Kerin, Workers creating Green Jobs.
23 http://www.guardian.co.uk/environment/2010/sep/07/Australia-green (viewed 10 September 2010).
24 Earthworker Cooperative, '100,000 Australians'.
25 Earthworker Cooperative, 'Media Release'.
26 Salleh, 'Green New Deal', p. 15.
27 ACTU, *The Greenhouse Effect*, inside back cover.
28 ibid., inside back cover.
29 ibid., pp. 25–26.
30 ACTU, 'Unions to pursue climate change solutions'.
31 ACTU, 'ACTU's Principles and Policy on Global Warming'.
32 ACF/ACTU, *Green Gold Rush*, p. 2.
33 ACTU, 'Campaign: Clean Energy Jobs'.
34 Burrow, p. 8.
35 ACTU, 'ACTU Congress 2009', pp. 1–2.
36 ibid., p. 3.
37 ibid., p. 5.
38 ibid., p. 7.
39 ACTU, 'Campaign: Clean Energy Jobs'.
40 ACTU/ACF, 'Creating jobs—cutting pollution'.
41 ACTU, 'Campaign: Clean Energy Jobs'.
42 ACTU, 'Campaign: Climate Change and Jobs'.
43 Burgmann, participant-observer notes.
44 'Uranium exploration a step in the right direction', 16 February 2012. http://www.awu.net.au/443735_5.html?H|0|443735|443735_52 (viewed 8 March 2012).
45 'Workers face year of living dangerously', 19 February 2012. http://www.awu.net.au/246519_2.html?H|A|246519| (viewed 8 March 2012).
46 ACTU, 'Campaign: Climate Change and Jobs'.
47 'Your Rights at Work—February Newsletter', 4 March 2011.

48 ACTU, *Climate Change.*
49 http://acfonline.org.au/articles/news.asp?news_id=3428 (viewed 5 October 2011).
50 CFMEU ME, 'Climate Change'.
51 T Maher, 'Crib Rooms and Climate Change', p. 1.
52 ibid., pp. 1–4.
53 http://www.cfmeu.com.au/index.cfm?section=29&Category=61&viewmode=content&ContentID=252 (viewed 8 July 2010).
54 CFMEU ME, 'Climate Change'.
55 T Maher, 'Crib Rooms and Climate Change', p. 4.
56 CFMEU ME, 'Climate Change'.
57 T Maher, 'Crib Rooms and Climate Change', pp. 1–6.
58 CFMEU ME, 'Climate Change Position Paper'.
59 CFMEU ME, 'Climate Alert!'.
60 CFMEU ME, 'Carbon capture and storage', p. 2.
61 CFMEU and WWF, 'Power to change', p. 1.
62 CFMEU, 'Response to the Carbon Pollution Reduction Scheme Green Paper', p. 2.
63 ibid., p. 9.
64 ibid., p. 9.
65 S Maher, 'Green jobs dopey—union leader'.
66 Baer, participant-observer notes.
67 Tony Maher, Speech to Australia Institute forum, 6 August 2010, reported in *Search News*, vol. 17, no. 4, August 2010, pp. 2–3.
68 Quoted in Alcorn, 'Coal industry'.
69 ibid.
70 Indymedia.
71 Quoted in *Australian*, 15 September 2009.
72 Rose, pp. 5, 8.
73 Butcher and Stilwell, p. 121.
74 AMWU.
75 NTEU.
76 QPSU, 'QPSU and ACF unite'; QPSU, 'About Climate Heroes'.
77 QPSU, 'QPSU and ACF unite'.
78 Reale.
79 Email from Union Climate Connector team, 2 September 2009.
80 Climate Connectors.
81 Reale.
82 Climate Connectors Jobs Action Kit.
83 Climate Change Conference—Saturday, 9 October 2010', assorted documents; Baer, participant-observer notes; Burgmann, participant-observer notes.
84 Burgmann, participant-observer notes.
85 ibid.
86 ibid.
87 UCAN, p. 2.

88 UCAN, email, 23 September 2011.
89 Burgmann, participant-observer notes.
90 ACTU Organising Centre.
91 Salleh, 'Green New Deal', p. 15.

Part III
Climate Politics From Below: the Grassroots Movement

10

Constructing the Climate Movement

In contrast to the middle-level sections of the climate movement, the movement that has emerged from below in Australia in recent years is less institutionalised—parts of it not at all—and its tactics tend to be confrontational rather than conventional: protest not pressure, more leafleting than lobbying. Yet there is considerable overlap between these layers: many members of middle-level organisations are involved to varying degrees in the grassroots movement; and many grassroots activists have connections with middle-level organisations. Hamilton has drawn attention to the renewed activism on the climate change issue of established ENGOS, campaigning alongside new groups and networks agitated about climate change and government inaction.[1] Their renewed activism is largely driven by the movement that has surged up from below. As Michael Green notes: 'This is a movement where the grassroots element is taking the lead and the NGOs are following, some of them faster than others.'[2]

The existence and rapid growth of the grassroots movement expresses ground-level scepticism about the willingness of institutionalised components of the climate movement to argue for radical mitigation measures and their capacity to apply sufficient pressure on governments to apply these. The mushrooming of this movement from below also indicates recognition that, even if institutionalised climate activists have the most honourable intentions and determination to succeed, it is only through ground-level mass mobilisations that middle-level negotiators carry any clout, given the reluctance of governments to make the tough decisions necessary to mitigate climate change. 'Governments and corporations would not be remotely green', as Tim Doyle points out, 'without the existence of the more non-institutional parts of the environment movement.'[3]

According to James Whelan, who runs the Change Agency, a Brisbane NGO that consults for activists, the climate movement is like none other. Perhaps overlooking the anti-conscription campaign during World War I and the anti-Vietnam war protests of the 1960s and early 1970s, he claims that:

> In the history of social movements in Australia, you can't find a parallel. There's nothing like it for its diversity, for its rate of growth, and for its inclusiveness. It includes coal miners. It's rural. It's urban. And it's a mistake for anybody to think the climate change movement is part of the environment movement. The climate movement is a much bigger beast. You can hold a public meeting in any urban centre in Australia now, and initiate one or more climate action groups.[4]

The grassroots movement is disparate in terms of participants, analysis of the sources of climate change, strategies for action, and proposed solutions for mitigation and adaptation.[5] It is diverse precisely because its constituency is large, bearing witness, as Green emphasises 'to a wellspring of concern rising from deep within the nation's psyche'.[6]

Individual activists and organisations often draw on a range of green philosophies and approaches in articulating their ideas about climate change: green social democracy, eco-socialism, deep ecology, eco-anarchism and eco-feminism are some of the most common. A dominant outlook is green social democracy: persuade politicians and corporate elites to prevent dangerous climate change through a government-supervised shift to renewable energy sources, facilitated by carbon pricing. Eco-socialists, also well represented within the movement, maintain that climate change mitigation ultimately requires the transcendence of capitalism, but because this cannot await the collapse of global capitalism, far-reaching reforms are necessary in the interim, such as: public ownership of utilities; expansion of public transport; minimisation of car use; ending of all coal-based energy sources; redistributive policies; greatly reduced consumption; perhaps a steep carbon tax, but one which adequately compensates low-income groups.

It is difficult precisely to pigeonhole organisations or even individuals: these categories exist on a spectrum where one merges into the other; and most organisations are heterogeneous and many individuals eclectic in their political world views. So left green social democrats are not easily distinguished from many eco-socialists; an organisation such as CEN or AYCC would have members of different political persuasions; and an individual climate activist might subscribe to a range of views. For example, Salleh's complex theoretical work has sought to reconcile eco-socialism with eco-feminism, an intellectual project that engages female climate activists frustrated with male domination of women and nature—and of the movement trying to prevent this destruction.[7]

It is much easier to identify climate justice perspectives wherever enunciated, whether focused on 'primary' or 'secondary' injustices in Eckersley's terminology. Concern with climate justice cuts across ideological and strategic divisions within the climate movement, but is least likely to be expressed within the ecomodernist paradigm. Green social democrats in the middle level of climate politics have expressed concerns about climate justice, as we have noted, but it is from below that the climate justice perspective is most strongly articulated, whether from within green social democrat, eco-socialist, eco-feminist or eco-anarchist viewpoints. While many grassroots activists prefer to focus on specific issues, such as renewable energy sources, many others call strongly and in broad terms for social justice for those who are least complicit and will be the most adversely affected by climate change, for instance by rising sea-levels (primary injustice) or rising electricity bills (secondary injustice).

Diversity of ideological and strategic leanings is characteristic of all significant social movements. This can be frustrating for activists attempting to cooperate together to effect a bigger political impact.

However, as Stilwell argues, 'breadth and diversity are also sources of strength in a politics of opposition', for there is 'a distinctively dialectical element' in the process by which radicals and reformists within a movement interact with each other.[8] Social movements, especially large and influential ones, are constructed, typically, out of the creative energies and even tensions within them. In this chapter we look at some examples of the disparate forces that have played a significant role in climate movement construction; and how diversity has facilitated this process.

Friends of the Earth and the Climate Action Research Group

Founded in California in 1969, FOE was established in Australia in 1971. FOE is an egalitarian, ultra-democratic organisation, the most radical of the principal green organisations, committed to ground-level activism. It argues that social and environmental issues cannot be separated: true environmental sustainability will not be realised without the establishment of socially equitable societies. Many of its activities are based in local communities, in keeping with its grassroots principles, but it has run an impressive nationwide Climate Justice Campaign since the turn of the millennium.[9]

FOE's Climate Justice Campaign is based on the notion that:

> Climate justice is a positive vision of the future, where all people enjoy equitable and sustainable access to the world's carbon resources. Climate justice also refers to the right to compensation and protection against the impacts of climate change for the peoples of the world who have not produced excessive amounts of greenhouse gas emissions.[10]

This Climate Justice Campaign works to bring about necessary changes in lifestyle, society, industry and policy to halt dangerous climate change.[11] FOE also runs a special-purpose website called CarbonEquity, which produces materials such as 'The two degree target: How far should carbon emissions be cut?',[12] and encourages people to form a Carbon Rationing Action Group, a group of people who act together to reduce their individual and collective carbon footprints.[13]

Climate justice to FOE also means providing particular assistance to those with the least financial means to adjust. It stresses the need to address the problems related to climate change faced by the seven million people of the twenty-two low-lying Pacific Island states, who have contributed least to causing the problem. FOE argues the 'ultimate' form of global warming adaptation will be the recognition of 'climate refugees' and urges the importance of 'considerable forward planning to put structures in place to assist people to move should global warming make their current existence untenable'.[14]

FOE produced 'The Citizen's Guide to Climate Refugees' and in early 2009 initiated a Climate Displacement Coalition, which campaigns for the Australian government to 'allocate adequate resources to assist with relocation and resettlement in climate affected communities and also to promote new migration pathways for people facing climate displacement', including 'water refugees' within Australia itself.[15] FOE is not just concerned with the low-lying islands nearby. In August 2009, it hosted an East Coast 'Big Melt' tour by two Nepalese visitors who held public forums at which they discussed the impact of climate change on Mount Everest and the Himalayas in the form of glacial melting, floods, and danger to local people. At the same time it published a report entitled 'Highstakes: Climate Change, the Himalayas, Asia and Australia', which was compiled by David Spratt and Damien Lawson.[16]

For Australia, FOE has demanded 'immediate deep cuts', arguing Australia should set an emissions reduction target of at least 90–95% from 1990 levels by 2050. It has recommended changes that federal and state governments should be implementing, including: supporting international agreements and setting national emission reduction targets; introducing carbon taxes or carbon rationing; promoting 'green energy' and renewable energy sources; replacing coal with 'clean energy' and supporting coal communities to make the transition to a green economy; promoting energy efficiency; developing a low-carbon transport system; introducing other controls on CO_2 emissions; and welcoming and supporting climate refugees to Australia.[17]

FOE is wary of market-based solutions to climate change, such as emissions trading and carbon offset schemes. In the case of the latter, it points to the example of the Bisasar toxic waste dump in

South Africa, kept open with World Bank funding in order to provide carbon credits through methane gas extraction. 'Both schemes rely upon a series of flawed assumptions, including that markets will enable us to trade our way out of pollution, when the very nature of human-induced climate change is rooted in the quest to generate profits.'[18]

Its open submission to the Garnaut Review expressed scepticism as to whether the design of the ETS would 'be able to withstand political pressures of fossil fuel industry and other polluters who will attempt to reduce liabilities under such a scheme'. FOE conceded an ETS could serve as part of the solution to emissions reduction, but only if it avoided the problems of previous trading schemes and only if it was 'tightly structured and managed'. Such an ETS would require that most sources of emissions be covered, the exclusion of forest and soil sinks, the auctioning of all permits, no granting of free permits to EITE industries, eschewal of offsetting schemes, and no borrowing of permits. FOE also recommended stabilisation of CO_2 concentrations at or below 350 ppm.[19]

In 2009 FOE and other climate organisations developed a special joint report entitled 'Plan B: An Agenda for Immediate Climate Action', to be implemented regardless of whether or not the CPRS was passed. This urged the government to adopt the following agenda:

> 1. saving energy by 'greening our homes and workplaces'; 2. fast-tracking the switch to renewable sources of energy; 3. driving a shift to low emissions vehicles, such as electric cars, and sustainable cities; 4. protecting forests and woodlands as a mean of sequestrating CO2 and encouraging more sustainable agricultural practices; and 5. developing a 'green job economy'.

The report spells out in detail the steps for carrying out Plan B, which include: the introduction of gross feed-in tariffs to ensure rapid industry development of all renewable energy technologies, such as occurred in Germany; an immediate moratorium on the construction of new coal-fired power plants; the development of 'sustainable transport', including public transport, cycling, and walking infrastructure; and ending the clearing of mature and regrowth forests, woodlands and grasslands by the end of 2011.[20]

As a decentralised organisation, FOE activities are based on local mobilisations. For example, FOE Melbourne has been active opposing the potential massive expansion of the fossil fuel industry in Victoria, with exploration licences being put forward for a range of coal seam gas (CSG), gas shale and coal projects. FOE Melbourne's Climate Justice website features news of a large range of campaigns, in keeping with its belief that climate justice is best achieved through action: 'Working with community groups to tackle climate change, campaigning for a moratorium on all new coal power stations and stopping the proposed HRL coal-fired power station in the Latrobe Valley, Friends of the Earth promotes a just transition to job-rich renewable energy solutions.'[21]

Operating in partnership with FOE is the Sydney-based Climate Action Research Group (CARG). CARG's objective is to research the emerging strategies of the climate movement, both through direct participation and through interviews with activists and onlookers. 'Our operating assumption is that answers to the problems of climate action are to be found in the process of mobilisation and contestation, at the "coal face", not in the abstract.' Its main focus has been to research the direct action Climate Camp, which has proliferated across carbon-intensive societies since the Dissent protest camp at the Gleneagles G8 meeting in 2005: 'As sites for strategising, popularising, and politicising, Climate Camps have defined a new genre in climate action. Creating an encampment for up to a week at the site of expanding carbon-intensive development, Climate Camp precisely poses the question of direct mass mobilisation on climate change.'[22] CARG focuses on four key dimensions: what gets people involved in climate activism and what keeps them there; how do activists bridge climate science and climate justice, to create a morality for climate action; how climate camps can be seen as an exercise in popular education, of creating mass counter-publics; and how Climate Camp expresses a new politics of direct climate action and whether this opens up new strategic pathways for challenging the political dispensation.[23]

CARG argues that, as a mass movement, climate action takes us well beyond the entrenched sectors and identities of social change.

Climate change affects everyone, so climate action is for everyone. It creates new social forces and new constituencies, politicising and radicalising people who are otherwise apolitical. This all-encompassing character of climate change means the climate movement is a political project that has now migrated far beyond environmentalism, into all facets of social change:

> As climate change subsumes society, the climate movement has the basis for subsuming and making possible all other movements. It is now widely recognised that there is no possibility of progressive social change without effective climate action. And ... there is no possibility of legitimate and therefore effective climate action without climate justice. In this context, it is becoming increasingly clear that the emerging climate movement may indeed be the long sought-after 'movement of movements'.

CARG believes that, despite the now widespread consciousness of the need for action, 'the climate movement still remains potential, not actual'. At best, we are seeing 'early signs of what is to come, pockets of strategising and experimentation'. As we approach the climate change 'tipping point', it asks whether there is a climate action 'tipping point': 'When will the climate movement emerge as a political force, capable of taking action, rather than deferring to the government of the day? When will we, as a society, set about dismantling the mines, infrastructures and industries that create the crisis?'[24]

As an organisation committed to climate justice, CARG differentiates between global and local climate justice imperatives, akin to Eckersley's primary/secondary injustice distinction. Where global climate justice focuses on the impacts of climate change on the poorest and least culpable societies, local climate justice in countries like Australia considers the impact of transition. Yet there are connections in the climate justice perspective. The demand for just transitions for coal-mining communities and climate justice for low-lying countries stems from the same principles: those who have profited from and contributed most to emissions should look after those who are the most vulnerable to both the effects of climate change and the drastic measures needed to mitigate those effects.[25]

To this end, CARG has developed a program for meaningful 'direct action' on climate change that recognises the urgency of the task in hand: achieving 350 ppm CO_2e by reducing global emissions by more than 85% below 1990 levels by 2050. The program recommends: 'the need to direct the economy and society to regenerative sufficiency, away from the productivist exploitation of natural resources; regeneration must be bound in with global climate justice, through climate debt repayments to address adaptation and mitigation needs in the Global South; localisation and social justice must guide the regenerative model, for example, decentralisation of energy supply can enable localisation of energy production and democratisation of provision; regulatory instruments must be deployed for large emitters, that is, minimum reductions should be announced for the 1000 companies listed in the CPRS legislation; decommissioning coal power and coal exports and the closure of fossil-fuel dependent power stations with a just transition program for affected communities and public investment redirected into a publicly owned renewables industry; public funds for de-carbonisation could be raised through more progressive direct taxes; and systematic expansion of sink capacity, through afforestation and changed agricultural practices, involving direct land regulation that would shift the meaning of land ownership.[26]

It is difficult to place FOE precisely within the layers delineated in this book. FOE is a long-established ENGO, and CARG utilises the talents of leading social scientists with access to university resources, so it operates to an extent within the middle level of climate politics, interacting with less radical ENGOs, such as the ACF. Yet, in its orientation and strategic disposition, FOE belongs with the grassroots movement. In short, it provides a valuable conduit between these two levels, bringing the intelligence and experience gained through grassroots activism to the ENGO sector and even the academy, and offering organisational resources and academic expertise to the grassroots climate movement, in which it plays a leading role.

The *Climate Code Red* team: David Spratt and Philip Sutton

Climate Code Red by David Spratt and Philip Sutton proposes a framework for climate change campaigning based on the argument that climate change mitigation is so urgent and immense it requires emergency measures like those adopted by the Allied powers during World War II. *Climate Code Red* was published initially in loose spiral binding by FOE in 2008 with the sub-title 'The Case for a Sustainability Emergency' then republished in expanded book form later in 2008 by Scribe with the sub-title 'The Case for Emergency Action'. (References are to the Scribe version.) Unlike other important climate movement texts, such as Diesendorf's books, *Climate Code Red* has been produced by climate movement 'organic intellectuals', not university academics. Spratt and Sutton were major figures in the grassroots climate movement before *Climate Code Red* and have become even more significant since.

Spratt is a longstanding peace and environmental activist. In 2008, with FOE Climate Justice Campaign Director Damien Lawson, Spratt established the Climate Action Centre, located in Melbourne Trades Hall, to strengthen the movement by developing, supporting and forging links between groups. The Centre provided informative materials, such as a '4 degrees hotter' primer in February 2011, which details the type of world humanity can expect if emissions are not adequately reduced, and a 'Talk Climate' reader for the Climate Action Summit in April 2011. It also served as an important physical space where CAGs and networks, including CEN and UCAN, could meet.[27] Spratt paid the Centre's rent until it closed in 2011. Along with Cam Walker, Spratt maintains FOE's CarbonEquity website. He regularly releases reports and commentaries on climate science and climate politics, often through CarbonEquity, for example 'The Big Melt: Lessons from the Arctic Summer of 2007' (October 2007).[28]

Sutton, who never completed the veterinary science degree he started at Sydney University in the 1970s, has been described as a 'green business advisor', having formed Environmental Management Systems to assist corporations in managing their environmental impacts.[29] He presents himself simply as a 'sustainability strategist'. By working unpaid for the climate movement over many years, Sutton effectively gave most of his inheritance to the movement.[30] He is Convenor of the Greenleap Strategic Institute, Convenor of CEN, co-founder of SCA and past-President of the Sustainable Living Foundation (SLF) and the Australia New Zealand Society for Ecological Economics. He was the architect of the *Victorian Flora and Fauna Guarantee Act 1988*, later adopted as a national model.

His work focuses on 'swift achievement of an ecologically sustainable economy and strategies for rapid social mobilisation'.[31]

In the same way that 'code red' signals to hospital workers that a patient needs advanced life-support, Spratt and Sutton's *Climate Code Red* argues that climate change sends a strong warning that societies and their leaders must declare a 'climate emergency', which transcends 'business and politics as usual' and indicates the will to prevent a global climate catastrophe. To create a safe climate, Spratt and Sutton contend that humanity needs to reduce the current global temperature to less than 0.5° C above the pre-industrial level and greenhouse gases to less than 320 ppm CO_2e.[32] Part I reviews the scientific evidence indicating the seriousness of climate change, Part 2 discusses debates about emissions reduction targets and proposes reductions consistent with a safe-climate future, and Part 3 calls for a rapid transition to sustainable economy.

Spratt and Sutton maintain that the United States, 'the most powerful economy of all time', can be harnessed to 'create a safe climate'.[33] Thus far, humanity has 'failed to build and maintain a system that has enabled modern society to ensure its own sustainability and that of other living species';[34] and the 'rich countries and classes' have contributed most to global warming.[35] In the case of Australia, federal and state governments have approached environmental issues cautiously so 'low expectations have become embedded in the relationships between lobbyists and government'.[36] In terms of climate change mitigation, governments have 'glued themselves to the fossil-fuel lobby and done almost nothing'.[37] They are deeply critical of neo-liberalism, under which the 'corporate agenda runs politics', ensuring that capitalism is environmentally unsustainable.[38] They envisage a green social democracy, a highly regulated capitalism, because governments, they maintain, have a crucial role to play in planning, coordinating, and implementing the transition to a safe climate.

Some business people, they insist, are open to concerted climate change mitigation and are part of the solution, because a 'safe-climate economy' that cuts emissions to nearly zero will require new technologies and products.[39] This would include highly efficient home appliances, zero-emissions vehicles, national energy efficiency programs, renewable energy plants, and shifting of long-distance road and air freight to rail. In addition to eco-efficient technologies, Spratt and Sutton propose: eco-taxation, a rationing system that sets limits on environmentally unsustainable practices by auctioning permits, the revenue from which could be used to create infrastructural changes; and the re-skilling of workers, communities, and industries in order that they move to a 'post-carbon world'.[40] Crucial to implementation of a climate emergency will be the energies and active support of a popular climate movement, involving community-action groups, unions, churches, neighbourhoods, schools, local government, political parties, even corporate elites.[41]

Climate Code Red has provoked constructive debate within the climate movement. For example, Diesendorf agrees there is indeed a climate emergency, but argues that climate activists must not delude themselves into thinking its implementation is as 'simple as making war'. He contends it would require 'radical transformations of the energy, manufacturing, agricultural and forestry sectors': much larger economic changes than wartime emergencies have entailed.[42] Socialist Alliance maintains it underestimates the extent to which corporations make or break political parties and politicians and the way that capitalism invariably causes climate change. The *GLW* review considered that Spratt and Sutton overlooked important questions such as: 'Can the planet be saved without first abolishing capitalism? Is a "state of emergency" compatible with democracy?'[43]

Because of the bold argument of *Climate Code Red* and the ensuing debates, it has helped to galvanise the Australian climate movement. CEN was formed, as we shall see, as a result of its publication. The book has not only provided a significant text for the movement but has enabled its co-authors, longstanding and relentless climate activists who have worked with a wide array of groups, to become even more useful leaders for the movement. They are both in high demand and speak frequently to local climate action meetings and large climate movement events, as well as to socialist and trade union conferences concerned with climate change.[44]

Rejecting the CPRS

Grassroots climate movement contempt for the CPRS was a significant factor in ensuring the Greens did not support it. To a Melbourne

University activist, 'It was the climate movement's opposition to the flawed CPRS and grassroots pressure that stopped the Greens voting for the CPRS in the Senate.'[45] At least sixty-six climate groups wrote to Rudd signalling opposition to the CPRS. The grassroots movement, according to Spratt, was 'uniformly angry' and appalled with ENGOs that supported it. Spratt stated: 'One of the great powers of the climate action movement is our capacity to withhold support from—and actively campaign against—government actions that are designed to fail, as the CPRS will ... And it is a tragedy that some people who should know better are warmly praising such a failure.'[46]

Stuart Rosewarne provided a powerful academic critique of 'this confidence in the capacity of the market to capture the cost of anthropogenic-generated emissions in a carbon price and for this to systematically drive change'. Exemplified in the Stern and Garnaut Reviews, such approaches are concerned primarily with maintaining economic growth; defining emissions reduction targets that will help avoid the potential for runaway climate change is a subordinate consideration. 'The *realpolitik* of the emissions trading system is a none-too subtle agenda to sustain capital accumulation.'[47] Rosewarne, a political economist, has worked consistently with James Goodman, another academic associated with CARG, to maintain rigorous criticism of market-based mitigation strategies.

CSIRO climate researcher Clive Spash also canvassed the futility of emissions trading. His article 'The brave new world of carbon-trading' critiques such schemes as ineffectual and a diversion from meaningful climate change mitigation. In 2009 this article was accepted for publication in *New Political Economy*, a prestigious peer-reviewed journal, but CSIRO management wrote to the journal withdrawing its authorisation to publish Spash's paper. This curtailment of his academic freedom prompted Spash to resign from the CSIRO on 2 March 2010—and focused even more climate movement attention on his article.[48]

Of the CPRS in particular, Goodman and Rosewarne argued it was unclear how it could possibly reduce emissions and, because the scheme accepted that economic growth must be sustained, it was presuming emissions would continue to increase.[49] After its demise in the Senate, Spratt described the mood in the movement:

> Since the CPRS was defeated in the Senate last week, there has been vigorous debate amongst activists and on a number of grassroots e-lists about this event, what it means, whether the decision of the climate action movement to oppose the CPRS as presented was wise, whether we will now get an 'even worse' scheme with Liberal support, and whether a better scheme was possible with Green–Labor negotiations ... So would the CPRS, even in its appalling form, be something we could support because it at least moves Australia 'in the right direction' (David Henry's favourite phrase) and starts to reduce emissions. The evidence is clearly to the contrary ... Incremental improvements and 'broad but shallow' advocacy that has dominated the climate scene for the past 10 years will not get us there ...The climate movement can't expect politicians and 'the system' in general to break out of failure-inducing incrementalism for as long as we in the movement are locked into the same model ... Part of what we must do is launch psychological warfare on the power elite.[50]

The CPRS debate spilled over into reflection on the Gillard government's carbon tax. Goodman and Rosewarne implied it was preferable to the CPRS, but reiterated their general critique of indirect solutions that rely on re-pricing carbon to shift incentive structures and thereby decarbonise. 'Re-pricing is said to internalise externalities, thus correcting market failure, to set us off on a new low-carbon growth trajectory.' Indirect market-based schemes conform to the dominant orthodox mythologies and are inadequate. Re-pricing emissions does not penalise the extraction of fossil fuels. The scale of re-pricing is insufficient to adequately incentivise renewables. Any resulting reduction in emissions could be offset by accelerated growth in overall energy demand, which has been the experience of countries with carbon taxation; only Denmark delivered significant reductions by directly spending tax revenues on energy efficiency and renewables, not by shifting incentives. The costs could be displaced to those least responsible for climate change and least able to pay. 'Revenue flows may be used to compensate low income consumers, but is unreliable and unlikely to keep pace with rising costs; revenue for corporate welfare, for the large emitters, is likely to be more reliable.'[51]

As Goodman and Rosewarne intimate, it is not just the ineffectiveness of market-based responses that is at issue but the 'secondary injustices' they create. This concern is articulated clearly by Ben Spies-Butcher's critique of Garnaut's neo-liberal approach to emissions reduction. The primary rationale for market-based responses is to minimise the cost of transition, but, as Spies-Butcher points out, carbon trading imposes a range of new costs that disproportionately affects particular industries, regions and income groups. He argues for a more 'holistic debate' about costs that is more than a cost-benefit analysis measured solely in the metric of GNP, maximising public engagement and the influence of pro-mitigation public opinion, open to complementary measures to overcome inertia and powerful opposing interests. None of this is likely to emerge from the dominant neo-liberal approach; but 'building this alternative is becoming increasingly urgent'.[52]

The concern with 'secondary injustices' is not simply a moral but also a strategic issue for the climate movement. Goodman and Rosewarne point to international disillusionment with market-based measures ('a healthy scepticism of climate policy that funds polluters and shifts the burden to consumers') as part of the reason for public opinion shifting back towards climate scepticism. 'If climate policy is captured for elite interests, where do the mass of people sit? If you are presented with a choice between the status quo and a patently unjust climate policy, that is in any case grossly inadequate, which would you choose?'[53] Since market measures are both deficient and unpopular, 'climate movements need positive transformative agendas' ready for the moment when market measure lose favour with policy makers.[54] The CPRS controversy and the wider debate about market-based mitigation strategies enabled the grassroots movement to clarify what it did not want and thereby to focus more sharply on alternative or at least complementary mitigation strategies.

Beyond Zero Emissions (BZE)

Established in 2006 by Matthew Wright and Adrian Whitehead, with Sutton's support, BZE is an independent, not-for-profit organisation run by volunteers, including many engineers. It focuses on achieving technological improvements and innovations in the stationary energy, transport, land use, housing and commercial buildings, industrial processes, and export revenue sectors of the economy. To facilitate 'the implementation of the social changes and technologies that will reduce the impacts of climate change and give our society and global ecosystems a chance of surviving into the future', it engages in: education; research; transition planning; climate science education; corporate education; political education; network building; community campaign building; and solutions development. In 2007 it commenced a weekly radio show, Beyond Zero, on Melbourne community radio station 3CR, interviewing specialists from around the world, such as James Hansen and Hans Joachim Schellnhuber.[55] It has recently commenced free monthly documentary nights at Kindness House, 288 Brunswick Street, Fitzroy, offering informal discussion 'and a chance to meet some of the BZE team'.[56]

As its name implies, BZE insists that greenhouse gases should not only be reduced immediately but be reduced to zero, because current atmospheric levels of greenhouse gases are already resulting in global climate disaster. 'If we allow greenhouse gas levels to increase we will see more fires, more storms, more extinctions and more death as a result of global warming and a changing climate.' It claims the broader climate action movement, climate scientists and many large environment groups now agree with this zero goal.[57]

BZE's *Zero Carbon Australia Stationary Energy Plan* (ZCA2020 Plan) has blazed a trail and galvanised climate movement support for this ambitious target. Even before its publication in 2010, BZE ideas connected constructively with 'just transition' strategies. For example, on 12 September 2009, BZE issued a media release in support of the Switch off Hazelwood protest: 'The Latrobe Valley, due to its existing infrastructure and skills base, is an ideal manufacturing hub for wind turbines, solar thermal mirrors and solar hot water systems.' It argued the federal government's $2.4 billion to support 'non-existent' 'clean coal' technology should be redirected immediately 'to supporting the rollout of renewable energy technologies in such a way as to maximise employment opportunities for Latrobe Valley power workers'.[58]

An enthusiastic audience of around 1000 people attended the launch of the BZE plan on 14 July 2010 at the University of Melbourne.

Speakers included Grattan Institute CEO John Daley, ANU Solar Thermal Group Leader Professor Keith Lovegrove and Pacific Hydro General Manager Lane Crockett. Matthew Wright stated Australians had been subjected to a sustained 'Can't Do' campaign in relation to renewable energy, mounted by the carbon lobby; BZE wished to counter this propaganda with a 'Can Do' campaign. A strong message from speakers representing 'green corporations' was that a carbon tax of at least $70 per tonne was essential to obtain finance from banks that needed to be persuaded renewable energy could compete successfully with coal-fired electricity sources; and that a carbon tax was simpler and more effective than an ETS.[59] More than 1000 people attended the launch at Sydney Town Hall on 12 August 2010. Keynote speakers here included Malcolm Turnbull, Bob Carr and Scott Ludlam. The Department of Political Economy at the University of Sydney endorsed the plan.[60]

The ZCA2020 Plan is motivated by the belief that 'The world stands on the precipice of significant change.' Climate scientists predict severe impacts from even the lowest estimates of global warming. Atmospheric CO_2 already exceeds safe levels. 'A rational response to the problem demands a rapid shift to a zero-fossil-fuel, zero-emissions future.' It outlines what it describes as 'a technically feasible and economically attractive way for Australia to transition to 100% renewable energy within 10 years'. The plan insists this immediate transition to a 100% renewable energy grid can be made based only on proven technologies already commercially available and used in large industries. 'Social and political leadership are now required in order for the transition to begin.'[61] The plan also proposes a 'transport revolution' that entails improving access to public transport and converting the vehicle fleet to renewable electricity, and the switch of other services currently provided by gas and oil to be supplied more efficiently, with renewable electricity. Because of this, under the ZCA2020 Plan, it is projected that 2020 grid electricity will be over 40% higher than today, from 228 terawatt hours per year up to 325 terawatt hours per year.[62]

The chosen renewable energy technologies are wind, concentrating solar thermal (CST), and solar. Wind generation is a key component because of its relatively low cost and industry maturity. Wind is proposed to meet 40% of the total grid-connected demand.

Large-scale CST with molten salt storage offers reliable electricity 24 hours per day. Solar power towers are proposed because of their technological maturity, higher operating temperatures and efficiency compared to other technologies. CST is proposed to meet 60% of the total grid-connected demand. Small-scale solar photovoltaic (PV) power has a role in reducing the demand for grid electricity during sunlight hours. This in turn allows the CST plants to accumulate more stored energy for release at night. For system backup, the plan recommends: existing hydroelectric generators have a useful role in providing first-dispatch backup and peaking power; and backup from crop-residual biomass for the CST plants offers energy security, protecting against occasions when a combination of low wind and low daily solar radiation occurs.[63]

The BZE plan reverses the normal tendency for capitalism to underuse labour resources and overuse environmental resources. It states that:

> ZCA2020 modelling shows that many more jobs are created with the construction of a 100% renewable energy grid than are lost with the phasing out of coal and gas from the stationary energy supply chain. The Plan will create up to 80,000 jobs from installation of renewable energy generation at the peak of construction, and over 45,000 jobs in operations and maintenance that will continue for the life of the plant ... If half the manufacturing for wind turbines and heliostats is done domestically, this would create up to 30,000 jobs in manufacturing ... The level of expansion in jobs would boost overall employment in the economy.[64]

The plan also confronts BAU investment patterns. The total investment required to implement the ZCA2020 renewable energy system, it calculates, is $370 billion, averaging $37 billion per year over the 10-year duration of the plan, equivalent to only 3% of GDP. 'Compared to Business-As-Usual, the ZCA2020 Plan is characterised by higher upfront investment, but lower ongoing costs. Under the Plan, by 2020, there would be $260 Billion more spent on the stationary Energy sector than Business-As-Usual.' If the ZCA2020 renewable infrastructure is paid for through electricity revenues, this would lead to a price increase of 6.5 cents per kilowatt hour, which would equate to only $420 per household per year for residential consumers or $8 per household per week.[65] The plan recommends the additional expenditure should be sourced from both public and private investment. It points out that, in due course, economic 'pay-back' occurs: around 2040 if the costs associated with oil imports and likely emissions costs are ignored; earlier if these oil and emissions costs are included.[66]

The ZCA2020 Stationary Energy Plan thus outlines 'a fully costed and detailed blueprint for transforming Australia's energy sources to 100% renewable supply ... achievable today, with no technical barriers to their deployment'. Implementing the proposed infrastructure in ten years, it insists, is well within the capability of Australia's existing industrial capacity. The required investment is the equivalent of an economic stimulus of 3% of GDP.[67] The plan concludes:

> 100% renewable energy in ten years is achievable and necessary, ensuring Australia's energy security, national security and economic prosperity for the future. Australia has some of the best renewable energy resources in the world, and should be positioning itself as a leader in the emerging renewable energy economy. What is required to make this happen is leadership from policy makers and society, with firm decisions made quickly that will allow this transition to occur.[68]

Salleh views the BZE scheme as a profound example of ecological modernisation, which might cut emissions but 'will do nothing to stop the extractive assault on the society–nature metabolism'.[69] The BZE plan has otherwise greatly inspired climate activists, its ideas permeating all levels of the movement. In a speech at the Sydney Institute in October 2008 Milne argued it was 'both necessary and possible for us to build a new zero emissions energy network, based on energy efficiency and renewable energy, in just a decade'.[70] The plan has been embraced warmly by trade unions promoting green jobs and by socialist organisations. BZE speakers are in high demand at climate movement events and trade union conferences on climate change.[71]

Socialists

For socialist groups, global warming reveals starkly the problematic nature of capitalism. The climate change issue thus offers them the prospect of engagement with people potentially responsive to the argument that global warming proves capitalism is unsustainable, who might therefore be persuaded to join a socialist party. Most socialist parties, especially Socialist Alliance and Solidarity, have become involved in the climate movement. Members of socialist groups are usually highly visible at climate rallies and conferences, displaying party placards and distributing literature, and are active within CAGs and networks. The climate movement has thereby gained dedicated and experienced activists who are sincerely concerned about the fate of the planet as well as seeing the climate change issue as an opportunity to argue for socialism or at least anti-market remedial measures.

Socialists press home the point that it is capitalism that has caused the problem and so capitalist market-based mechanisms cannot be relied upon to solve it. As the Socialist Alternative leaflet distributed at the anti-Hazelwood demonstration on 6 November 2010 in Melbourne's City Square put it: 'the capitalist system is to blame' and 'to save the planet, we need to go beyond it'.[72] They stress both the ineffectiveness and the secondary injustices caused by market-based 'solutions': that they are 'a way of collecting money from workers to bailout polluters' in the same way that the 'free market' has ensured that workers lose their houses while banks are given multi-billion dollar bailouts.[73] In a leaflet entitled 'Market solutions won't prevent a climate disaster', the Socialist Party Australia states:

> Even the best case scenario under the carbon tax requires a market-led death for the fossil fuel industry ... Instead, a socialist approach would be able to make a humane transition. Workers would be retrained for new roles, old energy sources could be strategically phased out, and the current holdings of energy companies could be rationally redeployed ... the needs of the people and the environment would be put before the needs of the fossil fuel industry.[74]

Solidarity has made climate change one of its principal campaign streams and regularly publishes articles on climate change and climate politics in its monthly magazine *Solidarity*, many written by Chris Breen. It has produced a pamphlet, *Climate Change is a Class Question*, which hopes that working-class industrial action—'green bans'—will in due course force polluting companies to cease production. 'This is why taking the arguments about climate change into the working class is so important for the climate movement.'[75] In December 2009, Solidarity published on its website the full text of Clive Spash's critique of carbon trading.[76] Solidarity has also distinguished itself by vehement opposition not only to emissions trading but also carbon taxation. Back in 2008, Breen argued that, while a carbon tax would be a simpler scheme than an ETS, corporations could nonetheless easily pass the expense onto consumers so there would be no incentive to cut emissions 'and our living standards would fall'.[77] In response to the Gillard government's carbon pricing mechanism, Solidarity maintains the polluters will pass on the costs incurred by a carbon tax to consumers and 'the climate movement will not be able to win the mass support needed for serious action by telling workers to take a cut in their living standards'.[78] Concern with secondary injustices is compounded by anxiety about recruitment.

Socialist Alliance is the most significant far-left organisation involved in the grassroots movement, because it produces *Green Left Weekly* (*GLW*), an important resource for the movement with its weekly articles on climate change impacts and climate politics, internationally and in Australia. In *GLW* and other publications, and at public events, Socialist Alliance argues for eco-socialist analyses and climate justice approaches to climate change. Members are highly active within the climate movement. For example, about forty attended the first Climate Action Summit in early 2009.[79] Resistance (the Socialist Alliance youth wing) also campaigns hard on the climate issue. For instance, on 16 July 2009 it organised a 'Billionaires for Coal' rally in Wollongong, which congratulated the NSW government for its pro coal-mining policies;[80] and the following month, it hosted Graham Brown, ex-coalminer climate activist, speaking at campuses in Melbourne about 'the urgent need to phase out coal without penalising the workers, and why the government needs to invest in a renewable energy industry to provide real green jobs'.[81]

Socialist Alliance has developed a Climate Charter, which stresses rich countries like Australia must take the lead. 'The rich countries have created the problem, so we must contribute the most to fixing it.' Australia should also accept a fair share of the environmental refugees displaced by climate change. 'This assistance is just part of repaying our ecological debt for our historical high emissions.' Its detailed Ten-Point Climate Action Plan proposes:

(1) emission reduction targets of 'at least 5% a year' to achieve zero emissions as soon as possible, and a target of '100% renewable energy by 2020';

(2) begin international negotiations aiming for 'a global target of at least 90% emissions cuts on 1990 levels by 2030' and 'Make cutting rich industrial nations' emissions a priority, and increase aid to poorer countries to help them to use clean energy for their development';

(3) transition to 'a zero-waste economy';

(4) require existing houses and commercial buildings, as well as all new buildings, to be energy efficient;

(5) bring power industries under public ownership and democratic control, immediate phasing out of coalmining and coal-fired power, jobs and retraining on full pay for affected workers, new sustainable industries in these communities, run 'maximum possible base-load power from existing natural gas and/or hydro power stations instead of coal until renewable energy is available';

(6) bring car industry under public control and re-tool it to 'manufacture wind turbines, public transport vehicles and infrastructure, solar hot water, solar photovoltaic cells';

(7) build solar thermal plants and wind farms and upgrade national grid 'to make it compatible with 100% renewable energy';

(8) ban old-growth forest logging and begin reforestation, carbon farming and biodiversity protection;

(9) phase out industrial farming reliant on fertilisers, pesticides, and fuel sourced from petroleum;

(10) make all public transport free, upgrade all public transport services, prioritise rail freight, and facilitate cycling.

This Climate Charter concludes: 'Change the System, not the Climate'. It argues the planet and the welfare of future generations must come before corporate profits. 'The greedy corporations that have resisted change have no right to take away our future.' To take on the greenhouse mafia, 'we need to strengthen a climate action movement that is independent of the big parties, which won't compromise on the steps needed to avert runaway warming'. The campaign must happen in the streets, workplaces, schools and universities to win wide public support and become an unstoppable force for sustainable change. 'Making those changes also means challenging the capitalist market, which has failed to protect future generations and cannot be allowed to stop us from averting climate disaster.'[82]

Socialist Alliance organises conferences to 'strengthen radical action to stop climate change', beginning with a three-day conference in Sydney on 'Climate Change/Social Change' 11–13 April 2008. Around 180 people attended. Keynote speakers included renowned American environmental sociologist John Bellamy Foster. Women were underrepresented as plenary and workshop presenters. The conference participants developed a statement that included a call for reduction to 300–325 ppm CO_2 and a climate sustainability policy consisting of five core elements: 'properly resourced public agencies to drive the sustainability effort, an international framework where the First World pays the vast bulk of the price of reversing global warming, an end to rampant consumerism, vastly strengthened campaigns for climate sustainability, and building a powerful political alliance for climate sustainability with social justice.'[83] A Climate Change/Social Change II conference was held in Melbourne

30 September – 3 October 2011, 'to urgently address the climate and social emergencies that we must overcome', again featuring John Bellamy Foster, plus Ian Angus, founder of the Ecosocialist International Network.[84]

Socialist Alliance has strongly criticised the Labor governments' market-based mitigation mechanisms, arguing direct government regulation and investment in renewables is preferable. *GLW* maintained a running cynical commentary on CPRS developments. For instance, in May 2009, it invited readers to check out the government's 'Greenhouse Challenge Plus' program and observed that the latest version of the CPRS included up to $24 billion to polluters over ten years, including $7 billion directly to coal-fired power plants.[85] The day before the CPRS first failed to pass the Senate, an editorial asserted the CPRS was 'worse than nothing', because it paid no attention to the climate science and would not cut emissions to safe levels. 'If passed into law, it will help keep the big polluters in business for longer.'[86] It was imperative that the climate movement 'help transform Australia's unions from a cheer squad for Labor's parody of a climate solution into a powerful force for campaigning against the biggest threat humanity faces'.[87] Workers, Socialist Alliance emphasises, have nothing to fear from socially just climate mitigation measures: 'transition to a post-carbon economy would entail job creation, with many new jobs in research, training, renewable energy, public transport, and manufacturing' and 'direct government investment in energy efficient public transport and public housing'.[88]

As for Gillard government initiatives, Socialist Alliance argued 'A carbon price does not equal climate action', because the prices discussed would 'simply stimulate a mass rollout of gas, extending Australia's commitment to fossil fuels at the expense of renewable energy'. A carbon tax could be 'a secondary or additional aspect' of climate policy,' but an effective policy must include 'large-scale public investment in a publicly owned and run renewable energy sector, a rollout of public transport, a shift to sustainable farming and other carbon abatement programs, as well as government regulation to phase out dirty industries'. Moreover, 'the government needs to stop fiddling with the market to cut carbon emissions and instead set emissions cuts targets based on the science'.[89] A *GLW* editorial 'Carbon price or climate action' on 1 June 2011 asserted that the 'worst thing about the Labor government's proposed carbon price scheme is that it's a diversion from real action on climate change'.

The Transition Decade Campaign

The Transition Decade campaign (T10) is committed within the 2010–2020 decade to creating a 'safe climate outcome', aiming to reduce the earth's temperature by between 0.3 and 0.8° C and reduce CO_2 in the atmosphere to between 280 and 320 ppm. The first two years of T10 focus on encouraging whole communities to say 'yes' to this objective, the subsequent eight years to obtain government commitment to the plan, and the last few years to adoption of an 'emergency transition mode' during which 'everyone gears up for physical rebuilding' and then engages in this process on a massive scale.[90] The 'Meta Strategy' of T10 thus envisages the transition decade having three stages: Getting National Commitment to transition; Post Commitment Planning for the transition; and Activating Intense Transition for the transition.[91]

The original concept of T10 was developed by Luke Taylor of the SLF; and SLF members developed the idea during 2009 and involved other organisations, in particular, BZE, CEN and FOE. Five 'sign on principles' were developed: 'Hold a vision for a sustainable world?', 'Recognise the climate emergency situation?, 'Want to restore safe climate conditions?', 'Grasp the scale of a national transition?' and 'Want to complete the transition in the next 10 years?'.[92]

On 14 February 2010, T10 was launched publicly to over 700 people at Melbourne Town Hall and another 500 around Australia in smaller events. At the Melbourne launch the speakers included: Governor David de Kretser; Will Steffen; Christine Milne; EV CEO Kelly O'Shanassey; Rob Adams, Director of Design and Urban Environment for the City of Melbourne; Uncle Bob Randall, Indigenous elder; Mark Ogge from BZE; and Philip Sutton.[93] In the lead-up to this launch, Andrew Bolt wrote an article 'The Governor becomes a far-left activist' on his blog on 6 February in which he asserted that de Kretser was abusing his non-partisan office and playing into the hands of 'activists from the radical green Left' and that there were even socialists in the climate movement. *GLW* responded: 'Is it a surprise that socialists are opposed to climate change? Is it a disgrace that the climate movement is not so prejudiced as to exclude people on the basis of their political opinion.'[94]

The core active groups of the T10 Alliance are SLF, BZE, CEN, Darebin Climate Action Now! and Yarra Climate Action Now! The T10 Alliance also has a very cooperative relationship with SCA; and the City of Melbourne has joined the Alliance as a full member.[95] T10 obviously promotes the 2020 plan of its member group BZE. A T10 postcard promoting it urged people to 'Talk it Up!' 'Promote the plan to your friends, family and people at work. Talk to your local members of Parliament, candidates and your local Council about adopting the plan.' It also recommended 'Get Active!' 'Join a local climate action group.'[96]

The T10 education and mobilisation campaigns aim in general to trigger extensive activism of two main kinds: political campaigning (mass organised, practical action to pressure change i.e. boycotts, rallies, letterboxing, letter writing, conventions, doorknocking); and community projects (small group localised action to support action, encourage, educate and raise awareness for change i.e. film nights, community market gardens, energy efficiency projects, food co-ops, school projects).[97] A recent leaflet explains:

> It is not possible to drift into emergency mode. A conscious social commitment is needed ... Without the commitment to a full package of emergency transition measures in the next few years, transition will not be achieved in the necessary time frame, and without action on a fully planned package of measures adequate to address the climate emergency, we will not achieve a safe climate.[98]

T10 brings together the significant climate movement forces discussed in this chapter. In the formulation of the T10 scheme and strategy we can glimpse FOE's commitment to significant lifestyle changes and grassroots mobilisation, *Climate Code Red*'s notion of emergency and extraordinary effort, wariness about market-based mechanisms such as emissions trading, BZE's ambitious reduction target and pathway to its implementation, and socialist ideas that capitalism must be strenuously restrained.

Notes

1 Hamilton, 'The social psychology of climate change', p. 192.
2 Green.
3 Doyle, p. xviii
4 Quoted in Green.
5 Whelan et al, p. 11
6 Green.
7 For an informative guide to Salleh's work, see Canavan et al.
8 Stilwell, p. 163.
9 Burgmann, *Power, Profit and Protest*, pp. 193–7; Doyle, p. 85.
10 Long, 'Market "Solutions" ', p. 20.
11 http://www.melbourne.foe.org.au/?q=cc/home (viewed 22 September 2011).
12 http://www.carbonequity.info (viewed 7 November 2011).
13 http://www.climateactioncentre.org/carbon-equity (viewed 7 November 2011).
14 Long and Walker, p. 26.
15 www.foe.org.au/climate-Justice/activities-and-projects/climate-refugees (viewed March 2010); Walker, 'Climate refugees in Australia', p. 12.
16 FOE Melbourne, 'Newsletter', Spring 2009, p. 3
17 http://www.foe.org.au/climate-justice/policy-position/federal-election-2007/foe-australia-campaign-priorities-on-climate-change-for-the-federal-election (viewed 17 September 2011).
18 Long, 'Market "Solutions" ', p. 20.
19 FOE, 'Open submission to Garnaut'.
20 http://www.foe.org.au/resources/publications/climate-justice/Plan%20B%20Report_Lores_FINAL.pdf/view (viewed 17 September 2009).
21 http://www.melbourne.foe.org.au/?q=cc/home (viewed 22 September 2011).
22 CARG, 'Research and action in the "movement of movements" ', p. 22.
23 ibid., pp. 22–3.
24 ibid., p. 22.
25 ibid., p. 23.
26 Goodman and Rosewarne, 'Climate Policy', p. 24.
27 Green; Baer, Burgmann, participant-observer notes; http://www.climateactioncentre.org/ (viewed 7 November 2011).
28 http://stuff.samassaveneessa.info/docs/2007BigMelt.pdf (viewed 7 November 2011).
29 Angel, p. 211.
30 Burgmann, personal information acquired from unnamed source.
31 http://www.green-innovations.asn.au/ps-bio.htm (viewed 9 February 2011; http://www.safeclimateaustralia.org/about-safe-climate-australia/ (viewed 6 October 2011).
32 Spratt and Sutton, *Climate Code Red* (Scribe, Melbourne, 2008), p. 6.
33 ibid., p. 112.
34 ibid., p. 151.
35 ibid., p. 138.

36 ibid., p. 158.
37 ibid., p. 161.
38 ibid., p. 216.
39 ibid., p. 212.
40 ibid., p. 213.
41 ibid., p. 234.
42 Diesendorf, *Climate Action*, p. 86.
43 Courtice, 'Climate change'.
44 Burgmann, participant-observer notes; Baer, participant-observer notes.
45 Ali.
46 Spratt, 'Suicidal sweethearts'.
47 Rosewarne, pp. 19, 47–8.
48 http://www.clivespash.com.org (viewed 28 September 2010).
49 Stuart Rosewarne and James Goodman, 'The ETS is a huge cost for almost no benefit', *Age*, 9 December 2009. http://www.theage.com.au/opinion/politics/the-ets-is-a-huge-cost-for-almost-no-benefit-20091209-kjpm.html (viewed 10 December 2009).
50 Email, 'CPRS emissions will be higher than baseline till 2035', 16 April 2009.
51 Goodman and Rosewarne, 'Climate Policy', p. 23.
52 Spies-Butcher, esp. pp. 67–8.
53 Goodman and Rosewarne, 'Climate Policy', p. 23.
54 ibid., p. 24.
55 http://beyondzeroemissions.org/about (viewed 1 August 2010).
56 http://beyondzeroemissions.org/events/film-nights (viewed 24 August 2011).
57 http://beyondzeroemissions.org/about (viewed 1 August 2010).
58 pressrelease@beyondzeroemissions.org, 'Grow jobs by ending coal', 12 September 2009.
59 Burgmann, observer notes.
60 Butler, 'Huge crowd'.
61 BZE, p. 2.
62 ibid., p. 3.
63 ibid., pp. 4–5.
64 ibid., p. 13.
65 ibid., p. 15.
66 ibid., p. 16.
67 ibid., p. 17.
68 ibid., p. 17.
69 Salleh, 'Climate Strategy', p. 126.
70 Milne, 'Christine Milne's Speech to the Sydney Institute'.
71 Burgmann, participant-observer notes.
72 Socialist Alternative, 'Why capitalism is costing us the planet', n.d., leaflet, 1p.
73 Jones, p. 21.
74 www.socialistpartyaustralia.org/archives/3030#more-3030 (viewed 17 September 2011).

75 Solidarity, p. 8.
76 http://www.solidarity.net.au/web/spash (viewed 3 March 2010).
77 Breen, 'Carbon trading', p. 22.
78 Editorial, 'Backing tax on workers no way to stop Abbott', *Solidarity*, no. 33, March/April 2011, p. 5.
79 White and Nichols, p. 2.
80 Socialist Alliance, *National Newsletter*, #3, 21 July 2009, pp. 2–3.
81 Resistance, 'Activist Calendar 09', leaflet, 2pp.
82 Socialist Alliance.
83 Baer, participant-observer notes.
84 NTEU, Climate Champions Update, Email 19 September 2011.
85 http://www.greenhouse,gov.au/challenge (viewed 15 May 2009); White, 'Rudd gambles with the planet'.
86 *GLW*, 12 August 2009, p. 3.
87 *GLW*, 19 August 2009, p. 7.
88 Emanuel.
89 Betzien.
90 Baer, participant-observer notes.
91 The Transition Decade, 'Strategic Framework', p. 1.
92 House, pp. 1–2.
93 Baer, participant-observer notes.
94 Courtice, 'Hot scoop'.
95 House, p. 2.
96 The Transition Decade, postcard.
97 The Transition Decade, 'Strategic Framework', p. 2.
98 CEN, 'A new paradigm of climate action'.

11
The Hard Work of Climate Movement Organisation

There are now hundreds of climate movement organisations across Australia. This chapter discusses the emergence of local climate action groups since around 2003 then traces the development of larger regional and nationwide climate movement groups and networks. Finally, it examines the role of the biggest conferences and summits in the growth and consolidation of the climate movement, helping to overcome what Whelan and others have described as a 'scattergun' approach to climate movement organising.[1]

Climate Action Groups (CAGs)

The growing anger on the part of those most concerned about climate change is manifested most obviously in the proliferation of CAGs. The origins of these probably date back to around 2003, certainly pre-dating the emergence of climate action networks. Figures on the number of CAGs in Australia vary, because their growth has been so spectacular, as Michael Green noted in December 2008:

> Grassroots climate action groups are appearing like white

blood cells at a wound. Over the past two years, an unprecedented, unreported and largely underestimated climate movement has sprung up throughout our cities and regions. Many of the members have dedicated decades to living simply and sustainably. The great majority though, are new. Groups start up so rapidly it is difficult to know their numbers.

Green cited Victorian Climate Action Centre estimates that Victoria had about fifty, most less than 2 years old; and that nationwide there were well over 200.[2] By this stage, according to Whelan and others, tens of thousands of Australians were active members of CAGs, many of whom had no previous experience as social activists, but were 'transforming themselves through their involvement in a collective effort that prompts them to reflect upon their efforts'.[3]

In March 2009, about a hundred CAGS were connected with the climate network website at ClimateMovement.org.au.[4] In December 2011, there were 138 groups listed.[5] This is a hub for CAGs around Australia 'to connect with each other, access resources, share success stories and collaborate to make a more powerful and effective climate movement'. It serves as: a collective blog for the community climate movement in Australia; a register of CAGs; a central calendar of events run by and for climate action groups; and a central directory of resources on climate science, solutions, campaigning and policy. It was an initiative of the Nature Conservation Council of NSW and has been supported since 2009 by the Climate Institute.[6]

There are more CAGs than listed on this site. An example is Climate Action Hobart. In 2010 it launched a document titled 'Ten Steps for a Safe Climate—Tasmania's Contribution to Preventing Dangerous Climate Change', which it had developed with input from industry experts, scientists, climate activists and the general public. The 'Ten Steps' it delineated are: carbon-neutral state by 2050; protecting Tasmania's carbon-rich native forests; reducing energy use by 20% by 2020; 100% renewables by 2020; sustainable cities and regions; investing in public and low carbon transport; just transitions to a low carbon future; local and sustainable production and consumption; closing resource loops and eliminating waste; and climate justice and education.[7]

Amongst the many Melbourne CAGS listed are two especially durable ones: Climate Action Moreland, formerly the Moreland Climate Group, and Darebin Climate Action Now! The latter, in inner northern Melbourne, aims to 'educate and inform the community about the urgency of the climate crisis' and 'urge Governments to recognise a climate emergency'.[8] It runs community information sessions, lobbies politicians, holds candidate forums before elections, and joins state-wide climate events. It currently has 70 supporters and holds meetings on the last Monday of each month at 7.30 pm.[9] Climate Action Moreland has the following objectives:

> 1. To put pressure on political leaders to take serious action on climate change via activities such as protests, attending rallies and engaging in discussion with local politicians, and to show that there is a strong community support for serious action;
>
> 2. To provide a focus and outlet for residents wanting to take action;
>
> 3. To provide community education and awareness on climate change and what we can do about it on individual, local and national levels; and
>
> 4. To create networks with other community-based organisations as part of a broader, more coordinated environment movement.[10]

It has links with other Melbourne CAGs, such as Darebin and Bayside, and ENGOs, such as FOE. Periodically, it holds public forums at the Edinburgh Castle pub in Brunswick. It meets biweekly, generally in the public library in West Coburg; it has a 'non-hierarchical structure, and decisions are reached democratically'.[11]

To indicate the activities of CAGs we offer two case-studies.

Case-Study: Climate Action Canberra

Climate Action Canberra (CAC) was formed in November of 2008, the seeds sown at a climate conference a few months earlier in June.[12] It describes its membership as 'a diverse group of students, scientists, workers, community activists and other Canberrans who are concerned about climate change and are helping to catalyse an emergency response to this threat to life on earth as we know it'.[13] By February 2010, CAC had over 400 members and has become one of the most effective CAGs.[14]

Internal documents reveal that CAC members debate and ponder the prospects of the wider climate movement. In one such memo, CAC identifies the reasons it believes climate change has been so difficult to act against:

> The crisis we face is that of a society where power is concentrated in the hands of mine owners, the owners of electricity generators, oil company owners and the owners of cement producers, the aluminium smelters, steel foundries, agribusinesses and coal export ports; in short, the owners of an economy entirely reliant on fossil fuels. These owners, simply to survive in a global market, are driven to be ruthless in their pursuit of profits—profits dependent on fossil fuels. We know their interests and we know their power; and although they do not own the entirety of the economy, they hold a commanding position within it.
>
> To prevent these powerful entities from pursuing their interests requires far more than a rational reason to change their behaviour. The modern corporation, as any look at recent events could tell us, is not a rational beast (least of all when it comes to environmental concerns). To defeat their interests their power must be broken.
>
> The breaking of the power of the polluters requires the use of a force more powerful than their own. Unfortunately, the force most people instinctively turn to—that of government—is on their side, not ours. Behind a facade of democratic and liberal slogans, we see Coalition and Labor politicians in every way beholden to the interests of the polluting few.[15]

In the same memo, CAC laments the slow progress that those trying to create a mass movement out of climate change action are experiencing. Unlike the successful social movements of the past, the climate change movement has not yet snowballed into a critical mass that has the power to influence the decisions of the world's leaders.

> The defining feature of this 'vacuum' is the absence of a powerful climate movement. All major social transformations owe their success to the common element of a social movement. The passage of the 1964 Civil Rights Bill in the United States owed much to Martin Luther King's 'I have a dream' speech, but that seventeen-minute speech would have never been heard in its entirety on national radio and television if not for the 300 000-strong March on Washington. With a movement comes attention from the broader public, and with that, the opportunity for change.
>
> The primary goal of Climate Action Canberra, and the movement at large, must be the building of our own movement. If we are to be in a position where we can present our program against that of Rudd and Abbott, our alternative to their inaction, it will be at the crest of a mass movement.
>
> But what sort of movement do we need now, and what do we do now in order to get it? This is the aspect of the crisis that we as activists most often face, and often get disheartened about. 2009 was a frustrating year in this respect, but 2010 may prove far different. Martin Luther King had had his teeth-grinding moments too, saying at Albany in 1962 '[we were] so vague that we got nothing, and the people were left very depressed and in despair'. The March on Washington occurred one year later.[16]

The memo then debates appropriate methods and the importance of identifying who exactly is the target audience. CAC believes a hands-on direct action approach will have a mobilising effect on sympathetic yet currently un-active people.

> In developing a strategic campaign we should seek, rather

> than avoid, confrontation ... Confrontation ... generates media attention and public debate, and while we want to be confrontational around easily understood issues, we should recognise that the only thing worse than being spoken about is not being spoken about. Just as anti-apartheid activists disrupted the Springboks tour in the 1970s and *The Chaser* team exposed the sham of security at APEC recently, we have to pick a fight. Sure conservative writers will sneer and liberal opinion-makers will disagree with our 'methods', but their audience is not our audience. If we (hypothetically) march in and occupy the Department of Climate Change, declaring that it is 'under new management', many people will think 'what a bloody disgrace' but our audience will think 'about bloody time'.[17]

CAC did not quite occupy the Department of Climate Change. However, during its protest outside its offices when the CPRS details were released and demonstrators were chanting and yelling slogans, a letter was sent to the staff inside:

> The purpose of the protest outside of your office this morning is to highlight the inadequacy of the ... CPRS as a response to climate change ...
>
> The Department of Climate Change is failing in its role of setting Australia on a path towards a sustainable economy. This is not a criticism of your own hard work. This failure is due solely to the fact that your Minister has no intention of achieving this goal. It is no doubt frustrating to work in such an intentionally hamstrung Department.
>
> As a public servant you are required to implement the policies of the government of the day, but as a private citizen you are not required to agree with them. In the absence of effective protection for whistleblowers we cannot ask you to do anything that could put your job at risk. We do however ask that you take action as a concerned citizen.[18]

Around this time, a CAC internal document entitled 'Countdown to Copenhagen' carefully considered the problematic nature of the CPRS, not just for the planet but for the movement. This document notes that the majority of people are generally supportive of the Rudd government's climate change measures and view any opposition to this stance as against climate action in general. This is problematic, because, without popular support for the more radical steps favoured by the movement, the government is not put under enough electoral pressure to break its close ties with the polluting industries:

> The most effective way of altering government and business operations is to make it politically impossible (i.e. profoundly unpopular) for the Rudd Government to continue with business-as-usual ... This means that the primary purpose of a climate action group must be to alter public opinion from a position of passive acceptance of the government's climate policies to a position of informed and passionate dissent.

This document also shows CAC's aspiration to shift the language used in the climate politics debate away from the politicians' and media obsession with the economy:

> The government currently sets the agenda, and has chosen to set the agenda firmly in the economic arena ...
>
> The most favourable ground for us is to debate climate change in moral and human terms. This would allow the harnessing of genuine moral outrage of the sort able to bring into question the very legitimacy of the government, as well as the positive counterpoint of real climate action as a heroic moral duty—the most just of causes, to bring out the best in ourselves.[19]

In the lead-up to the 2009 Copenhagen climate summit, an open letter was sent to their networks, organising a sit-in around Parliament House. It hoped this would be 'as peaceful and dignified an action as possible' but provided legal advice on the last few pages, because:

> One of the objectives of the protest will be to have as many people as possible arrested for peaceful, calm and dignified civil disobedience—Gandhian-style. As world leaders prepare for Copenhagen, we want to send the strong message that we are sick of our concerns about climate change being brushed aside. We would like to have a couple of hundred arrestees from across the movement and many walks of life showing Kevin Rudd that he can't continue ignoring our passion and concern.
>
> We also want many people who don't want to be arrested to come along in support. These people can participate in the protest until the point where the police warn people that unless they move they will be arrested.[20]

Indymedia reports that roughly 200 people attended, of which 130 were arrested, though none were charged:

> The protesters set up at 11 am and refused to leave the entrance as an act of civil disobedience. From midday they were gradually taken away by police. Greens Senator Christine Milne addressed the protesters and was received positively. Liberal Senator Bill Heffernan and Family First Senator Steve Fielding, both known as climate change sceptics, attempted to lecture the crowd, with Steve Fielding being laughed down. The police closed Parliament House to the public while detaining protesters.[21]

CAC's most notable success has been the 'Canberra loves 40%' campaign. Run in conjunction with other community groups, this campaign successfully lobbied the ACT government to legislate the most ambitious emissions reductions targets in the country. On 27 October 2010, the ACT Assembly passed a bill to establish an emissions reduction target for the ACT at 40% below 1990 levels by 2020.[22] The day after the bill passed into law, the website of the 'Loves 40%' campaign proudly displayed the heading 'We finally did it!' The rest of the article was celebratory but earnest about the continuing need for action.

> With the passing into law of our 40% emissions reduction target Canberra goes from a Nation leading city to a world-leading city.
>
> This is an ambitious target and there is no guarantee that it will be achieved—that is up to us—the community—both in terms of our own actions and in what we demand of the ACT Government. 40% is achievable and one thing is certain—if it is achieved our city will be transformed for the better.
>
> Setting this target makes us one of the first jurisdictions to accept responsibility for our emissions and commit to doing our share of what the science says is required globally for an effective response to climate change. While it is important for us to now weigh the costs and impacts of the options available for reaching this target, this isn't the time to hold ourselves back. This is an opportunity to make smart decisions, to deal with central concerns already facing the ACT in planning and growth, and to invest in our city in ways that will ensure the ACT is a prosperous, vibrant, sustainable community into the future.
>
> The vision is there, it is now time for the community, businesses and the Government to work together and all play our part to achieve it.[23]

In this collection of documents relating to the 'Canberra loves 40%' campaign, we see rosters drawn up to supervise community groups and their activities, ranging from the quite common (leafleting, lobbying outside parliament), to the more creative (organising to have '40 people doing something for 40%—40 dogs, 40 people on bikes, 40 trees planted?').[24] CAC's success in this instance was no accident. Much thought, debate and activity, typical of CAGs, has gone in to their organisational work.

Case-Study: Rising Tide (Newcastle)

Rising Tide was established in Newcastle in 2004. On its website, it states:

> We are a grassroots Newcastle group taking action against the causes of anthropogenic climate change and for equitable, just, effective, and sustainable solutions to the crisis. We are committed to the principles of Non-violent Direct Action. We are part of the global Rising Tide climate justice movement. We live in the biggest coal port in the cosmos.[25]

In the Upper Hunter region, twenty-six mines produce a total of 27 million tonnes of coal per year. Connor, Freeman and Higgenbotham describe the 'social field of coal-mine development in the Hunter Valley' as 'an intensely contested arena'. The opposition to coal mining is a diverse 'community of interest', consisting of local residents, wealthy vignerons, horse stud associations, 'ferals' or alternative people, and environmental groups.[26]

Rising Tide has mobilised the oppositional community of interest in the Hunter Valley. Along with Greenpeace it led the struggle to stop the Anvil Hill mine in October 2006, involving more than thirty local resident groups and local branches of other environmental and conservation groups. Building upon the concern of local people about environmental degradation, water shortage, pollution, and the threat to biodiversity from coal mining, Rising Tide and Greenpeace then introduced into this campaign 'a bold new rhetoric of climate change threat'.[27] Reflecting on the progress made since its formation, spokesman Steve Phillips told the *Newcastle Herald* in 2009 that the turnaround in public awareness about climate change and the problem of coal had been 'phenomenal'.[28]

Rising Tide manifests an explicit climate justice stance with a distinctive direct action emphasis. It argues: 'Climate justice is not just a theoretical goal. It is also a political practice founded on empowerment of marginalized communities, resistance to corporate control of global affairs, and to ending the growth addiction inherent in global capitalism.'[29] Its anti-capitalist perspective has no illusions about greening the state. Steve Phillips, as a Rising Tide representative, stated at a global warming conference at the University of Sydney on 2 March 2007: 'Coalmining companies have governments ...

wrapped around their finger, and anybody who dares call for a contraction of the coal industry is publicly caned as a dangerous extremist.'[30] Rising Tide emphasises the culture of consumption in contributing to greenhouse gas emissions and thus climate change. For example, on 1 April 2008, it unfurled a banner at the Westfield shopping centre in Newcastle that highlighted two issues: that Australian consumerism leaves Australia with a trade imbalance that is offset by expanding coal exports that prop up a non-sustainable local economy; and consumer demand for cheap goods from China and India accelerates the construction of new coal-fired power stations in those countries in order to provide the necessary energy and thus contributes to climate change.[31]

During 2009 Rising Tide started to attract attention. It generated national media coverage when its activists abseiled down the front of Parliament House on federal Budget day with a banner that read 'Carbon Budget Blowout', co-convened a huge summit for community climate action groups, stopped work at a notorious aluminium smelter for several hours with a non-violent protest, jointly organised a peaceful sit-in at Parliament House a week before the Copenhagen climate talks and staged a day-long blockade of the rail line into the world's biggest coal port at Newcastle. For this record, the ACF awarded it the 2010 Peter Rawlinson Award 'for tirelessly and fearlessly drawing attention to the looming climate crisis', commenting that 'As a small and agile group Rising Tide does things that bigger national groups like ACF simply can't do.'[32]

A particular focus of Rising Tide has been the role of Newcastle Harbour as 'the biggest coal-port in the cosmos'. It organises an annual 'People's Blockade of the World's Biggest Coal Port'.[33] For example, for the 'fun, effective and non-violent protest against the tripling of coal exports from Newcastle Harbour' on 28 March 2009, over 500 people paddled onto the harbour on canoes, surfboards, and homemade rafts, bringing the world's biggest coal port to a standstill for an entire day.[34]

These are some of the other Rising Tide protests in Newcastle Harbour. On 19 April 2008 around fifty people occupied the construction site of the third coal loader on Kooragang Island. The NSW government had approved the project in April 2007 to increase Newcastle's coal export capacity. Timed to coincide with the 2020

Summit in Canberra, the protestors asserted that the Rudd government needs 'to help this region move away from coal'.[35] On 26 September 2010, fifty Rising Tiders occupied all three coal-loading facilities, leaving twenty ships stranded offshore and caused millions of dollars costs to coal exporters. Rising Tide explained this was 'an emergency intervention' because Newcastle Harbour was 'exporting global warming to the world'.[36] In a return offensive against the Kooragang loader, Rising Tiders scaled the 15-metre high conveyor belt and suspended a banner 'We're sorry Somalia. Coal = climate change and starvation.' They explained scientists had that week made the link between human-made climate change and the deadly drought affecting more than ten million people across the Horn of Africa. Rising Tide spokesperson Naomi Hogan said: 'We are standing in the belly of the beast, here in the world's largest coal port, calling on the coal industry to reverse their plans to more than double Australia's coal exports. It's about time the coal giants took responsibility for their contribution to the dangerous impacts of climate change.'[37]

Rising Tide also stages cyber-attacks. On 9 November 2011, it released a rip-off website to counter the NSW Minerals Council's latest spin campaign, which defends the mining industry and tries to paint it as 'world class'. Rising Tide's parody website at www.worldclassminers.com.au imitated the layout and design of the original website, but pointed to the industry's poor track record in the areas of environment, community, economy, health and innovation.[38] As the ACF noted in 2010, 'a small, committed group of volunteers can make a big impact'.[39]

Climate Action Networks: Some Examples

Climate action networks connect existing groups into some cooperative arrangement with each other, avoiding duplication of efforts and, most significantly, combining efforts to produce greater political impacts.

Climate Emergency Network (CEN)

CEN is a Victorian-based network that emerged out of the workshop in February 2008 focused on Spratt and Sutton's *Climate Code Red*. Its leaflets quote Victorian Governor David de Kretser's comment on

17 July 2008 that climate change 'has reached the level of a state of emergency'. On 13 September 2008, CEN adopted a resolution outlining its rationale:

> All levels of all governments across the globe must recognise and work together to fulfill their responsibility to secure a safe climate.
>
> Underpinned by legislation, governments must lead a large-scale transformation to a post-carbon economy.
>
> Given the extreme urgency and enormous scale of transformation required, governments must acknowledge the climate and sustainability emergency, and implement a rapid response plan to urgently and effectively address it, whilst respecting basic human rights and freedoms.
>
> The community must be engaged in recognising and supporting the climate emergency. Therefore, we will work to engage citizens in taking responsibility for recognising and responding to the climate emergency.[40]

Describing itself as 'a network of community organisations and individuals campaigning for an urgent and effective response to a climate and sustainability emergency', CEN's objectives include: working towards 100% renewable energy in Australia by 2020 'or as fast as humanly possible'; no new fossil fuel power stations; protecting and strengthening the capacity of the earth's natural carbon sinks; and supporting carbon-neutral new technologies.[41]

In the wake of the 7 February 2009 Victorian bushfires, CEN produced a leaflet explaining the 'strong evidence that climate change is increasing the likelihood of conditions that favour uncontrollable fires'.[42] It has provided financial support for BZE, whose plan it vigorously supports, and to SCA to start its 'How Fast?' project.[43] More than twenty Victorian CAGs became affiliated with CEN.[44] Others have opted not to do so; the reasons for this are not entirely clear. No socialist groups belong to CEN, but individuals belonging to Socialist Alliance and Solidarity have been quite active in CEN.[45] There are internal debates about mitigation strategies, with some members supporting ambitious well-designed market-based schemes and others opposing them outright.[46]

Cutting across such differences is CEN's strong commitment to climate justice. At a general membership meeting on 22 May 2010, it passed the following resolution:

> Over the past decade or so, a climate movement has developed in many countries, both developed and developing. There has been a tendency for the climate movement in North America, Europe, and Australia to focus on ecological modernisation as a climate change mitigation strategy, that is one emphasising technological innovations such as renewable sources of energy and energy efficiency, while often downplaying the need to address social disparities emanating from the global economy, both within nation-states and between nation-states. While technological innovations constitute an important component of climate change mitigation, it is essential to recognise that the people of the world who have contributed the least to greenhouse gas emissions and anthropogenic climate change are suffering the most as a result of it and will continue to suffer from it.
>
> Climate Emergency Network resolves to make a commitment to promote climate justice both in Australia and abroad. While individual CEN members may disagree with some specific points of the People's Agreement produced at Cochabamba, as an act of solidarity with the international climate justice movement, CEN applauds its over-all spirit and commitment to climate justice.
>
> Furthermore, CEN encourages other groups involved in the Australian climate movement to more strongly address social justice issues and to recognise the role of the existing global economy in contributing to climate change, environmental degradation, and gross social disparities both at home and abroad. Climate injustice affects not only people in developing societies, such as in the South Pacific, sub-Saharan Africa, and Bangladesh, but also indigenous peoples and low-income groups, in developed societies. For instance, in the case of Australia, Aboriginals and Torres Strait Islanders are being adversely affected by

> rising sea levels and rising temperatures in the outback, and many rural people are being adversely impacted by a long-term drought related to climate change.[47]

As one of its major contributions to the work of T10, in 2011 CEN commenced work on a Safe Climate Economy campaign, 'to establish firmly in the public mind and the mind of community leaders that we need to complete the building of a safe climate economy by 2020 and to lead to a community demand that systematic rebuilding starts in a couple of years'. It recognises that most politicians think building a safe climate economy by 2020 is politically impossible, 'so we need to let the politicians know that the overwhelming majority of our campaign work is directed at getting a large majority of the community on-side—so that the campaign goal is eventually made politically feasible'.[48] On 23 July 2011, CEN launched its 'groundswell' for a safe climate, a proposed mobilisation across all of Australia, aiming to: build commitment in the Australian community to initiate the transformation of the Australian economy by 2020 to drive the restoration of a safe climate; and to catalyse the election of governments that accept the need for urgent action to restore a safe climate and are committed to take action on the timescale required.[49]

Climate Emergency Action Network (CLEAN)

CLEAN is a South Australian climate network that organises community actions, outreach and education in Adelaide and Port Augusta. Its website explains:

> CLEAN aims to help build a broad, popular climate movement and can assist Adelaide communities to start climate action groups, conduct presentations, or access information.
>
> CLEAN believes in a fast transition to a zero carbon economy, with a just transition for Australian workers in affected industries.[50]

It originated out of a conference in Adelaide 10–11 October

2008 on 'Climate emergency—no more business as usual', with keynote speakers including David Spratt, Mark Diesendorf and Barry Brook. About 250 people attended, many of whom felt inspired to establish an Adelaide-based network.[51] CLEAN held its inaugural meeting, organised by John Rice, Jenny Bain and others, on 23 October. Groups represented included Save Our Gulf Coalition, Adelaide Hills Climate Action Group, Eco-socialist Convergence, the 3D Environment Show, the Young Greens, Socialist Alliance, the Wilderness Society and the Conservation Council.[52]

Like CEN, CLEAN emphasises adoption of renewable energy sources, energy efficiency and technological innovations. CLEAN supports the BZE plan for 100% renewable stationary energy by 2020, opposes nuclear power plants as a purported mitigation strategy, favours a 60% cut in emissions by 2020 and supports Lock the Gate on CSG. It was critical of the CPRS and emissions trading in general. However, it does favour implementation of a '$40 plus carbon price as one step in the climate change challenge' and joined the 'Say Yes' to the carbon tax rally in Adelaide's Victoria Square on 5 June 2011.[53]

CLEAN emphasises climate justice issues even more strongly than does CEN, largely as a result of the key role of Eco-socialist Convergence members and the generally significant eco-socialist and also eco-anarchist presence in its membership. Its Positions Statement makes this clear:

> As the wealthy nations of the world have very largely caused the climate emergency and have unfairly exploited many of the poorest nations, people, and natural resources, the wealthy nations owe a very substantial duty to compensate materially poorer nations by providing them with financial aid to develop green energy infrastructure and to enable them to raise their standards of living to reasonable levels.
>
> Wealthy nations, due to their privilege and economic power, must play a leading role in mitigating climate change and be willing to make much more substantial cuts in their greenhouse gas (GHG) emissions, in absolute and relative terms, than poorer nations.
>
> In dealing with the climate emergency, all solutions to

> the emergency must be based on the principles of both ecological sustainability and social and economic justice between all nations and between individuals within nations.

One of its recommended 'specific measures' is: 'All workers in the fossil fuel industries are provided the opportunity for retraining on fair pay for suitable alternative employment in ecologically sustainable industries.'[54]

Australian Youth Climate Coalition (AYCC)

AYCC was created in November 2006 by lawyers Amanda McKenzie and Anna Rose. Members are easily spotted at climate movement events in their sky-blue and white AYCC 'Climate Action Now' t-shirts. AYCC claims to have 71 873 members and to function as a climate change coalition for about twenty-seven other youth organisations, including the Australian Medical Students Association, the Australian Student Environment Network, Engineers Without Borders, the Youth Food Movement, Law Students for a Just Community, the National Indigenous Youth Movement of Australia, the National Union of Students, the Sai Youth (Multi-Faith Youth Network), the Student Organised Network for Architecture and Youth Christian Students. The AYCC wants the Australian government to: (1) 'reduce Australia's emissions at home'; (2) 'invest in sustainability education', and (3) 'do our fair share internationally'.[55]

Anna Rose represented AYCC at the '2020 Summit' convened by Rudd in April 2008 to open 'a window on our democracy to let some fresh air in'. Instead, Rose noticed the disproportionate presence of coal industry representatives.[56] She recalled that 'I found myself in the climate stream with representatives of coal-mining companies including Xstrata and Shell, yet ... No-one from Friends of the Earth, the Australian Conservation Foundation, Greenpeace, Climate Action Network or any of the State Conservation Councils'.[57]

In mid-2009 AYCC and World Vision surveyed 37 432 young people: 97.5% favoured emissions cuts greater than those proposed by the government. AYCC also surveyed its membership base, asking how well they thought their elected representatives represented their concerns. Some 64.2% said 'not well' or 'not at all', compared to 5.6%

who selected 'very well'.[58] AYCC favours lowering of the voting age to sixteen. It takes its message to high school students through its 'Switched On Schools' initiative that links climate activists in schools across the country. Schools can sign up to a database that provides: resources for students and teachers to take action on climate change; contact with other switched-on schools; and information about climate campaigns. It provides a 26-page *Climate Action Kit. Manual for Schools* and offers speakers to educate about climate change and how to get involved in the movement. 'Switched On Schools is a dynamic program set to engage and empower the high school students of Australia. We provide the knowledge, skills and support that young people need to create a better world for themselves and future generations.'[59]

The regular conferences organised by AYCC are called 'Power Shift'. An especially large one, attended by about 1500 young people, was held at the University of Western Sydney 11–13 July 2009. The speaking list included: NSW Premier Nathan Rees, South Australian Premier Mike Rann, Christine Milne, Nick Xenophon and Tim Flannery. While Rees was speaking, a high-school student donned a green plastic hard hat and yelled 'What about the world's largest coal port?' Rann was heckled by anti-uranium activists; and the conference gave a standing ovation to the Aboriginal woman who spoke against Rann's uranium drive and described the impact of nuclear testing on her community.[60] Observer Troy Henderson was impressed by the diversity and dynamism of this conference:

> It was a snapshot of a movement in motion and the image is one that excites. While most climate change events leave people feeling exasperated, powerless and depressed, Power Shift sent 1500 young people back to their homes with new friends, new knowledge and the message that they can change the world. Power Shift was cool, inspiring and fun—three words young people don't readily associate with politics. It drew new people into the movement, some of whom may well blockade a power station one day.[61]

However, other sections of the climate movement criticised the conference sponsorships. Resistance noted that, in addition to

GetUp!, WWF-Australia and Climate Institute support, conference organisers accepted: $10 000 from Rann, who was 'promoting construction of a large uranium mine'; $10 000 from Pacific Hydro, a renewable energy company that 'earns money from carbon offset scams'; and $5000 from Westpac, the largest shareholder in BHP Billiton. Representatives from the two companies also spoke at the conference.[62] Resistance commented on these three speakers: 'We do not believe these people are part of the climate change movement. They are climate criminals and should not be provided with more space to tell young people that their "business-as-usual" organisations are part of stopping climate change.'[63] Solidarity likewise argued this conference constituted a 'platform for some of Australia's worst polluting politicians and businesses to spout their greenwash'.[64]

Guy Pearse has also criticised AYCC, to which AYCC has responded, insisting it had not supported the CPRS and that there were other false allegations:

> The AYCC is in favour of switching to 100% renewable energy as fast as humanly possible, and reducing our emissions in line with what the science says, to get us below 350 ppm CO_2 in the atmosphere. We do not support clean coal. We do not support dodging our emission reduction targets by using offsets from overseas.[65]

AYCC did support the Clean Energy Future legislation. When it finally passed the Senate on 8 November 2011, the AYCC Facebook page was jubilant: 'We did it!!! The Australian Parliament said YES to introducing the carbon price at 12.30 pm today! ... Let's keep building on our success: Today, a carbon price. Tomorrow, 100% renewable energy!'[66]

Climate Action Network Australia (CANA)

CANA is 'an alliance of over seventy-five local, state, national and international environmental, development, research and advocacy groups from throughout Australia'. It was formed in 1998 as the Australian arm of the global Climate Action Network International, with branches in over seventy countries. CANA's 'core principle' is that 'a cohesive, mutual and cooperative civil society is an essential pre-requisite if Australia is to effectively and justly meet the challenges posed by climate change'. CANA sees its role as 'providing the behind-the-scenes work necessary to ensure that such cooperation is possible'; its 'foundational work' allows for symbiotic cooperation between members 'in the complex ecosystem of social change and advocacy'.[67]

The affiliated groups on its website include: NGOs and institutes such as Greenpeace, WWF-Australia, ACF, Climate Institute, Wilderness Society, Women's Environmental Network Australia, 350.org Australia, AYCC and GetUp!; aid organisations; the Liquor, Hospitality and Miscellaneous Union; faith-based organisations such as Australian Religious Response to Climate Change; renewable energy and energy-efficiency organisations; legal and research organisations; state-based ENGOS such as EV and the Conservation Council of Western Australia; regional environmental organisations; and CAGs, such as Rising Tide.[68]

It brings its constituent organisations together in regular conferences. For example, 11–12 May 2009 it held a conference on 'Australia's climate movement in 2009: the turning point', to develop collaborative relationships among CANA member groups. On 24–25 February 2010 it held a conference on 'Climate action 2010: rebuilding capacity and strategy post-Copenhagen', taking the opportunity to discuss tactics at that gloomy moment. Its website provides valuable information and resources about climate change for these organisations as well as the general public.

CANA has recommended much more drastic emissions cuts than envisaged by governments. During the Rudd government, its policy paper called for emissions to peak by 2010, drop by at least 40% by 2020 and by at least 95% by 2050 (from 1990 levels). It strongly criticised the Garnaut Review for low targets, willingness to allow the coal industry to continue functioning and advocacy of CCS.[69] In March 2011, CANA developed three key tests for the Gillard government's proposed carbon price: 'The current parliament must ensure that Australia's emissions peak and begin to decline by 2013'; 'The price mechanism must be part of a package that ensures transformation of the entire economy, including policies that promote energy efficiency, prevent new coal-fired power stations and encourage large-scale investment in land-sector abatement and large-scale renewable technologies'; and 'The package must ensure that Australia can meet its international obligations to reduce emissions by 25% by 2020 and provide its share of climate finance.' Its analysis concluded that the Clean Energy Future package 'stacks up against these tests'.[70]

Conferences and Summits

Networks, as we have seen, frequently organise conferences, and these are important in building the climate movement. There are also many smaller regional conferences that have taken place across the country in the past few years, for example, the 'Earth Matters Climate Change Forum' on 8 August 2008 in Ballarat, addressed by David Spratt, Mark Diesendorf and Richard Denniss. There are also community-based conferences held within the largest cities, for instance, the 'Activate: community organising training for climate solutions' conducted by the Nature Conservation Council of NSW on 17 October 2009 in Sydney, addressed by Sydney Alliance Director Amanda Tattersall and Unions NSW Campaigns Director Mary Yaager.[71] These kinds of gatherings indicate the need so many people feel to get together to discuss their concerns—an important aspect of any social movement.

Conferences and summits assist climate movement consolidation. In this next section we describe some of the most significant *general* gatherings of the climate movement and their contribution to its development. These larger meetings enable activists within CAGs and networks to meet each other, literally, and form further linkages. Indeed, it is often at conferences and summits that networks have been established.

Climate Movement Convergence 2008

Important in the recent spate of conferences was the Climate Movement Convergence that met at Northcote High School in Melbourne on 9 February 2008, organised primarily by the SLF, FOE, the Greenleap Strategic Institute, and Sustainable Business Practices. Supporting organisations included BZE, CarbonEquity, the Central Victorian Greenhouse Alliance, EV, Greenpeace, the Moreland Energy Foundation, the Western Region Environment Centre, Zero

Emissions Network, and an array of suburban CAGs. The AMWU, the Australian Services Union and the Australian Education Union sent representatives.[72]

The conference handbook stated:

> In planning today, we have simply attempted to create a space where the many and very diverse elements of the 'climate change movement' can get together, share info, educate and inspire each other, and hopefully decide if we do, in fact, have an emerging movement. And if so, to consider the question—what are we aiming to do about it, and how can we work together in future in the most effective ways.[73]

Around 250 people attended. A CANA representative, Owen Pascoe, gave an opening address on 'The Current Landscape', in which he discussed new opportunities for climate action in light of a new ALP government that had ratified the Kyoto Protocol and committed itself to creation of an ETS. This 'convergence' contained within it the varieties of approach typical of the climate movement: speakers and workshop organisers proposed strategies that sought to mitigate climate change within the parameters of 'green capitalism' while others were committed to direct actions and moving well beyond 'business-as-usual'. A highlight was the launching of *Climate Code Red: The Case for a Sustainability Emergency* report, the original edition sponsored by FOE, which led to the formation of CEN.[74]

Climate Action Summit 2009

Even before the government's CPRS White Paper was released, various organisations, including the Bega-based group Clean Energy for Eternity, had been planning a summit in Canberra to herald the first 2009 session of federal parliament. In its rationale for the forthcoming summit, the organisers stated:

> Kevin Rudd was elected to government in the world's first climate change election ... we are still waiting for the urgent action needed to peak our greenhouse gas emissions and begin this urgent transition to renewable energy

> ...We need to ensure the Australian government goes to Copenhagen with the strongest commitment to reducing greenhouse pollution and averting runaway climate change.[75]

The ensuing Climate Action Summit held at the ANU on 31 January to 2 February 2009 was organised by activists from the Australian Student Environment Network, Rising Tide, Greenpeace, FOE, CAC, and other individuals.

In part the Summit was designed to coordinate the numerous CAGs across the country. In the *Climate Reader* prepared for this summit by the Victorian Climate Action Centre, Damien Lawson delineated ten lessons for the climate movement: Changing government does not mean a change in policy; Continuous mobilisation; If we are not frightened then no one else will be; Knocking on doors is as important as climbing smokestacks; Alliance building is more than box-ticking; Propose solutions that will work; Stop talking about the reef and start talking about people; But it is the economy, stupid?; We are activists, not policy advisors; Our movement is and must be global.[76]

Over 500 attended: representatives of around 150 CAGs; members of the organisations involved in staging the summit; climate activists from other organisations, such as AYCC; eco-socialists from Socialist Alliance, Solidarity and Socialist Alternative; and various eco-anarchists and other individual climate activists. Socialist Alliance commentators described the diversity of people present:

> From vegan to vegetarian groups to local sustainable technology businesses, from professors of environmental science to high school students, from local Climate Action Group activists to organic farmers, from trade unionists to environmentalists active around water and forests, from Greens to socialists and people who 'hate politics', the summit was a representative potpourri of the growing army of Australians who are alarmed about global warming and determined to act against it.[77]

The summit consisted of several plenary sessions and numerous concurrent workshops. Meals consisted of catered vegan food.[78]

In his opening speech on 31 January, Clive Hamilton noted: 'I am sure that, like me, many of you are asking yourself how we could have allowed ourselves to be seduced into believing that a Rudd Government would face up to the reality of climate change and do what is needed'.[79] Hamilton observed that 'there is an unholy alliance between government and industry to defer and delay action, to deny the true implications of global warming, and to hoodwink the public into sharing their view that protecting the old energy industries must come first. Except for the Greens, the main political parties have been captured by the fossil fuel lobbyists and climate sceptics.' He went on to argue, 'The only hope lies in a campaign of radical activism aimed at confronting the powerful in government and industry with their failure at every turn ... A campaign of radical activism must confront the public with their complacency, because until the public demands it far-reaching action simply will not happen.'[80]

Eighty-one recommendations were put forward at the summit. Amongst the nine that were rejected were ones supporting nuclear power, CCS and biofuels. The seventy-two resolutions passed included demands that the Australian government should: participate in an international effort to reduce global levels of carbon dioxide to 300 ppm no later than 2050; reduce Australia's CO_2 emissions by at least 60% by 2020 and 90% by 2030 (from 1990 levels); enact a policy of 100% renewable energy by 2020; create 'green jobs'; establish a moratorium on all new coal- and gas-fired power plants immediately; revolutionise energy efficiency; promote public transport; foster agricultural biological resilience; and create a moratorium on native forest logging.[81] Although no consensus was achieved on the structure for a national climate network, the Summit agreed on four key mass actions during 2009: 27 March when the CPRS was expected to be presented to parliament; 13 June on World Environment Day; September actions over an expected arctic melting; and 9 December nation-wide demonstrations before the Copenhagen Conference.[82]

Participants at the summit had different views on emissions trading in general, but agreed the CPRS was a flawed proposal, with its low reduction targets and free permits and compensation to polluters. The summit unanimously opposed the CPRS legislation.[83] Yet the tone of the conference was more accepting of ecological modernisation than fundamental economic and social transformation to mitigate climate change. According to Goodman, 'Nowhere in the list of eighty policy proposals approved by the Summit was there mention of economic growth or consumerism as a driver of emissions, and as something to be addressed.'[84] Commentary from socialist groups suggests the eco-socialists did not have the numbers. Chris Breen complained in *Solidarity* that, in such a large gathering, decision-making on resolutions was all too often orchestrated by summit organisers and insiders:

> Proposals were drawn up in detail by working groups that involved a small minority of summit participants and then presented as a virtual fait accompli to the whole summit. Furthermore the consensus process meant there was little free-flowing discussion about proposals, with participants only able to raise strenuous objections ... Around 80 policy recommendations were voted on individually, without debate, at a plenary on Sunday (the policy was available before the conference, and there had been debate in small policy sessions beforehand, but only a minority participated).[85]

While the Greens were not sponsors of the summit, Milne spoke at it and Brown addressed its concluding rally at Parliament House on 3 February. Summit organisers claimed around 2500 attended this rally; the *Canberra Times* could see only about 1250.[86] Also on 3 February and the following day, about forty summit delegates lobbied around thirty-five MPs to support the summit's policy recommendations.[87]

Climate Action Summit 2010

Around 300 attended the Climate Action Summit 13–15 March 2010 at the ANU, fewer than at the previous year's summit. Delegates represented a range of climate groups from every state and territory, and joined with officials from union, religious, and community organisations. Plenary speakers included David Karoly, Ian Fry (Chief Climate Change Negotiator for Tuvalu), Christine Milne, Damien Lawson, Mark Ogge (BZE), Walden Bello and Clive Spash. With a new national Community Climate Network formed to implement summit strategy and coordinate climate groups across Australia, this summit website announced:

> Our governments are not providing real climate solutions. Whether it is Rudd or Abbott's plan or Copenhagen we deserve better. This three-day conference will look at strategy, policy and actions for the grassroots climate movement.
>
> There is a growing community movement in Australia that is building people power for immediate action on climate change. Our vision is to work together at emergency speed to restore in a just way a safe climate in time for all people, all species and all generations.[88]

The summit was structured around different campaign streams: 100% renewables, coal, vote climate/election, climate emergency, and green jobs.[89] The plenaries, sessions and workshops focused on a huge variety of topics, including ones critical of ecological modernisation, such as 'critiques of carbon trading, technology, and the growth economy' and 'alternatives to the CPRS' and ones concerned with 'climate justice' and ethical issues, such as 'food choices and climate change'.[90] The summit adopted World Environment Day on 5 June 2010 as a national day of climate action revolving around the themes of renewable energy sources, cessation of coal production, a safe climate transition, and green jobs.

The summit set out key policy demands on clean renewable energy, green jobs and the need for an immediate carbon levy; and mapped out a strategy for the coming election, including a focus on key marginal seats, particularly Green–Labor swing seats. A spokesperson for the summit stated that, with this new strategy, the climate movement would be a major force in the coming election. 'We will harness the strong feeling in the community that after three years Kevin Rudd has not done nearly enough on climate change.'[91]

Climate Action Summit 2011

The third consecutive Climate Action Summit occurred 9–10 April 2011 at the University of Melbourne. Some 300 people, representing over 100 CAGs, attended to hear various climate activists, scholars, and climate change experts discuss and debate climate action strategies. From discussion in the panels and in the halls, it appeared most participants regarded the Gillard government's proposed carbon price as an improvement on the CPRS, while still regarding all market-based strategies as insufficient. Many felt the movement should support the carbon tax, then campaign against it rolling over into an ETS.[92]

In a communiqué released to the media, the summit declared that 'current climate policy options in Australia are inadequate and bear little relationship to actions demanded by the science of climate change'. The summit called on the government to:

> develop and implement a comprehensive national climate policy framework that includes a national plan for transitioning to a zero emissions economy, with substantial reductions in energy use; make polluters responsible by ceasing subsidies for fossil fuels use immediately, and not compensating polluting industries; redirect subsidies from animal agriculture to sustainable plant-based farming and to just transition programs for affected workers and communities; provide generous support providing a just transition for affected workers and communities via income redistribution and/or direct assistance or other measures; and place a much greater focus on so-called 'complementary' measures.[93]

At the plenary session that debated this communiqué, women activists drew attention to the serious under-representation of women in leading positions within the movement and asked those who shared this concern to stand, which everyone did.[94]

The mood at this conference was cautiously optimistic about the opportunities opened up by the 'balance of power' position in which the Greens found themselves, but sombre about the limited room for manoeuvre allowed by parliamentary politics.[95] At the social evening on 9 April, Sydney climate activist Jenny Curtis hosted a hilarious parody of the Oscars, dispensing awards for various forms of climate movement heroism to about a dozen surprised activists. One of these, Terry McBride, who earned his living installing television antennae, had become concerned about climate change by listening to climate scientists on ABC radio as he drove between jobs. Another, Michael Morphett, was an elderly musician, who had become engaged by radical politics during the anti-capitalist moment at the turn of the century.[96]

One of the highlights of this conference was the session late on 10 April featuring veteran Queensland environmentalist Drew Hutton speaking about the Lock the Gate Alliance, opposing coal-seam-gas exploration. In order to work with Queensland farmers, who still wanted to 'string him up' for preventing their tree-clearing proclivities, Hutton had resigned from the Greens, despite being a founding member. He predicted that the Lock the Gate Alliance, forging a powerful coalition of farmers and environmentalists for the first time ever, would become the biggest social movement in Australia history.[97] This emergent force is discussed in the next chapter.

Notes

1 Whelan et al, p. 11.
2 Green.
3 Whelan et al, p.15.
4 http://www.climatemovement.org.au (viewed 21 March 2009); 'Australian climate action groups'.
5 http://www.climatemovement.org.au/groups/ (viewed 8 December 2011).
6 http://www.climatemovement.org.au/about/ (viewed 11 November 2011).
7 Austin.
8 http://www.climatemovement.org.au/groups/darebin-climate-action-now/ (viewed 13 May 2010).
9 http://www.climatemovement.org.au/groups/darebin-climate-action-now/ (viewed 25 October 2011).
10 http://www.climateactionmoreland.org (viewed 13 May 2010).
11 Baer, participant-observer notes.
12 'Taking Action in an era of Climate Change: Wednesday June 18 [2008]', leaflet for conference at ANU, Clive Hamilton speaking, CAC Papers.
13 'Climate Action Canberra: Who we are', leaflet, CAC Papers.
14 'Climate action Canberra submission to national capital to solar capital: options for an expanded act: electricity feed-in tariff scheme', 26 February 2010, CAC Papers.

15 'The essential ingredients: a strategic primer for 2010', p. 2, CAC Papers.
16 ibid., pp. 2–3.
17 ibid., p. 6.
18 'To the staff of the Department of Climate Change', letter, CAC Papers.
19 'Countdown to Copenhagen', CAC internal document, p. 1, CAC Papers.
20 Letter headed 'Dear Friend', CAC Papers, pp. 3, 1.
21 'Climate civil disobedience action in Canberra for 40% carbon reduction target', 23 November 2009, Indymedia, available online at: http://indymedia.org.au/2009/11/23/climate-civil-disobedience-action-in-canberra-for-40-carbon-reduction-target (viewed 8 December 2010).
22 'ACT sets ambitious emissions target', ABC News, 27 October 2010. http://www.abc.net.au/news/stories/2010/10/27/3049154.htm (viewed 7 December 2010).
23 'We finally did it!', 28 October 2010, Canberra Loves 40% website. http://love40percent.org/news/ (viewed 7 December 2010).
24 'Canberra hearts 40%, Agenda, Monday 24 May, 2010', CAC Papers.
25 http://www.climatemovement.org.au/groups/rising-tide-newcastle/ (viewed 11 November 2011).
26 Connor et al, pp. 10, 19–20.
27 ibid., p. 14.
28 Thompson, 'Switched on'.
29 Rising Tide North America, 2009, quoted in G Evans, p. 202.
30 Quoted in G Evans, p. 211.
31 G Evans, p. 213.
32 http://farmer.cobaltmedia.com.au/articles/news.asp?news_id=2888 (viewed 11 November 2011).
33 G Evans, p. 211.
34 http://www.theherald.com.au/news/local/news/general/rising-tide-climate-protesters-take-to-newcastle-harbour-in-battle-with-coal-ships/1788214.aspx (viewed 11 November 2011); Alcorn, 'World's biggest coal port'.
35 Alcorn, 'Workers, environmentalists and the fossil fuels industry'.
36 G Evans, p. 199; 'World's biggest coal port shut down,' *GLW*, 6 Oct 2010, p. 1.
37 Branley.
38 http://risingtide.org.au/ (viewed 11 November 2011).
39 http://farmer.cobaltmedia.com.au/articles/news.asp?news_id=2888 (viewed 11 November 2011).
40 http://climateemergencynetwork.org/about_cen/#Charter (viewed 9 September 2009).
41 CEN, 'Membership Application'.
42 CEN, 'Bushfires', p. 1.
43 CEN, 'Roundtable', p. 1.
44 http://climateemergencynetwork.org/about_cen/#Affiliates (viewed 8 December 2011).
45 Information obtained by Baer while serving on CEN Facilitation Committee 2009–2010.

46 Baer, participant-observer notes.
47 ibid.
48 CEN, 'Roundtable', p. 3.
49 CEN, 'groundswell'.
50 http://www.cleansa.org.au/ (viewed 7 November 2011).
51 Baer, participant-observer notes.
52 Ratcliffe, 'Climate Emergency Action Network'.
53 http://www.cleansa.org.au/ (viewed 7 November 2011).
54 http://www.cleansa.org.au/Position%20Page%20nov%202009.html (viewed 7 November 2011).
55 http://www.youthclimatecoalition.org (viewed 22 October 2010); http://www.asen.org.au (viewed 22 October 2010); http://www.aycc.org.au (viewed 22 March 2011, 8 November 2011).
56 *GLW*, 30 April 2008, p. 10.
57 Quoted in Rintoul, '2020'.
58 A Rose.
59 http://www.switchedonschools.org.au (viewed 26 May 2010).
60 Henderson, pp. 45–6.
61 Henderson, p. 47.
62 *GLW*, 29 July 2009, p. 1.
63 Resistance.
64 Supple, 'A power shift'.
65 http://aycc.org.au/2011/09/12/response-to-article-in-the-monthly/ (viewed 8 November 2011).
66 http://aycc.org.au/ (viewed 8 November 2011).
67 http://cana.enviro-geek.net/general-info/cana-who-are-we (viewed 11 November 2011).
68 http://cana.net.au/hot-topics/cana-member-organisations (viewed 11 November 2011).
69 CANA, 'Garnaut's Supplementary Report'.
70 http://cana.enviro-geek.net/news/test-news (viewed 11 November 2011).
71 Baer, participant-observer notes.
72 Bolton, 'Australia's first climate convergence'.
73 *Climate Movement Convergence Handbook*, Melbourne 2008. http://www.sustainabilityconvergence.org.au (viewed 17 September 2008).
74 Baer, participant-observer notes.
75 http://www.foe.org.au (viewed 22 December 2008).
76 Lawson, 'Looking back, moving forward'.
77 White and Nichols, p. 2.
78 Baer, participant-observer notes.
79 Hamilton, The time for radical activism, p. 3.
80 ibid., p. 6.
81 Baer, participant-observer notes; Butler, 'Climate Action Summit'.
82 Baer, participant-observer notes.
83 Spratt, 'Suicidal sweethearts'.
84 Goodman, 'Responding to climate crisis', p. 145.
85 Breen, 'Climate summit', p. 27.

86 *Canberra Times*, 4 February 2009, p. 1.
87 Baer, participant-observer notes.
88 http://www.climatesummit.org.au/national-climate-summit-maps-out-election-plan (viewed 26 May 2010).
89 Saunders.
90 Climate Action Summit 2010, Program.
91 http://www.climatesummit.org.au/national-climate-summit-maps-out-election-plan (viewed 26 May 2010).
92 Baer, participant-observer notes.
93 Australian Climate Action Summit communiqué, 'Not good enough!', April 2011.
94 Burgmann, participant-observer notes.
95 Burgmann, participant-observer notes.
96 Conversations with Burgmann, 9 April 2011.
97 Burgmann, participant-observer notes.

12
Demos and Direct Action

More than any other social movement in history, the climate movement is involved in a conflict that will not be resolved without major social change. There is widespread understanding in the movement that demands upon governments must be backed up by strong ground-level political mobilisation to pressure governments otherwise reluctant to confront business in the interests of climate change mitigation. Demonstrations and disruption through direct action are both mobilisation in themselves but also tactics to mobilise yet more people by drawing attention to the serious threat of climate change.

Here we describe the major demonstrations of the past few years, actions targeting coal and the very new campaign against coal-seam gas. On-line activism has been invaluable in advertising and explaining these events—and hassling politicians and corporations. GetUp!, 'a movement of over half a million Australians trying to make our nation a better place', has been an especially important electronic avenue—so much so that, early in November 2011, a Coalition Senator angrily referred to GetUp! as 'the Hitler Youth wing of the Greens' political movement'.[1]

National Demonstrations

The climate movement organises regular demonstrations around the country. Often these are state-based events, targeting state government policies. For example, a Climate Emergency Rally in Melbourne on 5 July 2008, addressed by Bob Brown, demanded of the Victorian government: 'No desalination plant, phase out coal, no new freeway tunnel, no bay dredging; and yes to renewable energy, public transport and urgent action to stop global warming.'[2] Amongst nationwide mobilisations, the most leisurely are undoubtedly the annual Walks against Warming. On 11 November 2007, 50 000 people walked in Melbourne, 30 000 in Sydney (including a 5-metre effigy of Peter Garrett with a piece of coal around his neck); and tens of thousands in more than fifty other cities and towns.[3] These have continued successfully, often drawing in people unwilling to attend more militant climate movement events.

By late 2008, it became apparent that climate movement frustration had reached boiling point. When Rudd announced the 5–15% interim emissions reduction targets on 15 December 2008, climate groups across the country organised angry protests, at Rudd's electoral office in Queensland, Wong's office in Adelaide, Garrett's office in Sydney and elsewhere. Three female protesters were wrestled out of the National Press Club for interrupting Rudd's announcement by shouting:

> No, that's not good enough. It's spineless, totally spineless. This will mean runaway climate change. This will kill the Great Barrier Reef; this will wipe out the pacific islands. You're siding with polluting industries against the people. There's still time to change, we need leadership, give us some hope. We will not walk away from climate change, unlike Kevin Rudd and Penny Wong.

A joint statement from the Australian Climate Movement denounced the interim targets as 'a total failure of climate policy, a demonstration of the Rudd Government's capitulation to pressure from the big polluters, and simply unacceptable'.[4]

After more spontaneously angry rallies outside ALP headquarters in Sydney on 10 March 2009 and the federal government offices in Melbourne on 11 March 2009, the climate movement conducted a national day of action against the CPRS at federal parliamentarians' offices on 27 March 2009. Among those targeted were Rudd, Garrett, Lindsay Tanner (MP for Melbourne), Jenny Macklin (MP for Jaga Jaga and Minister for Aboriginal Affairs) and Maxine McKew (MP for Bennelong, who captured John Howard's former seat in the 2007 election). Demonstrations also occurred at the federal Department of Climate Change in Canberra and the Commonwealth Parliamentary Offices in Adelaide.[5]

On 13 June 2009, simultaneous Climate Emergency Rallies occurred in Adelaide, Brisbane, Canberra, Hobart, Melbourne, Perth, Sydney and Wollongong. An estimated 11 000 rallied around Australia.[6] John Rice, a member of CLEAN in South Australia, addressed some 1500 people in Adelaide.[7] In response, Penny Wong told the media: 'What many of these people are calling for simply can't be done.'[8] A few months later, in the build-up to Copenhagen, climate activists around Australia participated in the International Day of Climate Action on 24 October 2009, organised by 350.org, demanding reduction of global emissions to 350 ppm CO_2.

Direct action protests against fossil fuel targets, discussed below, absorbed climate movement energies for much of the next few years. By 2011 national demonstrations focused on support for the anticipated carbon price legislation. Preceded by a 10 000-strong pro-carbon price rally at Gillard's Melbourne office in Treasury Place on Saturday 12 March 2011, synchronised 'Say Yes Australia' rallies were held on Sunday 5 June 2011 in Sydney, Melbourne, Adelaide, Brisbane, Perth, Hobart, and Canberra.[9] Say Yes Australia was the collaborative effort of the ACF, ACTU, the Climate Institute, Greenpeace, WWF-A, CANA, GetUp!, AYCC, and EV (on behalf of Australia's Conservation Councils). It called for protesters to 'say yes to a price on pollution' in order to 'pave the way for Australia to join the booming clean energy market, to create thousands more jobs, to have healthy, clean air and to protect our beautiful environment for generations to come'.[10] GetUp! emailed its supporters: 'We have one thing the opponents of a carbon price don't: people power. But people power is like a muscle—if you don't use, you lose it.'[11]

An estimated 45 000 protesters attended these Say Yes Australia rallies, with about 5000 in Canberra, 10 000 in Melbourne, 10 000 in

Sydney, 3000 in Perth, 4000 in Adelaide, and 4000 in Hobart. At the Sydney rally, Simon Sheikh, GetUp!'s National Director, stated: 'Our politicians are negotiating a carbon price. But we say to them, "make it ambitious. Make sure it invests in clean energy and energy efficiency because we are ready to power our economy with 100% renewable energy".'[12] Climate movement activists critical of the carbon price either distributed leaflets at the various rallies, explaining their opposition, or, in the case of Ben Courtice, opted not to attend at all, to avoid joining the ALP 'just say yes' cheersquad.[13]

Anti-Coal Protests: Some Highlights

Many climate activists became so tired of waiting on the government to adopt even moderate mitigation strategies they decided to embark upon a series of direct actions focusing on the coal industry. Noting that 160 people in Australia were arrested taking direct action on climate change in 2008, Greenpeace Climate Change Campaign Coordinator John Hepburn explained: 'More people are starting to realise that we need to take direct action, including civil obedience actions that physically stop the expansion of the coal industry and start the phase out of coal and its replacement with renewable energy.'[14]

In the Howard government days of anti-coal campaigning, Greenpeace actions led the way. For example, on 27 July 2005, Greenpeace's *Rainbow Warrior*, bearing a sign saying 'Quit Coal', dropped anchor in the channel leading to the Kooragang and Carrington terminals in Newcastle harbour. In addition, teams of activists occupied a 2.5-million tonne coal stockpile and one of four huge coal loaders. Greenpeace spokesperson Ben Pearson explained the action:

> We are here today to expose Australia's dangerous addiction to coal export dollars and its significant contribution towards global climate change. Newcastle sends 80 million tonnes of coal to the world each year, and every tonne causes worse droughts, bushfires, storms and floods. Every hour that we are here, we stop coal that could emit thousands of tonnes of greenhouse gases from leaving.[15]

Another Greenpeace stunt coincided with the start of the 2007 APEC meeting hosted by Howard in Sydney and attended by President Bush, Russian President Vladimir Putin, Chinese President Hu Jintao and other Pacific-rim leaders. At dawn on Sunday 2 September, twelve Greenpeace activists painted the message 'Australia Pushing Export Coal' in two-metre high letters on the side of a coal ship, *Endeavour*, and unfurled a large banner in Chinese calling on China to be cautious of Howard and Bush's attempts to sabotage Kyoto. The message aimed to expose Howard's agenda to protect Australia's coal export industry by undermining the Kyoto Protocol. Pearson stated: 'Australia's climate policy is to "Push Export Coal" and to hell with the consequences for the planet.'[16]

The following day, Monday 3 September, activists wearing orange 'This is Real Action on Climate Change' t-shirts shut down part of the Loy Yang power station in Victoria's Latrobe Valley to highlight inaction on climate change by APEC member governments. Four people were arrested. At the subsequent media interview outside the station, a spokeswoman for the group said 'we need to see our political leaders supporting the move to the renewable energy future'.[17]

Anti-coal actions gained rather than lost momentum after 2007 with the election of federal Labor governments. Several of the direct actions against coal have been part of the Camps for Climate Action movement, which was especially active during 2009 when there were four Climate Camps. On its website Climate Camp 09 delineated three core principles:

> 1. To provide quality information from a range of people like former coal-miners, scientists, journalists, field experts and traditional owners so you can ask questions, discuss topics to make up your own mind as well as come-up with brilliant ideas and solutions.
>
> 2. To empower individuals, families and groups who are legitimately concerned about climate change by connecting them with each other over the three days in an atmosphere that is fun, creative, diverse, safe and which practises sustainable living.

3. Explore, celebrate, learn and practise non-violent direct action as a necessary, responsible, legitimate way to demand a rapid and just transition to a low-carbon economy.[18]

The Climate Camp movement in Australia commenced in 2008. To coincide with climate camps in Europe and North America, climate activists affiliated with FOE, Rising Tide, the Australian Student Environment Network and other groups, organised a Camp for Climate Action Australia on 10–15 July 2008 in Newcastle. Already the world's largest coal port, protesters were reacting to expansion plans: the NSW government had approved a plan on the part of the Newcastle Coal Infrastructure Group, led by BHP Billiton, to expand its annual export capacity from 80 million tonnes to 211 tonnes of coal. In the words of the organisers, what was needed was a 'just transition from coal to clean—not an expansion of the coal industry'.[19] This camp sought to take direct action to 'prevent the expansion of fossil fuel industries that are endangering the earth'.[20] The camp attracted about 1000 people. Apart from forming a human sign in Wickham Park spelling out 'Cut carbon now or never', the campers also succeeded in shutting down the coal-port train line for an entire day. They climbed a fence, and either locked themselves onto the tracks, chanting for the end of coal exports, or boarded coal trains; thirty-seven were arrested.[21] Camp spokesperson Georgina Woods said the protesters wanted to show their opposition to coal exporting and spread their message to the world. 'We wanted it to be an empowering show of community opposition to the expansion of the coal industry ... and it has been exactly that, we're really pleased.'[22] Straight after the camp, some 150 people, including camp participants, local Newcastle residents, and union members, picketed the office of NSW Treasurer Michael Costa, calling for green jobs and opposing plans to privatise electricity provision.[23]

In November 2008, three protesters entered the Tarong coal-fired power station, near Kingaroy in Queensland. They succeeded in shutting it down for five hours. Strangely, no convictions were recorded against the protesters.[24] In response, according to Philip Dorking in the *Canberra Times* on 15 December 2008, federal Energy Minister Martin Ferguson convened a meeting of federal and state energy ministers, which called on Australia's attorneys-general to raise legal penalties for interference with infrastructure important for national security, such as coal mines and coal-fired power stations.[25]

In the early morning of 20 May 2009 about fourteen climate activists shut down a coal digger at the Hazelwood Power Station in Victoria's Latrobe Valley.[26] This was one of the first of many direct action protests at Hazelwood, which opened in 1964 and burns locally mined brown coal. EV reports it is Australia's 'dirtiest' power station, responsible for 16 million tonnes of greenhouse pollution each year, the largest single source of emissions in Australia (3% of Australia's and 15% of Victoria's). Owned by British company International Power, the plant was scheduled to close down in 2005 until its life was extended to 2031 by the Bracks government. Numerous studies have shown it is possible to replace Hazelwood with a combination of renewable energy, energy efficiency and peaking gas plants. The jobs in construction and operation of these power stations would far exceed the 500 jobs that currently exist at Hazelwood.[27] During 2009, the company was embroiled in an industrial dispute with the CFMEU ME, prompting union activist Dave Kerin to tell 3CR radio early in August 2009 that the union would flout anti-secondary boycott laws if necessary: 'a call will go out from the community asking that people get onto buses and come down and assist in the closure of Hazelwood'.[28]

The poster advertising Switch off Hazelwood, Switch off Coal 12–13 September 2009 (one of the four Climate Camp 09 events) invited people to: 'Join a peaceful community protest and mass civil disobedience ... Camp on Saturday night. Action Sunday.'[29] Although Victorian law protects 'critical infrastructure' sites against threats of sabotage, with penalties of up to 25 years' gaol, ACF spokesman Dave Sweeney told the *Age* on 12 September that 'conventional methods of lobbying against coal-fired power had proved fruitless, and the urgency of climate change meant "non-violent, direct action" was necessary'.[30]

A few days beforehand, climate groups sponsored a public forum in Morwell on transition options for the Valley, addressed by Gippsland Trades and Labour Council representative John Parker, Latrobe Valley resident Denise McKenna, BZE renewable energy expert Mark Ogge, and FOE's Damien Lawson.[31] About twenty

Hazelwood workers showed up for the meeting and did not disagree with the proposition that, if they could get good jobs in the renewable energy industry, they would be happy to take them. Nor did they take issue with those who stated that 'clean coal' would not come online in time to save their jobs, given most predictions give it twenty years to come into fruition, if it works at all.[32]

On Sunday 13 September, some 500 protestors, about half of whom had spent the previous night camping nearby, gathered for the rally. This started with a two-kilometre march to the power station, with entertainment provided en route by groups such as Wombat Warriors, Carbon Cops, Radical Cheerleaders, Ministers for Energy Resources and Silly Walks, and the Climate Clowns. David Spratt and Scott Ludlam were among the speakers at the rally outside the plant. A police helicopter buzzed low over the crowd and a temporary fence had been erected in front of the parking lot outside the main gate, but more than twenty people, including two wombats, managed to scale the fence to put a 'Community Decommission Order' on the plant before they were tackled by police. Other groups of protesters repeatedly tried to push over the temporary fence.[33] After twenty-two people were arrested, FOE climate campaigner Louise Morris told the ABC a mass show of civil disobedience was needed to draw attention to climate change.[34] The day concluded, according to Indymedia, with protestors forming a human sign in the shape of a wind turbine 'to symbolise their hope and determination to see dinosaurs like Hazelwood shut down' and replaced with renewable energy that will provide both clean jobs and power. 'The vibe was definitely up and empowered by the days end with people well aware that many more such actions will be needed to achieve sane Climate Change policies in Australia.'[35] The *Age* reported: 'Protestors expressed concern at the level of police force, particularly the tactic of charging police horses into crowded groups of protesters who were approaching the fence.'[36]

The Port Augusta Climate Camp on 24–27 September 2009 was held on the edge of the Flinders Ranges in South Australia, about 60 kilometres from the rally site near two Port Augusta coal-fired power stations. The radical temper of this camp was exemplified in its assertion that it constituted 'a place for anyone who wants to take action on climate change; for anyone who's fed up with empty government rhetoric and corporate spin; for anyone who's worried that the small steps they're taking aren't enough to match the scale of the problem'.[37] Workshops and talks were conducted on topics such as the role of capitalism in generating climate change, the history of union action in defence of the environment, the Angela Pamela uranium mine near Alice Springs, and renewable energy sources.[38]

On 26 September 2009, around fifty protestors—CLEAN members, students and others—assembled at the rally site, demanding the power stations' closure and replacement with renewable energy operations. The Northern and Playford B plants are powered by low-grade brown coal, were privatised in 2000, are owned by Babcock and Brown Power, and supply 20–30% of South Australia's electricity. The Playford B plant operates only 15–20% of the time, to meet peak demands. Although the plants employed about 1000 workers during the 1980s, the present workforce is around 200. Authorities designated the power stations a 'protected area' under the *Protective Security Act 2007* for the period of the camp. The protestors marched two kilometres into this 'protected area' but were met by around seventy police, including twelve on horseback. Speakers at a temporary fence addressed the protesters and the police permitted eight individuals to ride in a van to the main gate to deliver a Community Decommission Order and 350 flowers. In the lead-up to the camp, activists had discussed taking non-violent direct action against the train line that runs from the Leigh Creek mine to the power stations, the longest coal train in the world. They advised Flinders Power to suspend coal-train operation over the weekend, which it did, eliminating any need for direct action against the train. The train restarted on 27 September, but derailed due to a technical problem.[39]

About 200 people attended a climate camp in Helensburgh, a small coal-mining community between Sydney and Wollongong from 9–11 October 2009. Workshops at the camp discussed topics such as the environmental impacts of long-wall coalmining on the southern coalfields and the inadequacies of the CPRS.[40] In a workshop on 'Beyond coal—green jobs and the transition to clean energy', speakers included retired coalminer Graham Brown and NSW Greens MP John Kaye.[41] On the last day of the camp, about 600 protestors marched to the entrance of the colliery owned by Peabody Energy, Australia's oldest operating coalmine, forcing coal trucks to be cancelled. A contingent of local residents conducted a counter-rally opposing the protesters' demand that the mine be closed. The protesters had alienated locals by downplaying the fact that the mine produces coking coal, used in making steel. Helensburgh is a divided community: many of the residents feel dependent on the mine; others oppose it and its recently approved expansion, particularly those relatively new to the area and who commute to work in Sydney.[42]

From 17–21 December 2009 around 200 people attended Camp for Climate Action (Western Australia) near Collie, 200 kilometres south of Perth, which has several mines and coal-fired power stations in the area. The camp aimed to create 'the foundations of a movement, capable of taking all action necessary to bring about an end to carbon industries and begin our sustainable future'. It placed emphasis on reaching out to the workers and community in Collie, so there was a community forum in the lead-up to the camp, to encourage local support. The camp included educational workshops covering climate science, non-violent direct action and green alternatives to carbon industries. It culminated in a peaceful community picket of the Muja coal-fired power station, which involved 100 people. Most power station workers happily stopped and heard about protesters' desires for renewable energy and a 'just transition' away from coal, which would protect the livelihood and community of workers now reliant on carbon-intensive industries. During the camp the response from workers and the community was largely positive, despite fear-mongering statements from the police and some early negative press. Some Collie residents even joined the camp. According to *GLW*, useful links were made with trade unions, climate activists, and the community and was a great basis for the climate action movement to build on.[43]

Departing from Newcastle, a bus-load of climate activists took a 'Just Transition Tour' through coal communities in the Hunter Valley and western coalfields of NSW, 21–26 November 2009. The aim, according to Rising Tide, was 'to learn about coal, its contribution to climate change and its impacts on communities and environments. The tour is also about promoting the potential for a just transition to a renewable energy future that guarantees jobs and protects communities.' It welcomed 'anyone who is worried about climate change, and wants to find out more from communities on the ground about how we can work together to make a fair and fast transition to renewables in NSW'.[44] *GLW* reported that the message of the tour was that 'a "just transition" is needed—a move away from burning coal for energy, but done in a way that protects the people who work in the industry'.[45]

The tour met coalminers and their families in mining towns, strengthened ties with local environmental groups and visited farming communities fighting to stop big coal companies from degrading prime agricultural land. Retired coalminer and Hunter Valley climate activist Graham Brown served as the tour guide. Of the people in the communities visited, he stated: 'To a man and a woman they all said coal was destroying their lives and our country's future. What we're seeing now from all those people is a joining together of forces that we haven't seen before.' Tour participant Bronwyn McDonald said: 'An overarching memory that will stay with me is the women in these communities ... they talked about being so tired from fighting a faceless bureaucracy about the toxins in the air ... and their children's health.' Tour co-organiser and Socialist Alliance activist Simon Cunich explained:

> We wanted to go out to communities where a lot of people are employed in the coal industry and say that there are alternatives. The science of climate change demands that we do transition away from coal, but it must be done in a way that doesn't leave people without employment and doesn't leave communities without livelihoods.

Cunich said building alliances with workers and farmers in affected communities was 'a non-negotiable requirement for the climate movement'.[46]

A tour participant subsequently wrote about the effects of coalmining on the communities visited:

> It's true that coal creates jobs. But the streets of Muswellbrook and Singelton should be paved with gold, and they most certainly aren't. Childhood diseases and

social disfunction are a poor prize for back-breaking labour and the loss of the bucolic lifestyle that a local pensioner described to me as life in Singelton before mining.[47]

A 30-minute documentary film, *There's no future in coal*, was made about the tour. It features interviews with people around NSW confronting the expansion of the coal industry, exploring issues of coal dust pollution, impacts of mining on rivers, the loss of community, campaigns against expansion and the potential for a just transition away from coal to renewable energy.[48]

On the other side of the continent, LD Operations, Vasse Coal and Southwest Coal announced plans in early 2010 to open a coalmine in the Margaret River area south of Perth. In response, a few days later, some local residents formed NO COAL!tion, with the slogan 'Margaret River—Too Beautiful to Undermine'.[49] NO COAL!tion organised a Walk Against Warming rally in August 2010, which attracted about 1000 people, and activists another rally in Perth on 9 September 2010. As a result of the rallies and inundation with over 750 submissions, the Environmental Protection Authority delayed its decision whether to approve the project.[50] In March 2011, British comedians Ben Elton and Michael Palin lent their support to NO COAL!tion, which is still working to 'save the beautiful environment of Western Australia's South West Capes region from disruption and pollution by proposed coal mining'. It emphasises this 'pollution of the air, soil and water with the waste products of coal mining, including carcinogenic heavy metals and bio-destructive acids and salts' would 'damage the region's pristine international tourism brand and wine industry' and the proposed mine would be 'the thin edge of a coal-mining wedge that would be driven through the heart of the Capes Region from Dunsborough to Augusta'.[51] On 7 February 2012 the state government finally rejected the proposal for a coalmine at Margaret River, a major victory for activism against powerful Western Australian mining interests.[52]

Timed to coincide with an International Day of Climate Action on 10 October 2010 (10/10/10), another 'Switch off Hazelwood, Switch on Renewable Energy' protest was sponsored by about twenty different groups, including various CAGs, FOE, Greenpeace, CEN, the Victorian Climate Action Centre, BZE, 350.org, AYCC and EV. The flyer urged people: 'Take direct action against Australia's dirtiest coal-fired power station: Join the peaceful protest for renewable energy.'[53] The mood was festive, though the 250 protesters were fewer than in 2009, while police numbers were up to about 100 officers, dozens on horseback. Authorities had erected a new fence designed to prevent repetition of the previous year's successful attacks from the western side. Cam Walker and Sean Murray from FOE welcomed the protesters and the Legal Support Team warned that the state government had passed laws in 2009 to suppress 'eco-terrorists' and asked protesters to report incidents of police harassment to the team.

On the two-kilometre walk to the main gate of Hazelwood, marchers chanted: 'Coal. Don't dig it! Leave it in the ground. Get with it!' Penelope Swales opened up the festivities at the gate with her protest songs and Rod Quantock served as MC. Samantha Dunn, a local Greens candidate, read a statement from Christine Milne: 'The Greens hear you and are calling for 100% renewables. Coal-fired power stations must be closed, and coal companies must pay for pollution.' Mark Ogge explained it was technically difficult to produce electricity from coal, and referred to the tremendous amount of water required to generate electricity in this way. 'Climate Emergency Service' workers installed a five-metre high wooden structure with the words 'Solar Power 24/7' inscribed on it, then protesters clustered together holding aluminium-foil panels to create a 'human model solar thermal power plant', which was officially opened by activists dressed up as a politician (who delivered an hilarious speech, dubious about the sun as a continuing source of energy), a union official and a renewable energy CEO. An overhead plane took a picture of the construction, and festivities followed into the afternoon.[54]

Conversations were conducted with fellow-protesters about their motivations in attending such a long-distance demonstration. Jane from Ripponlea, who appeared to have a wind turbine coming out of her backpack, came 'to promote renewable energy—it's a much better alternative to the dirtiest power station in Australia we have here on our right'. It was her first time on such a demonstration, but she wanted to 'get it shut down' and felt the government was 'not moving fast enough'. Will from Ballarat, carrying a sign 'Our atmosphere is our survival', was there 'to try and shut down Hazelwood and get the government to act on climate change as soon as possible'.

Peter from Emerald, a veteran of last year's protest, explained: 'Basically we need to do something quick about coal power. It's a major driver of climate change and global warming. The opportunities and technologies are there and available for transition, so we need to put pressure on politicians to do the right thing … there's a lack of political will.' Since last year, he felt 'the urgency has increased'.[55]

Shortly after this protest, Labor Premier Brumby told ABC radio: 'it's important to close down Hazelwood because it's the dirtiest and most polluting power station in Australia'. Sniffing possible victory, on 6 November, a well-attended 'Rally to Replace Hazelwood' met at the City Square then marched to Parliament House. GetUp! advertised the rally: 'Imagine replacing Hazelwood—the symbol of everything that's wrong with climate policy in this country—with clean, renewable energy. It can be done.' In ill-fated words, EV urged Victorians to make Hazelwood an issue in the upcoming state election. 'If ever a major victory was in reach for the Australian climate movement this is that time. We can make history by ensuring that Hazelwood is Australia's first coal-fired power station closed for climate change reasons. And we can do it properly by looking after workers and communities in the process.'[56]

In mid-2011 the Baillieu Coalition government slashed EV's state government funding, singling it out as the only advocacy group to be defunded—to the tune of $1 million.[57] EV had released a report titled *Turning It Around: Climate Solutions for Victoria* which seeks to demonstrate how Victoria can cut its greenhouse gas emissions 54% by 2020. It identifies four areas in which emissions could be cut: (1) the promotion of sustainable products and consumption, (2) increased energy efficiency, and (3) the creation of a low carbon energy supply, and (4) an increased shift to public transport.[58] In 2009, EV had made numerous recommendations to the Victorian government, including cutting emissions by at least 50% by 2020 from 1990 levels, switch from coal to natural gas as a 'transition fuel', the creation of new gas and renewable energy manufacturing jobs.[59] (In May 2010 EV caused controversy within the Victorian climate movement, by releasing a report advocating natural gas as a possible 'transition fuel'.)[60]

However, the key to EV's defunding is that it has been prominent, with FOE, Greenpeace and Victorian CAGs, in opposing the EPA's approval of the first stage of a new coal-fired power station for Victoria, the first in twenty years. A few weeks before this de-funding, a rally had gathered at the Victorian State Parliament on 24 May 2011 to object to this approval.[61] HRL, a Chinese firm, had resubmitted its Works Approval Application in September 2010, to build this 600-megawatt facility near Morwell. HRL markets this as a 'clean-coal' project, simply because it will dry brown coal with gas-fire so that it is only as dirty as black coal. If constructed, it would increase Victoria's emissions 3–4.2 million tonnes annually, use three billion tonnes of water annually, create only 350 new jobs and uses '$150 million of public funds that should be diverted into renewable energy'.[62]

HRL is indeed backed by federal government funding of $100 million and state government funding of $50 million as innovative technology.[63] However, as EV points out, all that technology does is make brown-coal burning about as polluting as black-coal burning. EV states:

> there are better, cleaner and safer ways for Victoria to produce its electricity. Ways that will put us on the technological cutting edge, help us create new green jobs and secure a safer climate. Every dollar we spend on old, polluting technology is a dollar wasted. We could be spending it on clean energy instead.

EV considers it's bad enough for Victoria to house Australia's dirtiest power station, Hazelwood, but 'the idea of having mass expansion of the coal industry in the age of climate change is pure insanity'. It is appealing the EPA decision in court on the grounds that the HRL plant does not meet best practice standards for electricity generation, as required by Victorian law, and that the EPA failed to consider non-coal-based technologies in its determination of best practice. 'If we win, no new polluting coal-fired power stations would ever be built in Victoria again—a lasting legacy for our environment. But if HRL goes ahead it well keep Victoria in the technological dark ages—addicted to coal and shunning clean energy and the jobs that come with it.'[64] In conjunction with Greenpeace, EV organised a petition and 'Dirty Money' postcards to the members of the House of Representatives:

> At a time when Australia is taking historic steps to act on climate change, this coal-fired power station will pump millions of tonnes of greenhouse pollution into our atmosphere each year. We believe in a clean, renewable energy future for Victoria not new polluting coal. We the undersigned call on the House to withdraw all federal funding for this project and invest in a clean, renewable energy future for Victoria instead.[65]

The legal challenge to this HRL project proceeds.

Amongst these larger anti-coal protests, Greenpeace stunts have continued. For instance, in August 2011, five Greenpeace activists were arrested and pleaded guilty to charges related to blocking a BHP Billiton coal train from Mt Arthur coal mine in the Hunter Valley, one of the most polluting mines in the country. They were demanding BHP Billiton pay their fair share towards helping Australia develop clean energy for our future. Activist Trish Harrup stated: 'BHP Billiton makes billion in profits every year from Australian resources. A carbon price is loose change to them just 0.6% of annual revenue, but a big deal for a clean future in Australia. Instead of standing up and paying for their pollution, BHP Billiton has been busy pressuring the Government to exempt them from the tax.'[66]

The Lock the Gate Alliance

BZE and many other groups in the climate movement fear that the Gillard government agenda is a shift to gas, not concerted transition to renewable sources of energy, including the especially contentious coal seam gas (CSG), which the climate movement argues poses serious risks to water, agricultural land, biodiversity, climate and health. Combet has indeed stated that the carbon price will drive a transition to gas. There are enormous reserves of CSG in eastern Australia; and governments have approved exploration for CSG sites in Queensland, Western Australia, Victoria, South Australia, Tasmania and the Northern Territory. In NSW, CSG exploration licenses cover most of the state and the government has exempted the industry from royalty payments for five years. Companies such as ExxonMobil, BP, Shell and Chevron, the same fossil fuel companies whose profits are at risk from meaningful climate action, dominate the gas industry.[67] To extract CSG, huge amounts of water are removed from natural aquifiers to force gas out of the ground, then pumped back into the earth filled with industrial contaminants (a process known as 'fracking'). Gas can escape, contaminating water supplies. People get sick. Agricultural land is spoiled and, once an aquifer is contaminated, there is no known way to fix it.[68]

In a history-making, path-breaking alliance, farmers and graziers have joined with climate activists in the Lock the Gate movement. The Lock the Gate Alliance is a national coalition of 122 community groups and hundreds of individuals concerned about CSG.[69] Drew Hutton is national president. The tactic of Lock the Gate is that landowners refuse to negotiate about exploration on their land, though, after 20 days, refusal is against the law. The strong anti-CSG sentiment in rural communities has been assisted by the US documentary film *Gasland*.[70]

At his speech to the Climate Summit on 10 April 2011, Hutton described the 'massive imposition on the landscape' of the Darling Downs, how it 'looks like a pincushion', with a well every 60 acres, making normal landowner activity almost impossible. The rhetoric of governments, he explains, is that gas is a transition fuel en route to renewables, but this is crap: it will stop at gas for 30–40 years. It might be only 75% of coal-based emissions, but its ecological footprint is hideous, it will damage the Great Artesian Basin for centuries and it diverts resources away from renewables. Initially, Hutton had called the movement Shut the Gate until 'cockies to the right of Genghis Khan' had said 'Not Shut the Gate, Lock the Gate!' With increasing numbers of landowners locking the gate and putting up signs telling CSG companies not to enter, Hutton sees the movement as a significant conduit between rural people and urban climate activists.[71]

In August 2011, GetUp! informed its list that CSG was 'spreading like a cancer across the eastern seaboard, with over 700 new wells in Queensland alone'.[72] With reports flooding in of illnesses caused by the drilling of gas wells and carcinogenic chemicals discovered in nearby water bores, in September it organised a national petition 'to urgently halt the unbridled expansion of the coal seam gas industry until we have fully understood the impacts the industry is having, and we have proper regulations and safe-guards in place'.[73]

Local rallies against CSG have proliferated along with the wells themselves.[74] With Victoria facing a wave of exploration licenses for CSG, and huge shows of community opposition from Bacchus Marsh to the far west, FOE Melbourne called on the Victorian government in September 2011 to ban CSG. It compares the Victorian Coalition government invidiously with the support given the Lock the Gate Alliance by federal and other state Coalition politicians.[75] In early October it hosted Hutton on a 'CSG Roadshow', touring western and central Victoria, to highlight the threat posed by CSG, coal and shale gas across rural Victoria.[76]

The Lock the Gate website exudes energy, with 4880 stories currently posted and a long list of sites and related links.[77] This extraordinary alliance of normally conservative rural communities with largely urban, politically radical climate activists could become the significant force that Hutton predicts and certainly an increasingly important part of the larger climate movement. It indicates the potential of climate movement activism to involve new constituencies in reaction to shortsighted, inadequate and dangerous 'mitigation' strategies such as CSG.

Notes

1 Simon Sheikh, National Director, GetUp!, Email, 4 November 2011.
2 Baer, participant-observer notes.
3 Ker and Moynihan, 'The great Australian walkabout'.
4 http://www.climatemovement.org.au/2008/12/climate-protests-erupt-across-the-country/ (viewed 17 September 2010).
5 *Climate Action Summit, Fortnightly Bulletin* #1, 2009; Baer, participant-observer notes.
6 *GLW*, 17 June 2009, p. 3.
7 *GLW*, 1 July 2009, p. 10
8 Quoted in *GLW*, 24 June 2009, p. 13.
9 http://www.greenlivingpedia.org/ (viewed 25 October 2011); http://www.getup.org.au/campaign/ClimateActionFund (viewed 19 April 2011).
10 http://www.greenpeace.org/australia/en/what-you-can-do/events/Say-Yes-to-cutting-carbon-pollution/ (viewed 4 June 2011).
11 GetUp! email, 19 May 2011.
12 *GLW*, 15 June 2011, p. 2.

13 Courtice, 'Don't just "say yes" '.
14 Hepburn, p. 14.
15 http://www.greenpeace.org/international/en/press/releases/australiacoalexportportaction/ (viewed 15 November 2011).
16 http://takvera.blogspot.com/2007/09/australia-greenpeace-activists-arrested.html (viewed 15 November 2011); Vromen, Gelber, and Gauja, p. 319.
17 realactiononclimatechange.blogspot.com (video viewed 18 April 2010).
18 www.climatecamp.org.au (viewed 21 November 2009).
19 *Camp for Climate Action Reader*, 2008, p.1. http://www.climatecamp.org.au (viewed 8 September 2009).
20 http://www.climatecamp.org.au (viewed 21 November 2009).
21 Breen, 'Camp success'.
22 Cuneo.
23 Parker.
24 ABC News, 23 December 2008, cited in Kevin, p. 7.
25 Cited in Kevin, p. 7.
26 'Green activists protest at Australian power plant'; http://www.reuters.com (viewed 26 May 2010).
27 http://www/environmentvictoria.org/replacehazelwood (viewed 12 May 2010).
28 Manning, 'Protests threat to power station'.
29 Image available at http://www.switchoffhazelwood.org (viewed 10 September 2009).
30 Ker, 'Hazelwood protest'.
31 FOE Melbourne, *Newsletter*, Spring 2009, p. 4.
32 Courtice, 'Who is really protecting power industry jobs?'
33 Breen, 'Hazelwood protest'; Yu et al; http://www.switchoffhazelwood.org (viewed 7 May 2010); Baer, participant-observer notes; Cam Walker, email, 13 October 2009.
34 http://www.abc.net.au/news/stories/2009/09/13/2684396.htm (viewed 7 May 2010).
35 http://indymedia.org.au/2009/09/09/switch-off-hazelwood-coverage (viewed 7 May 2010).
36 Ker, 'Police tactics'.
37 http://www.climatecamp.org.au (viewed 1 October 2009).
38 Clarke, 'Climate Camp'.
39 Ratcliffe, 'SA Climate Camp'; Baer, participant-observer notes.
40 2009 Program—Helensburgh, Climate Camp 09; http://www.climatecamp.org.au (viewed 1 October 2009).
41 C Williams, 'Australia's oldest coalmine'.
42 Cavagnino.
43 Peterson and Geilman.
44 http://www.risingtide.org.au/justtransitiontour (viewed 16 November 2011).
45 Butler, 'Bus tour'.
46 ibid.

47 'To protect the planet, coal must stop', 4 February 2010. http://transitiontour.wordpress.com/ (viewed 16 November 2011).
48 http://transitiontour.wordpress.com/ (viewed 16 November 2011).
49 Iqbal, 'Residents fight'.
50 Iqbal, 'EPA delays decision'.
51 http://www.nocoalitionmargaretriver.com/ (viewed 9 September 2011).
52 Iqbal, 'Margaret River community'.
53 http://www.switchoffhazelwood.org/supporters (viewed10 May 2010).
54 Burgmann, participant-observer notes; Baer, participant-observer notes.
55 Burgmann, conversations with fellow-protesters.
56 info@nowwethepeople.org, email, 5 November 2010; GetUp!, email, 29 October 2010; Burgmann, participant-observer notes.
57 EV, 'An urgent note to our most important supporters', 20 August 2011, Leaflet.
58 http://www.encironmentvictoria.au/halving-our-emissions (viewed 18 January 2010).
59 EV, The People's Climate White Paper.
60 EV, Fast-Tracking Victoria's clean energy.
61 http://www.foe.or.au (viewed 4 June 2011).
62 FOE Melbourne, *Newsletter*, Spring 2010, p. 4; Darebin Climate Action Now, 'Dirty tricks, Mr Brumby', leaflet, 2010, 1p.
63 *Age*, 23 October 2009, p. 1.
64 Kelly O'Shanassy, CEO, EV, 4-page letter.
65 Greenpeace/Environment Victoria, Petition and Postcard, authorised by Kelly O'Shanassy, n.d.
66 http://www.greenpeace.org/australia/en/mediacentre/media-releases/climate/Greenpeace-activists-face-court-for-blocking-BHP-Billiton-coal-train/ (viewed 4 October 2011).
67 Moore, 'Carbon price'.
68 Parliament of Australia Parliamentary Library Background Note: The development of Australia coal seam gas resources, 28 July 2011; National Water Commission, *The Coal Seam Gas and Water Challenge: National Water Commission Position*, December 2010; Ross Dunn, Australian Petroleum Producers and Explorations Association, *Lateline*, Coal Seam Gas Debate, 2 August 2011; Doctors for the Environment Australia, Submission to the Rural Affairs and Transport References Committee Inquiry into management of the Murray Darling Basin—impact of mining coal seam gas; Senator Heffernan questioning Mark McFarlane of Santos in Federal Senate Inquiry, ABC Radio, 9 August 2011, cited in GetUp!, email, 16 August 2011.
69 http://lockthegate.org.au/groups/ (viewed 18 November 2011); Ausburn, 'Activists organise'.
70 Thomas, 'Rural communities'.
71 Burgmann, participant-observer notes.
72 GetUp!, email, 16 August 2011.
73 http://www/getup.org.au/campaigns/coal-seam-gas-petition/stop-

coal-seam-gas?t=dXNlcmlkPTMyNzUwMixlbWFpbGlkPT15NA (viewed 14 September 2011).

74 Williams, 'Thousands'; Ausburn, 'Protest halts work'.

75 FOE, 'Call for Victorian government to ban coal seam gas mining', Media Release, 16 August 2011. http://www.melbourne.foe.org.au/?q=node/983 (viewed 22 September 2011).

76 http://Melbourne.foe.org.au/?q=node/908 (viewed 14 September 2011).

77 http://lockthegate.org.au/news/ (viewed 18 November 2011).

Afterword: Towards a Safe Climate and Climate Justice

While globalising capitalism has resulted in impressive technological innovations, it is a system fraught with contradictions, including an incessant drive for economic expansion; growing social disparities; authoritarian, militarist, and imperialist practices; depletion of natural resources; and environmental degradation, resulting in global warming and associated climatic changes. Even more than in earlier stages of capitalism, transnational corporations make or break governments and politicians. In Michael Hardt and Antonio Negri's words, they 'construct the fundamental connective fabric of the biopolitical world' and 'tend to make nation-states merely instruments to record the flows of commodities, monies and populations that they set in motion'.[1]

As Diesendorf argues: 'the dominant economic system is destructive to the majority of people as well as the environment. It is impoverishing the majority of poor countries and transferring wealth from the poor to the rich countries. Within poor countries, it is making the poor poorer and the rich richer.'[2] Inequality has increased significantly during the past few decades of corporate globalisation. Back in 1992, in Australia the average income of a top CEO was 27.2 times greater than the average annual money wage of an employed

Australian; by 2002, the average remuneration of a top CEO was 98.4 times the annual wage of an Australian worker.[3] The maldistribution of wealth is even more pronounced than that of income. In 2010 the wealthiest 20% of Australian households owned 62% of total wealth and the bottom 60% of households owned a mere 8% of wealth.[4]

Globalising capitalism is not only increasing inequalities amongst people everywhere but has embarked humanity upon a trajectory to 'fry the planet'. It has become increasingly clear that human societies, including Australia, will have to adapt to the reality of climate change in a variety of ways, including technological innovations, reliance upon renewable energy resources, significant expansion and improvement in public transport systems, more efficient forms of heating and cooling, the development of buildings and dwelling units that are more energy efficient, the redesign of cities to control their energy demands and heat outputs, restoration of degraded environments, reforestation, protection of biodiversity, and less reliance on airplanes and private motor vehicles as forms of travel. As important as these and other strategies will be in mitigating climate change, they might well be insufficient if not part of a longer-term effort to transcend or at least significantly restrain global capitalism. As Tokar insists: 'With climate chaos looming on the horizon, this transformation is no longer optional. Our very survival now depends on our ability to renounce the global status quo and create a more humane and ecologically balanced way of life.'[5]

Democratic Eco-Socialism: a Vision for the Future

If societies were more egalitarian, there could be less emphasis on economic growth, so harmful to the environment. If the cake were shared more equally, there would not be the widespread obsession with making the cake larger each year to protect or improve most people's material circumstances. The existing global economy has manifestly failed the planet. In the phenomenon of climate change the left could arguably rest its historic case; but the left also carries uncomfortable baggage. In the nineteenth century, various revolutionaries and reformers imagined alternatives to capitalism. Efforts to create alternatives at the national level started out with the Bolshevik Revolution in Russia in 1917 and included subsequent revolutions in other countries, including China in 1949, Vietnam in 1954, Cuba in

1959, and Nicaragua in 1979. Scholars have spilled much ink and used many printer cartridges trying to determine whether these societies constituted examples of 'state socialism', 'state capitalism', 'new class societies' or whatever. Suffice it to say these ill-fated attempts to transcend capitalism failed to achieve authentically democratic socialist societies and contributed to climate change as they embarked upon ecologically unsustainable forms of industrial development in competition with Western countries.

Since the 1920s many people on the left have acknowledged the failings and undesirability of such societies but it has only been in the past few decades that the left has become acutely sensitive to the environmental travesties that occurred in them. As a result, scholars of the left have sought to develop ideas about democratic eco-socialism.[6] Democratic eco-socialism rejects a statist, growth-centred or productivist ethic and recognises that humans live on an ecologically fragile planet with limited resources that must be sustained and renewed as much as possible for future generations. The creation of democratic eco-socialism would entail the following components: an increasing movement towards public ownership of productive forces at the local, regional, national and international levels; the development of an economy oriented towards meeting social needs, such as basic food, clothing, shelter, and health care, and towards environmental sustainability rather than profit-making; a blending of both representative and participatory democratic processes; the eradication of health and social disparities and the redistribution of human resources between developed and developing societies and within societies in general; the curtailment of population growth that in large part would follow from the eradication of health and social disparities; the conservation of finite resources and the development of renewable energy resources, such as wind, solar and geothermal energy; the redesign of settlement and transport systems to reduce energy demands and emissions; the reduction of waste through recycling; and resisting the reigning culture of consumption.[7]

In this concluding chapter, we offer a vision of a safe-climate Australia that draws on the concept of democratic eco-socialism as well as the insights of climate activists concerned with climate justice. In part our thinking is guided by Erik Olin Wright, who has called for 'real utopias', which challenge capitalism but also serve as transitional steps to achieving a better world.[8] As Stilwell observes, a 'socialist alternative must embody inspirational ideals while also offering down-to-earth proposals for dealing with the problems of contemporary Australian capitalism'.[9] After recommending some down-to-earth proposals, we consider the vexed question of strategies to achieve these, in ways that do not place undue faith in market-based solutions.

A Steep Carbon Tax at the Site of Production

An emissions tax or carbon tax is fraught with danger, because of the tendency of corporations to find loopholes and pass the burden of the tax onto working-class people: compounding the primary injustice of climate change with secondary injustices. However, a carbon tax designed to prevent such unfair outcomes would assist emissions reduction. Climate scientist James Hansen and Yale economist William Nordhaus maintain that carbon taxation is a crucial measure to be employed in mitigating climate change.[10] This is especially so if the tax is extracted at the point of production. Hansen and others have proposed a 'carbon tax to be implemented at the wellhead, mine gate or port of entry', given the flaws in other market mechanisms, particularly emissions trading schemes and carbon offsetting.[11]

Indicative of the potential effectiveness of carbon taxation is the fact that vested interests are much more hostile to it than emissions trading. In 1993, the Clinton administration's proposal for a British thermal unit-based tax on energy fuels met fierce opposition and was defeated in the Senate Finance Committee. Political parties in many countries, especially conservative parties, have continued to oppose carbon taxation and prefer emissions trading.[12] Obviously, carbon taxation would be difficult to implement, given the greater resistance to it. However, climate-related regulation and climate-related tax codes do already exist in various countries.[13] Norway has a carbon tax of about US$50 per tonne of CO_2, Sweden one of about US$40 per tonne CO_2 (with industry only paying about half of the tax), and Denmark one of about US$14 per tonne of CO_2.[14] France taxes chemical industries a modest tax for emitting nitrous oxide. Unfortunately, efforts to impose a carbon tax in the EU met with failure, because certain member states, such as the UK, capitulated to various industries' claims that they would not be able to compete with non-EU industries.[15]

To prevent secondary injustices, carbon taxation schemes must incorporate measures to ensure poor consumers do not suffer. Diesendorf argues that revenues 'from taxes (and from types of emissions trading that involve auctioning of permits) can fund compensation for low-income families that are vulnerable to higher energy prices'.[16] Dick Nichols delineates several constructive arguments in favour of a carbon tax: a rising carbon price is 'a powerful incentive for capitalists to move to less carbon-intensive production methods'; 'a properly designed carbon tax would make possible generous compensation schemes for working people and people on welfare faced with rising fuel and electricity prices'; and, if combined with the elimination of the GST, increases in welfare payments and cuts to income tax for the low-paid, it could provide a 'tax mix' that 'puts people and the environment before polluting corporations'.[17]

Greening Energy and Employment

Despite the fact that Australia is well positioned to derive much of its energy from renewable sources, such as solar, wind, hydro, geothermal and wave power, its reliance on such sources has decreased over the past two decades. The proportion of Australia's electricity output provided by renewable forms of energy declined from 10.2% in 1990 to 7.7% by 2006.[18] In 2009 the sunburnt country of Australia had a solar generation capacity of a mere 105 megawatts, while cloudy Germany had a solar photovoltaic capacity of 9785 megawatts, Spain 3386 megawatts and Japan 2633 megawatts.[19]

As we argued in Chapter 5, Australia's poor record in renewables is explicable in large part to the privatisation of electricity provision in many states in the recent past which reversed earlier initiatives in developing renewables, in particular solar power, during the 1970s and 1980s when these crucial utilities were all publicly owned. Historically, public ownership of productive forces in Australia has not been a particularly radical idea. Even Robert Menzies stated in 1943: 'Few people would have any quarrel with government control of railways, or tramways, or water supply, or such other great public utilities.'[20] Climate change mitigation strategies to increase energy efficiency and discourage energy consumption would be much easier to instigate if energy supply was publicly owned. Reverse privatisation as it is known, is not uncommon in response to crisis. After 9/11 the United States government nationalised the privatised airport security systems; and governments have taken control of some private financial institutions since the global financial crisis, such as the Dutch government's nationalisation of Fortis and the UK government's partial nationalisation of the Royal Bank of Scotland. In other instances, reverse privatisation has occurred simply when public interest arguments manage to win the day, for example the resumption of total state control of New Zealand's railways and rolling stock in 2008. Looming climate crisis and public interest point clearly to the need to reverse the privatisation of electricity.

Public ownership of other important productive forces would contribute further to climate change mitigation policies by removing the profit motive from the provision of goods and services, because the imperative of profit maximisation obviously encourages consumption and cheaper production processes regardless of ecological consequences. Ideally, publicly owned companies should involve workers and consumers in decision-making to create a 'participatory economy' in an 'authentically democratic society', in ways envisaged by Brian Dominick.[21]

However, even the development of a renewable energy infrastructure under public ownership will require metals, such as steel and aluminium. At present, steel production relies on coking coal; but CSIRO is investigating the use of charcoal derived from the oil in mallee trees and other native hardwoods. Renfrey Clarke argues that 'biochar technology represents an enormous plus for the effort to sharply cut Australia's carbon emissions'.[22]

Such cutting-edge developments require labour. In Chapter 9 we examined union pressure for green jobs, for the obvious reason that green jobs utilise labour rather than exploiting resources. For example, Diesendorf points out that wind power in Australia employs two to three times, and bioelectricity three to four times the number of job-years per kilowatt-hour of coal power, including the associated coal mining.[23] Labour movement enthusiasm for green jobs is mirrored in the climate movement. For instance, in February 2010, FOE, the Total Environment Centre, the Boomerang Alliance, EV and the ACF called for a comprehensive green jobs package from the Victorian government, which they calculated had the potential to provide 20 000 new jobs in 'green industries' within the next 5 years: in energy and water efficiency, 6900 new green jobs by 2014; in public transport, 6650 new jobs by 2014; in renewable energy, 4000 new green jobs immediately; in recycling, 3100 new green jobs by 2014; and in solar water heating, 1500 new green jobs by 2020.[24] Green jobs are a crucial aspect of 'just transition' plans that seek to alleviate the 'secondary injustices' of climate change: the burden of mitigation strategies.

While both renewable energy sources and improving energy efficiency can play an important role in climate change mitigation, they are not in themselves panaceas, as some climate movement activists believe. Elmar Altvater points out that 'the transition to renewable energy requires appropriate technologies, but requires even more appropriate social institutions and economic forms'.[25] In the nineteenth century, Jevons observed that increasing the energy efficiency of the steam engine led to an increase in coal consumption. According to what has become known as the 'Jevons paradox', conserving energy by improving efficiency within the context of a capitalist system actually increased its use. The electric car might encourage continuing private vehicle usage rather than a shift to more sustainable public transport. Green technology does not spontaneously determine ecologically sound outcomes.

Resisting and Reducing Consumption

A growing number of people in developed societies, including Australia, are opting to 'downshift', reducing their levels of consumption in myriad ways, though the numbers are not yet adequate to make a significant contribution to climate change mitigation.[26] Like most climate change mitigation strategies, reduced consumption can contribute also to human well-being. Frank Stilwell maintains that, while capitalism has proven impressive as a means of producing a vast array of commodities, many people have come to recognise that 'our principal sources of satisfaction—personal security, social recognition and interesting work—are often destroyed in the preoccupation with consumption'.[27]

From an eco-socialist perspective, Magdoff and Foster argue:

> An economic system that is democratic, reasonably egalitarian, and able to set limits on consumption will undoubtedly mean that people will live at a significantly lower level of consumption than what is sometimes referred to in wealthy countries as a 'middle class' lifestyle (which has never been universalized even in those societies).[28]

In one of his popular on-line commentaries, Immanuel Wallerstein, the principal architect of world systems theory, delineates three overarching obstacles to overcoming climate change: the 'interests of producers/entrepreneurs' who act as the purveyors of the capitalist treadmill of production and consumption; the 'interests of less wealthy nations', like China and India, which are emulating the developed countries; and the 'attitudes of you and me'.[29] While much more than mere individual efforts are required to mitigate climate change, Wallerstein is right that, for many of us, climate change mitigation starts at the individual level, for climate justice requires that the affluent of the world, including millions of Australians, scale back their consumption of material goods. While many working-class Australians turn to consumerism as a compensation for alienation in the workplace and in everyday social life, consumerism correlates with class position: the wealthier the Australian, the greater the consumption, the greater the emissions.

So, in preaching reduced consumption, there are dangers for the climate movement if the climate justice perspective is ignored. Jonathan Neale cautions climate activists not to talk without discrimination about 'sacrifice', because 'they will find themselves without the support of ordinary people'.[30] Rather, as Neale continues, it is imperative that climate activists make social justice a central theme in the climate movement to 'mobilise the people of the world to survive, share, help each other' and reserve the notion of sacrifice for those whose consumption threatens the survival of all—'to make the established powers of the world sacrifice'.[31]

Shorter Working Hours

Reduced working hours would help mitigate climate change in two main ways. First, it decreases transport to the extent that trips to and from employment are curtailed. Second, it challenges the culture of consumption by inviting people to spend time rather than money, possibly including time to provide more of their own foodstuffs at home or in community gardens. In relation to the culture of consumption, Magdoff and Foster maintain:

> A simpler way of life, although 'poorer' in gadgets and ultra-large luxury homes, can be richer culturally and in reconnecting with other people and nature, with people working the shorter hours needed to provide life's essentials. A large number of jobs in the wealthy capitalist countries are non-productive and can be eliminated, indicating that the workweek can be considerably shortened in a more rationally organized economy.[32]

Despite the historical fact that Australian workers pioneered the 40-hour week in the mid-nineteenth century, most full-time employed Australians in various occupations and professions are working over 40 hours a week, often well over 40 hours a week. In 1991 radical Harvard economist Juliet Schor identified the phenomenon of the 'overworked American'.[33] The land down under has in recent years become a country of 'overworked Australians'. By 2005, only 28% of employees were working a standard week of between 35 and 40 hours. In the case of men in full-time jobs, 42% were working more than 45 hours a week, more than 30% were working more than 50 hours a week and nearly 15% were working more than 60 hours a week. Women in paid work were catching up fast on overworking: between 1978 and 2004 the proportion of women working more than 45 hours a week more than doubled, from 12% to 25%, and in the same period the proportion of women working more than 50 hours a week doubled, to more than 15%.[34] Casualised workers could be working even longer hours than those recorded officially in full-time employment.

Overwork is encouraged by the culture of consumption but it has also been necessitated for many by decreased hourly rates of pay at the lower end of the labour market, associated with decreased unionisation rates. Internationally, there is a correlation between shorter working hours and strong trade unions—and Australia is now drifting towards the American pattern of longer working hours and weaker trade unions. Scandinavian countries provide a contrast. In 2005 Norwegians, with a unionisation rate of 53%, worked on average 39 hours per week; Swedes, with a 78% unionisation rate, worked on average 36 hours per week. Australians in 2005, with unionisation halved in the previous quarter century to around 23%, worked 46 hours per week, only slightly less than American workers, even less unionised, who put in 46.2 hours per week.[35] As part of reducing transport usage and resisting the culture of consumption, a return to the land of the long week-end would assist climate mitigation. If unions were able to embrace this challenge and mobilise successfully for reduced working hours, with increased hourly rates of pay where necessary, they would in the process also continue their historic role of improving the well-being of workers.

Sustainable Settlement and Transport

Modern cities have evolved following, in large part, the dictates of capital with its need for manufacturing, financial, commercial, distribution and communication centres, as well as state bureaucracies. Australian cities developed as regions of urban and suburban sprawl, committed to the notion that each household have a quarter-acre block and one or more cars to transport its occupants through the resulting sprawl. This spatial arrangement has made commuting and moving around in the larger cities increasingly difficult, even where elaborate public transport infrastructure exists, because much more money has been poured into road-building than developing and upgrading public transport, further encouraging car use. Car-dependency is especially acute in outer suburban areas, so an immediate priority for governments should be to improve public transport services in these areas to match the quality found in inner and middle suburban areas.[36]

The Rapid and Affordable Transport Alliance, which consists of seventeen environmental, health and union groups, including the ACF and the ACTU, published a report on 23 June 2009 entitled *Investing in sustainable energy: our clean, green transport future*. This report recommended that the federal government devote two-thirds of transport funding to public transport and only one-third to roads during the next decade. It noted that between 2004 and 2009 the federal government had spent $1.2 billion on rail but $14 billion on roads.[37] Within and between urban centres, people and freight need to be able to utilise energy-efficient forms of transport. According to Peter Newman: 'The biggest challenge in an age of radical resource-efficiency requirements will be a way to build fast rail systems for the scattered car-dependent cities.'[38]

Canberra, designed by American Walter Burley Griffin with the emerging automobile culture in mind, is emblematic of this unsustainable vicious circle in transport and urban planning. A national capital city with no rail link to any state capital except one (and very slow) and no internal rail network is clearly world worst-practice. In 2004 Hans Baer asked an ACT urban planner why the ACT had not developed a rail system and was informed that the idea had been considered but the powers-that-be deemed it too expensive. In the meantime, in the ACT as elsewhere, more and more money is spent on constructing roads and maintaining the existing ones, which invite people to use them to the point where congestion and eventually gridlock occur. In cities where significant public transport infrastructure exists but has not been developed and upgraded sufficiently to discourage car use, the proliferation of cars also reduces the efficiency of road-based forms of public transport. In Melbourne, for example, buses and even trams are slowed greatly by congestion caused by cars, in contrast to the much speedier light-rail stretches between the city and both Port Melbourne and St Kilda, and ordinary rail services, which are now inadequate to meet the increased demand. Every car user has a rational interest in improved public transport rather than improved roads.

In response in large part to looming climate crisis, a new urbanism that seeks to make cities more environmentally sustainable, and also thereby more liveable, has emerged around the world. Richard Register, for example, maintains that cities should be designed for people, not for cars. He proposes the notion of pedestrian cities in which people will not need cars and will be able to walk, cycle, or take public transport to get around.[39] Community gardens and family unit gardens are viewed increasingly as part of the effort to address environmental problems and mitigate climate change. Such experiments have been going on or undergoing a process of rejuvenation in cities such as Dar es Salaam, Havana,

Hanoi, Caracas, Paris, Chicago, Philadelphia, Melbourne and Brisbane. These endeavours are encouraged by what some psychologists have termed *eco-psychology*, the notion which stresses the need for people to have contact with the natural environment and sources of food supply.[40] The eco-village and transition towns movements supported by Ted Trainer offer food for thought about urban design for climate change mitigation.

Such ideas have fortunately permeated urban planning in Australia. For instance, Melbourne's Director of City Planning and Design, Rob Adams, proposes 'saving the Australian dream' but by creating a very different paradigm from that in which the original dream was constructed: one that contains future development and infrastructure within the current city boundaries to the greatest extent possible, while achieving greater efficiencies and affordability. Adams' plan specifies that all the existing and future major public transport corridors need to be clearly identified and adequately funded, the appropriate level of building development in specified 'urban corridors' be four to eight storeys with all developers required to provide 25% affordable housing in any residential development, and streets everywhere should be modified to favour rapid public transport, bicycles and pedestrians over cars.[41]

Cars that remain with us need to be more energy-efficient and as small as possible. Large four-wheel drive vehicles should be banned from urban areas, where their use has become not just an ecological problem but an increasing hazard for other road users. Air travel for business purposes should be restricted via intelligent use of technological alternatives to face-to-face meetings. More sustainable forms of holidaying would entail more trips closer to home, ideally by means other than private motor vehicle, and fewer journeys to distant lands.

Sustainable Agriculture and Forestry

With emissions from our agricultural practices especially significant, an important form of climate change mitigation would be a decided shift in food production from a heavy reliance on meat, particularly certain forms of meat. This need not entail complete vegetarianism for everbody, but excessive meat consumption is bad for both the planet and people's health. According to Bert Metz, 'changing to a vegetarian diet can avoid N_2O emissions from grasslands, CH_4 emissions from livestock and manure, CO_2 emissions from fossil fuel use, and free land for other purposes.[42] A vegetarian diet requires about 80% less land than that needed to feed a person on a meat-based diet—and some animals are more injurious than others. The amounts of grain required for each kilogram of flesh are 7 for beef, 3.5 for pork, 2 for poultry and 1.2 for fish.[43] Drastic reduction of current forms of meat consumption, perhaps with increased consumption of kangaroo as a lean and relatively healthy red-meat alternative, and rabbit as a white-meat alternative, would greatly decrease emissions from food production—and health problems for humans.

Where animal production continues, methane from animal digestion could be reduced through practices such as vaccinations and chemical inhibitors; and changes in tillage practice and vegetation cover. It may also be possible to reduce NO_2 emissions from fertilisers through fertiliser management, soil and water management, and fertiliser additives.[44] Last but not least, bioregionalism in terms of food distribution and consumption could potentially prevent millions of 'food miles' creating emissions by the unnecessarily long-range distribution and longer-term storage of food products.

Without exception, Australia's forests and particularly old-growth forests must be preserved for their carbon-absorption capacities. Since European settlement, forests on the east coast and Tasmania have been reduced by 30–50% and much of the woodland in the south-west has been decimated due to clearing for farm land.[45] Excessive tree-clearing in Queensland and elsewhere has even occurred in recent times, despite awareness about the need to reduce carbon pollution. Forest plantations, as Lester Brown argues, can reduce pressures on the earth's remaining forests, including those of Australia, but only as long as they do not replace old-growth forest.[46]

Sustainable Population Growth

Diesendorf argues that 'population growth must be ended as a matter of priority in countries with high per capita emissions and, in the longer term, in all countries'; in less developed countries, 'the main policy needs for reducing births are empowerment and education of women, the provision of contraception and the right to use it, and economic development that provides social security for the aged'.[47] Demographers have often observed that the greatest rate of population growth tends to occur among the poor, especially in developing countries, where people in both rural and urban areas regard children as an economic asset, because they are productive on the land or sell products or services in the cities. A more equitable distribution of economic resources would help slow population growth, particularly if improvements in the economic conditions of households are accompanied by policy changes that give women more control over fertility decision-making.[48] At international level, reversing the trend towards increasing economic inequality between and within countries, and improving the rights and status of women, clearly constitute the best means to reduce population growth.

In Australia, however, the population issue is inextricably linked with the problem of affluence rather than poverty, as Diesendorf's prescription acknowledges. Most Australians are consuming many, many times more resources than the vast majority of people in developing countries. Therefore it is essential that any climate movement arguments around the issue of Australian population growth acknowledge this fact. When the Rudd government announced its projection that the Australian population could increase from its present 22 million to 35 million by 2050, some environmentalists contended that such a population would not be sustainable in a country whose resources are being strained by both urban growth and drought in various regions, particularly the south-eastern portion of the country.

If the overall level of consumption in Australia at the present time were to continue or even grow, this argument would undoubtedly be true. Mark O'Connor and William Lines' *Overloading Australia* implies that population reduction is desirable because it would permit affluence-as-usual. They admit that, 'if the world's resources were ever shared equally, *no one* would have a first world lifestyle', but if the world's population could somehow someday be reduced to two billion 'then there might be enough resources for everyone to live a European material standard of living (standard of consumption) which is about half that of the USA'.[49]

Given that Australian population growth is based more on immigration than natural replacement, affluence-as-usual reasoning nourishes prejudice about immigrants. An alternative Australia committed to a simpler way of living along the lines envisaged by Ted Trainer, and redesigned for sustainable living, could support a larger population and gainsay the present widespread xenophobia about immigrants and particularly refugees. Various climate movement spokespeople have rightly sought to counteract hostility towards refugees by arguing for increasing the humanitarian component and decreasing the business component of our immigration program. As climate change creates its own refugees, many in our immediate vicinity, a climate justice perspective demands that Australia plays its part in offering climate asylum to peoples of the world considerably less complicit than average Australians in causing climate change.

From Here to Sustainability

Achieving a democratic eco-socialist Australia would have to be part of a larger global effort in which a greatly enlarged climate justice movement works in alliance with other anti-systemic movements, such as the global justice movement and the new Occupy movement. The Australian climate movement could more fully join in the struggle for social justice and environmental sustainability both globally and at home if it were to place greater emphasis on climate justice and less on ecological modernisation, which has a role to play in climate change mitigation but will not counteract the capitalist treadmill of production and consumption and its drive for continual economic expansion.

Christine Frank imagines an 'uncompromising environmental movement led by working people in alliance with other oppressed groups', one that is imbued with 'ecosocialist principles that go beyond the maintenance of capitalism and its suicidal and genocidal policies, and instead advance a zero-growth, zero-waste, steady-state, democratically planned socialist economy that puts planetary and human needs before profits'.[50] Frank's hopeful imagination stands in stark contrast to that of John Foster, who notes that:

> For a long time many people thought, or hoped, that proper attention to our environmental responsibilities would have to mean an end to this system, which had so plainly generated ecological crisis ... But the overthrow or

voluntary abdication of capitalism on environmental grounds was never remotely likely ... Concerned people of all shades of green simply need to face this fact ... For good or ill, the global market economy is where we have to operate to save the planet: it's there, or nowhere.[51]

Foster's assumption that the global market economy is capable of saving the planet is questionable; his insistence that there is no alternative closes down the utopian impulse that motivates and mobilises people to act in hope of a far better world than such an economy offers.

While the notion that democratic eco-socialism may be implemented in Australia or any developed society may seem utterly ridiculous, history tells us that social changes can occur very quickly once economic, political and social factors have reached a tipping point. Magdoff and Foster suggest such a time is nearing: 'Everywhere radical, essentially anti-capitalist, strategies are emerging, based on other ethics and forms of organization, rather than the profit motive: ecovillages; the new urban environment promoted in Curitiba in Brazil and elsewhere; experiments in permaculture, and community-supported agriculture, farming and industrial cooperatives in Venezuela, etc.'[52]

We two authors agree on the goals to reach but disagree about how to achieve them. Hans Baer believes that the implementation of the proposals we have outlined requires the formation of an authentically new left party that represents ordinary people rather than the corporate interests. It might constitute an 'unholy' alliance of various progressive groups and individuals—disaffected left-wing ALP types, many Greens, Socialist Alliance members, Solidarity members, independent eco-socialists and eco-anarchists. He admits that, to long-time Australian progressives, the notion of such a party may harken back to earlier unsuccessful initiatives such as the New Left Party that formed in the early 1990s, the Rainbow Alliance that was active between 1987 and 1993 and more recently the Progressive Labor Party that eventually imploded.[53] He hopes a rejuvenated new left party would learn from the pitfalls of those parties which fell by the wayside. Verity Burgmann has been there and done that, and is convinced such an effort would again fail or produce yet another small socialist sect. In any case, she believes that it is up to people in struggle, in this case climate mitigation struggle, to invent their own forms of organisation as occasion arises and in the process of mobilisation—and that to lay down in advance any organisational blueprints is authoritarian and elitist.

However the vision might be implemented, we both agree that vision is crucial, for inspiration and hope are essential to mobilisation. Tokar argues that the climate justice outlook embraces the 'counter-hegemonic aspects of environmentalism' in its promise that reorienting societies towards a renewed harmony with nature can help spur a revolutionary transformation of our world.[54] 'For many involved in movements for social change', as James Goodman observes, 'it is axiomatic that there is no action without hope.' Acknowledging there might be little to hope for in terms of avoiding runaway climate change, he points to how activists, and through them movements, are producing their own foundations and motivations for action: 'activists can despair at the built-in logic of climate change, and at the same time vest hopes in mass action to create new possibilities ... Climate activists are self-consciously engaged in this intellectual process of generating new visions and new models for action in the context of climate change.'[55] As the existing capitalist world system continues to self-destruct due to its socially unjust and environmentally unsustainable practices, democratic eco-socialism provides a vision to mobilise Australians to prevent their country becoming even more sunburnt and committed instead to a safe climate and climate justice.

Notes

1 Hardt and Negri, p. 31.
2 Diesendorf, *Climate Action*, p. 120.
3 Meagher and Wilson, p. 227.
4 ABS, p. 3. Socialists have over the years engaged in intense debates over what sort of wage differentials should exist in a socialist society. Stilwell argues that a ratio of 3:1 from the highest to lowest incomes would be a tolerable standard for a socialist society. Others argue for a somewhat larger income differential on the grounds that some individuals may choose to work many more hours than others, while others prefer to work fewer hours in order to pursue other activities, favouring the intrinsic rewards of intellectual or physical stimulation over material rewards.
5 Tokar, p. 52.

6 For example, Panitch and Leys; Kovel; JB Foster.
7 Boswell and Chase-Dunn; Baer, Singer, and Susser; Baer and Singer.
8 Wright.
9 Stilwell, p. 134.
10 Brohe, Eyre and Howarth, p. 292.
11 Frank, p. 36.
12 Brohe, Eyre and Howarth, p. 155.
13 Lohmann, p. 334.
14 Metz, p. 229.
15 O'Riordan and Jordan, p. 84.
16 Diesendorf, *Greenhouse Solutions*, p. 296
17 Nichols, 'Five questions'.
18 Christoff, p. 38.
19 Buckman, p. 83.
20 Quoted in Wettenhall, p. 428.
21 Dominick, pp. 385–6.
22 Clarke, 'Can biochar help stop climate change?'
23 Diesendorf, *Climate Action*, p. 52.
24 FOE, 'Victoria must invest in green jobs now', Media Release, 24 February 2010.
25 Altvater, p. 54.
26 Hamilton and Mail; Hamilton and Denniss, pp. 153–177.
27 Stilwell, p. 56.
28 Magdoff and Foster, p. 26.
29 Wallerstein, p. 1.
30 Neale, p. 45.
31 Neale, p. 46.
32 Magdoff and Foster, p. 26.
33 Schor.
34 Hamilton and Denniss, p. 86.
35 Tiffen and Gittins, pp. 82, 86.
36 Matthews, 'How do we fix public transport?'
37 Matthews, 'Too many cars'.
38 Newman, p. 108.
39 Register.
40 L Brown, pp. 160–162.
41 R Adams, pp. 119–131.
42 Metz, p. 241.
43 ibid., pp. 46, 235.
44 Rickards and Tucker, p. 93.
45 Steffin et al, p. 32
46 L Brown, p. 195.
47 Diesendorf, *Climate Action*, p. 119.
48 Bodley, p. 218.
49 O'Connor and Lines, p. 28.
50 Frank, p. 43.
51 Foster, *The Sustainability Mirage*, p. 13.

52 Magdoff and Foster, p. 25.
53 Stilwell, pp. 165–170.
54 Tokar, p. 52.
55 Goodman, 'Climate movement', pp. 20–21.

Bibliography

AAP, 'Australia "could lead way in green jobs" ', 3 November 2009. http://www.adelaidenow.com.au/news/breaking-news/australia-could-lead-way-in-green-jobs/story-e6frea7l-1225794113169 (viewed 7 May 2010).

AAP, 'Ratify Kyoto to shape future', 30 October 2007. http://news.ninemsn.com.au/article.aspx?id=311340 (viewed 14 July 2010).

ABC, 'The Carbon War', *Four Corners*, 19 September 2011.

ABS, Household Wealth and Wealth Distribution, Australia, 2009–10, ABS, Canberra, 2012.

ACF and other organisations, *Australia Responds: Helping our Neighbours Fight Climate Change*, n.d. [mid 2010]. http://www.acfonline.org.au (viewed 3 June 2010).

ACF, 'Cut carbon pollution—set effective targets', Policy Brief 2.1, September 2008, 10pp.

ACF, 'Projected financial impact of expanded compensation under the Carbon Pollution Reduction Scheme (CPRS)', Research note, 15 October 2009, 6pp.

ACF, 'What is an emissions trading scheme?', 7 October 2007. http://www.acfonline.org.au/articles (viewed 26 October 2011).

ACF/ACOSS/Choice, Energy & Equity, Preparing Households for Climate Change: Efficiency, Equity, Immediacy, n.d. http://www.acfonline.org.au (viewed 3 June 2010).

ACF/ACTU, 'Creating Jobs—Cutting Pollution. The Roadmap for a Cleaner, Stronger Economy', n.d. http://www.acfonline.org.au (viewed 3 June 2010).

ACTU Organising Centre, 'Changing unionism through values campaigning', 27 April 2010. http://www.actu.org.au/Images/Dynamic/attachments/6918/B5%20Changing%20Unionism%20Through%20Values%20Campaigning.pdf (viewed 8 July 2010).

ACTU, 'ACTU Congress 2009. A Fair Society. Environment and Climate Change Policy', 2–4 June 2009. http://www.actu.asn.au/Tools/print.aspx (viewed 10 May 2011).
ACTU, 'ACTU's Principles and Policy on Global Warming', 17 February 2007. http://www.actu.asn.au/AboutACTUPublications/ACTUsPrinciplesandPolicyonGlobalWarming.aspx (viewed 19 May 2009).
ACTU, 'Campaign: Clean Energy Jobs', 2010. http://www.actu.org.au/Campaigns/CleanEnergyJobs/default.aspx (viewed 8 July 2010).
ACTU, 'Campaign: Climate Change and Jobs', 2011. http://www.actu.org.au/Campaigns/ClimateChangeJobs/default.aspx (viewed 12 May 2011).
ACTU, 'Unions to pursue climate change solutions including carbon trading scheme, renewable energy sources and green building codes', 16 February 2007. http://www.actu.asn.au/Media/Mediareleases (viewed 19 May 2009).
ACTU, Climate Change is Union Business, ACTU, Melbourne, 2011.
ACTU, The Greenhouse Effect: employment & development issues for Australians, ACTU, Melbourne, 1992.
ACTU/ACF, 'Creating Jobs—Cutting Pollution', n.d. [2009–2010]. http://www.actu.org.au/Images/Dynamic/attachments/6971/ACF_Jobs_reporrt_190510.pdf (viewed 8 July 2010).
ACTU/ACF, *Green Gold Rush*, ACTU/ACF, Melbourne, 2008.
Adams, Peter, 'A gleam of green', *Age*, 7 October 1989, Extra p. 2.
Adams, Rob, 'Transforming Australian cities', in H Sykes (ed.), *Climate Change on for Young and Old*, Future Leaders, Sydney, 2009, pp. 119–131.
AIGN/BCA, Carbon Pricing Forum Program, Parliament House, Canberra, 23 March 2011.
Alcorn, Zane, 'Coal industry job scare baseless', *GLW*, 13 July 2011, p. 10.
Alcorn, Zane, 'Greens put carbon tax on the agenda', *GLW*, 3 February 2010, p. 10.
Alcorn, Zane, 'Workers, environmentalists and the fossil fuels industry', *GLW*, 30 April 2008, p. 9.
Alcorn, Zane, 'World's biggest coal port brought to a standstill', *GLW*, 25 March 2009, p. 11.
Ali, Jasmine, 'Post-Election 2010: Where to for the Climate Movement?', *Climate Action Working Group Melbourne Uni Zine*, n.d. [2010], 4pp.
Alison, Ian, et al, *The Copenhagen Diagnosis: Updating the World on the Latest Climate Science*, Climate Change Research Centre, University of New South Wales, 2009.
Altvater, Elmar, 'The social and natural environment of fossil capitalism', in L Panitch and C Leys (eds), *Coming to Terms with Nature—Socialist Register 2007*, Merlin Press, Monmouth UK, 2006, pp. 37–59.
AMWU, 'AMWU members want action on climate change', 24 July 2008. http://www.amwu.org.au/read-article/news-detail/154 (viewed 3 June 2010).
Anderson, James, 'Afterword: only sustain ... the environment, "anti-globalization" and the runaway bicycle', in J Johnston, M Gismondi, and J Goodman (eds), *Nature's Revenge: Reclaiming Sustainability in an Age of Corporate Globalization*, Broadview, Toronto, 2006, pp. 245–273.

Angel, Jeff, *Green Is Good: An Insider's Story of the Battle for Green Australia*, ABC Books, Sydney, 2008.
ARG, Climate Institute Marginal Electorates Election Campaign Poll, ARG, Melbourne and Sydney, November 2007.
Aronowitz, Stanley, 'The situation of the left in the United States', *Socialist Review*, vol. 23, no. 3, 1994, pp. 5–79.
Arup, Tom, 'Plant owners open to price talks on shutdown', *Age*, 11 July 2011, p. 6.
Arup, Tom, 'We're not pissants: Garnaut', *Age*, 1 June 2011, p. 2.
Arup, Tom, and Adam Morton, 'Emissions surge to record level: Labor pushes new plan on carbon tax', *Age*, 31 May 2011, p. 1.
Arup, Tom, and Adam Morton, 'Garret pulls plug on solar', *Age*, 10 June 2009. http://www.theage.com.au/environment/energy-smart/garrett-pulls-plug-on-solar-20090609-c28d.html (viewed 10 June 2009).
Ashley, Michael, 'No science in Plimer's primer', *Australian*, 9 May 2009. http://www.theaustralian.com.au/news/ian-plimer-heaven-and-earth/story-e6frg8no-1225710387147 (viewed 12 May 2009).
Ausburn, Kate, 'Activists organise to lock out coal seam gas mining', *GLW*, 22 June 2011, p. 11.
Ausburn, Kate, 'Protest halts work on gas rig', *GLW*, 12 July 2011, p. 4.
Austin, Susan, 'Ten steps for a safe climate', *GLW*, 2 February 2010, p. 11.
Australia Institute, 'A clean energy future? The Australia Institute's analysis of the government's carbon tax announcement', 11 July 2011. https://www.tai.org.au/index.php?q=node%2F19&pubid=877&act=display (viewed 9 October 2011).
Australia Institute, 'A win for voluntary action', 11 July 2011. https://www.tai.org.au/?q=node/335 (viewed 9 October 2011).
Australian Business & Climate Group, 'Stepping Up: Accelerating the Deployment of Low Emission technology in Australia', August 2007. http://www.santos.com.au (viewed 19 September 2009).
Australian Government and Australian Consumers' Association, *The Energy Guide*, Australian Government Publishing Service, Canberra, 1991.
Australian Government, *Carbon Pollution Reduction Scheme: Australia's Low Pollution Future—White Paper*, vol. 1, Australian Government, Canberra, December 2008.
Australian Government—The Treasury, *Climate Change Mitigation Policy Modelling*, Australian Government, Canberra, October 2008.
Australian Greenhouse Office, *The National Greenhouse Strategy*, Commonwealth of Australia, Canberra, 1998. http://www.greenhouse.gov.au/pubs/ngs (viewed 29 September 2011).
Australian Greens, 'Policy: Climate Change and Energy, Final', November 2009. http://www.greens.org.au/about/policy (viewed 20 July 2010).
Australian Greens, 'Rudd's white flag of surrender on climate change', 16 December 2008, posted on blog, John Kaye, Greens NSW MP. http://www.johnkaye.org.au (viewed 6 April 2009).
Baer, Hans A, 'The impact of the war machine on global warming and health: a political ecological perspective', in M Singer and GD Hodge (eds), *The War*

Machine and Global Health: A Critical Medical Anthropological Examination of the Human Costs of Armed Conflict and the International Violence Industry, AltaMira Pres, Walnut Creek CA, 2010, pp. 157–177.
Baer, Hans A, Merrill Singer, and Ida Susser, *Medical Anthropology and the World System: A Critical Perspective*, Praeger, Westport CT, 2003.
Baer, Hans, 'Global warming as a by-product of the capitalist treadmill of production and consumption: the need for an alternative global system', *Australian Journal of Anthropology*, no. 19, 2008, pp. 58–62.
Baer, Hans, and Merrill Singer, *Global Warming and the Political Ecology of Health: Emerging Crises and Systematic Solutions*, Left Coast Press Inc., California, 2009.
Bailey, Ian, 'Australian climate politics', *Arena*, no. 110, 2011, pp. 32–35.
Bandt, Adam, 'Melbourne the newest green electorate', *Green. Magazine of the Australian Greens*, issue 32, Summer 2010, pp. 14–15.
Barnett, Jon, 'The Geopolitics of Climate Change', *Geography Compass*, vol. 1, no. 6, 2007, pp. 1361–1375.
Bartlett, Andrew, 'The rise of the Greens ... so far', *Green. Magazine of the Australian Greens*, issue 32, Summer 2010, pp. 11–13.
BCA, 'Strategic Framework for Emissions Reduction', April 2007. http://www.bca.com/content/101042apex/acc (viewed 9 April 2009).
Bean, Clive, and Ian McAllister, 'The Australian election survey: The tale of the rabbit-less hat. Voting behaviour in 2007', *Australian Cultural History*, vol. 27, no. 2, 2009, pp. 205–218. http://eprints.qut.edu.au/31530 (viewed 9 November 2011).
Bean, Clive, Ian McAllister, and John Warhurst (eds), *The Greening of Australian Politics: The 1990 Federal Election*, Longman Cheshire, Melbourne, 1990.
Beder, Sharon, *Global Spin: The Corporate Assault on Environmentalism*, revised edn, Scribe Publications, Melbourne, 2000.
Benedek, Paul, 'Rinehart grabs for media control', *GLW*, 15 February 2012, pp. 3, 12.
Bennett, Ebony, 'The election that was', *Green. Magazine of the Australian Greens*, issue 32, Summer 2010, pp. 4–5.
Bennett, Ebony, National Campaign Coordinator, Australian Greens, 'Top 10 reasons to vote Green this Saturday'. http://www.greens.org.au (viewed 20 August 2010).
Benton, Ted, 'Marxism and Natural Limits', *New Left Review*, no. 178, November/December 1989, pp. 51–86.
Betzien, Jody, 'Carbon price could delay real action', *GLW*, 23 March 2011, p. 10.
Black, John, 'Wealthy Greens the new DLP', *Weekend Australian*, 5–6 June 2010, Inquirer p. 2.
Blair, Tim, 'Life's a blast in Geothermia', *Daily Telegraph*, 3 March 2007. http://www.dailytelegraph.com.au/news/opinion/lifes-a-blast-in-geothermia/story-e6frezz0-1111113087567 (viewed 5 August 2010).
Bodley, John H, *Anthropology and Contemporary Human Problems*, 5th edn, AltaMira Press, Walnut Creek CA, 2008.

Bolt, Andrew, 'Global warming catastrophist Tim Flannery is at it again', *Herald Sun*, 21 December 2006. http://www.dailytelegraph.com.au/news/opinion/lifes-a-blast-in-geothermia/story-e6frezz0-1111113087567 (viewed 5 August 2010).
Bolton, Sue, 'Australia's first climate convergence a success', *GLW*, 20 February 2008, p. 4.
BOM, 'Record warmth for Victoria and Tasmania', 3 May 2010. http://www.bom.gov.au/announcements/media_releases/ho/20100503.shtml (viewed 17 September 2010).
BOM, 'Special climate statement 17: The exceptional January – February 2009 heatwave in south-eastern Australia', 9 February 2009. http://www.bom.gov.au/climate/current/statements/scs17c.pdf (viewed 7 July 2009).
BOM, 'Special Climate Statement 19: A prolonged spring heatwave in central and south-eastern Australia', first issued 24 November and updated 2 December 2009. http://www.bom.gov.au/climate/current/statements/scs19b.pdf (viewed 17 September 2010)
BOM, 'Special climate statement 27: An exceptional summer heatwave in greater Sydney and the Hunter Valley', 8 February 2011. http://www.bom.gov.au/climate/current/statements/scs27.pdf (viewed 9 September 2011).
BOM, *Annual Australian Climate Statement 2009*, 5 January 2010. http://www.bom.gov.au/announcements/media_releases/climate/change/20100105.shtml (viewed 17 September 2010).
BOM, *Annual Climate Summary 2010*. http://www.bom.gov.au/climate/annual_sum/2010/index.shtml (viewed 9 September 2011).
Boswell, Terry, and Christopher Chase-Dunn, *The Spiral of Capitalism and Socialism*, Lynne Rienner, Boulder CO, 2000.
Boucher, Geoff, and Matthew Sharpe, *The Times Will Suit Them: Postmodern Conservatism in Australia*, Allen & Unwin, Sydney, 2008.
Bradford, Sue, 'Green greed is no good answer', *GLW*, 20 July 2011, p. 5.
Bradstreet, Katherine, 'Meltdown as climate emergency hits', *GLW*, 11 February 2009, p. 3.
Bramble, Tom, 'Labour movement leadership', in R Kuhn (ed.), *Class and Struggle in Australia*, Pearson Longman, Sydney, 2005, pp. 73–89.
Branley, Alison, 'Rising Tide protesters hit Kooragang conveyor belt', *Newcastle Herald*, 5 August 2011. http://www.theherald.com.au/news/local/news/general/rising-tide-protesters-hit-kooragang-conveyor-belt/2249641.aspx (viewed 11 November 2011).
Breen, Chris, 'Camp success a chance to renew movement', *Solidarity*, issue 6, August 2008, p. 10.
Breen, Chris, 'Carbon trading and market solutions: why they don't halt climate change', *Solidarity*, issue 5, July 2008, pp. 20–23.
Breen, Chris, 'Climate summit provides direction for movement', *Solidarity*, issue 12, February 2009. http://www.solidarity.net.au/12/climate-summit-provides-direction-for-the-movement/ (viewed 21 March 2009).
Breen, Chris, 'Hazelwood protest needs fight for jobs at its centre', *Solidarity*, issue 17, August 2009, p. 7.
Breen, Chris, 'Victorian Labor backs new coal-fired plant', *Solidarity*, issue 6,

August 2008, p. 10.
Breen, Chris, and Amy Thomas, 'Climate movement needs to target CPRS', *Solidarity*, issue 14, April 2009, p. 9.
Brett, Judith, 'Fair share. County and city in Australia', *Quarterly Essay*, no. 42, Black Inc., Melbourne, 2011, pp. 1–90.
Brohe, Arnaud, Nick Eyre, and Nicholas Howarth, *Carbon Markets: An International Business Guide*, Earthscan, London, 2009.
Brook, Barry, 'Nuclear Ammonia—a sustainable nuclear renaissance's "Killer App"?', 4 October 2011. http:// www.bravenewclimate.com (viewed 10 October 2011).
Brown, Bob, and Peter Singer, *The Greens*, Text Publishing Company, Melbourne, 1996.
Brown, Bob, *Memo for a Saner World*, Penguin, Melbourne, 2004.
Brown, Lester R, *Plan B 4.0: Mobilizing to Save Civilization*, WW Norton, New York, 2009.
Brownhill, Leigh, and Terisa E Turner, 'Women and the Abuja Declaration for Energy Sovereignty', in A Salleh (ed.), *Eco-Sufficiency & Global Justice. Women Write Political Ecology*, Pluto and Spinifex, Melbourne, 2009, pp. 230–248.
Buckman, Greg, 'Weaknesses and reform of Australia's renewable electricity support', *Journal of Australian Political Economy*, no. 66, 2010/2011, pp. 71–97.
Bunting, Andrea, The social shaping of wind power in Australia, PhD thesis, School of Social Sciences, Media and Communication, University of Wollongong, 2003.
Burgmann, Meredith, and Verity Burgmann, *Green Bans, Red Union: Environmental Activism and the New South Wales Builders Labourers' Federation*, UNSW Press, Sydney, 1998.
Burgmann, Verity, 'Moloch's Little Mate: The Ruling Class and the Australian Labor Party', in N Hollier (ed.), *The Power, Privilege and Politics of the New Ruling Class*, Australian Scholarly Publishing, Melbourne, 2004, pp. 48–69.
Burgmann, Verity, *Power and Protest: Movements for Change in Australian Society*, Allen & Unwin, Sydney, 1993.
Burgmann, Verity, *Power, Profit and Protest: Australian Social Movements and Globalisation*, Allen & Unwin, Sydney, 2003.
Burgmann, Verity, with Colin McNaughton and Jennifer Penney, *Unions and the Environment*, ACF/ACTU, Melbourne, 2002.
Burrow, Sharan, 'Combating global warming is a shared task', *Australian Options*, no. 56, 2009, pp. 8–9.
Burton, Bob, 'Nice names—pity about the policies: industry front groups', *Chain Reaction*, no. 70, January 1994, pp. 16–19.
Butler, Simon, 'Abandon Affluence—25 years on', *GLW*, no. 849, 15 August 2010. http://www.greenleft.org.au/node/45108 (viewed 2 December 2010).
Butler, Simon, 'Australia's dirty role at Copenhagen', *GLW*, 20 January 2010, p. 11.
Butler, Simon, 'Australia—world's biggest climate vandal', *GLW*, 21 January 2009, p. 4.

Butler, Simon, 'Bus tour takes on coal industry', *GLW*, 2 December 2009, p. 17.
Butler, Simon, 'Carbon trading promise is misleading', *GLW*, 29 June 2011, p. 3.
Butler, Simon, 'Climate Action Summit unites new environment movement', *GLW*, 11 February 2009, p. 6
Butler, Simon, 'Climate crisis calls for complete social change', *GLW*, 6 April 2011, p. 7.
Butler, Simon, 'Emissions trading—how to avoid action on climate change', *GLW*, 4 March 2009, p. 3.
Butler, Simon, 'Gillard: same old spin on climate', *GLW*, 7 July 2010, p. 11.
Butler, Simon, 'Green groups take dirty money', *GLW*, 22 June 2011, p. 6.
Butler, Simon, 'Huge crowd at energy forum', *GLW*, 18 August 2010, p. 2.
Butler, Simon, 'The false hope of "clean coal" ', *GLW*, 16 September 2009, p. 12.
Butler, Simon, 'The renewable energy scheme that isn't', *GLW*, 26 August 2009, p. 3.
BZE, *Zero Carbon Australia Stationary Energy Plan. Synopsis*, BZE/University of Melbourne Energy Research Institute, Melbourne, 2010.
CAC Papers, Collection of documents relating to Climate Action Canberra, in possession of Anne O'Brien.
Cahill, Damien, and Stephen Brown, 'The rise and fall of the Australian Greens: the 2002 Cunningham By-election and its implications', *Australian Journal of Political Science*, vol. 43, no. 2, June 2008, pp. 259–275.
CANA, 'Garnaut's Supplementary Report: Targets and trajectories', 5 September 2008.
Canavan, Gerry, Lisa Klarr, and Ryan Vu, 'Embodied materialism in action: an interview with Ariel Salleh', *Polygraph*, no. 22, 2010, pp. 183–199.
CARG, 'Research and action in the "movement of movements" ', *Chain Reaction*, no. 107, November 2009, pp. 22–23.
Carney, Shaun, 'Rudd lost in policy maze', *Age*, 28 April 2010. http://www.theage.com.au/opinion/politics/rudd-lost-in-policy-maze-20100427-tpxy.html (viewed 28 April 2010).
Castells, Manuel, *The Information Age: Economy, Society, and Culture*, Blackwell, Malden MA, 1997.
Cavagnino, David, 'Lessons to learn from local response to Helensburgh climate camp', *Solidarity*, issue 19, November 2009, p.12.
CEN, 'A new paradigm of climate action', n.d. [c.2011], 1p.
CEN, 'Bushfires and the Climate Emergency', CEN, Melbourne, 2009, 2pp.
CEN, 'Groundswell for a safe climate Launch', 23 July 2011, Melbourne, 1p.
CEN, 'Membership Application', n.d., 1p.
CEN, 'Roundtable to explore social & political strategy for a Campaign for a Safe Climate Economy by 2020', 12 February 2011, Melbourne, 6pp.
CFMEU and WWF, 'Power to Change. Carbon Capture and Storage', 2009. http://www.cfmeu.com.au/storage/documents/Power_to_Change_%20CCS_LowRes.pdf (viewed 21 July 2010).
CFMEU ME, 'Carbon Capture and Storage: making it happen. A CFMEU position paper', December 2007. http://www.cfmeu.com.au/storage/

documents/CCS_report_dec07.pdf (viewed 19 May 2009).
CFMEU ME, 'Climate Alert! Briefing Note', November 2007. http://www.cfmeu.com.au/storage/documents/Climate_5points_1nov07.pdf (viewed 21 July 2010).
CFMEU ME, 'Climate Change Position Paper', September 2007. http://www.cfmeu.com.au/storage/documents/Climate_pol_sep07.pdf (viewed 19 May 2009).
CFMEU ME, 'Climate Change', 2009. http://www.cfmeu.com.au/index.cfm?section=29&Category=61&viewmode+content&ContentID=252 (viewed 19 May 2009).
CFMEU, 'Response to the Carbon Pollution Reduction Scheme Green Paper', 10 September 2008. http://www.cfmeu.com.au/storage/documents/SUBCFMEU_CPRSGreenPaper12SEP08.pdf (viewed 21 July 2010).
Chandler, Jo, 'The end of the world is nigh? It's just a matter of degrees, says one expert', *Age*, 13 July 2011, p. 6.
Christoff, Peter, 'If not now, then when?', in H Sykes (ed.), *Climate Change on for Young & Old*, Future Leaders, Sydney, 2009, pp. 29–45.
Christoff, Peter, and Robyn Eckersley, 'The Kyoto Protocol and the Asia Pacific Partnership on Clean Development and Climate', in T Bonyhady and P Christoff (eds), *Climate Law in Australia*, Federation Press, Sydney, 2007, pp. 32–43.
Clark, Brett, and Richard York, 'Carbon and metabolism: global capitalism, climate change, and biospheric rift', *Theory and Society*, no. 34, 2005, pp. 391–428.
Clarke, Renfrey, 'Can biochar help stop climate change?' *GLW*, 1 April 2009, p. 11.
Clarke, Renfrey, 'Climate camp targets Port Augusta power', *GLW*, 12 August 2009, p. 7.
Clarke, Renfrey, 'How to crash a planet: just follow Garnaut', *Adelaide Courier*, 17 September 2008, p. 9.
Climate Action Summit 2010, Program. http://climatesummit.org.au/network/ (viewed 15 October 2010).
Climate Connectors Jobs Action Kit. http://www.climateconnectors.org/images/phocagallery/ucc_pics/jobs_kit.pdf (viewed 22 July 2010).
Climate Connectors, 2010. http://www.climateconnectors.org/index.php?option=com_content&view=article&id=10&Itemid=7 (viewed 22 July 2010).
Climate Institute, 'Climate change top tier issue in deciding 2007 election', 26 November 2007. http://www.climateinstitute.org.au/index.php?option=com_content&view=article&catid=39:media-releases&id=128:climate-change-top-tier-issue-in-deciding-2007-election (viewed 14 July 2010).
Climate Institute, 'Passing the pollution policy test: Australia enters the global clean energy economy', Policy Brief, 13 July 2011. http://www.climateinstitute.org.au/images/reports/tci_policybrief_passingthepollutionpolicytest_july2011.pdf (viewed 8 October 2011).
Climate Institute, 'World's first climate changed election? Climate Institute

releases final party score card', 23 November 2007. http://www.climateinstitute.org.au/index.php?option=com_content&view=article&catid=39:media-releases&id=127:worlds-first-climate-changed-election-climate-institute-releases-final-party-score-card (viewed 14 July 2010).
Climate Institute, *Action Plan on Pollution and Climate Change: A Policy Roadmap for Australia*, Climate Institute, Sydney, July 2010.
Climate Institute, *Climate of the Nation. Australians Attitudes to Climate Change and its Solutions*, Climate Institute, Sydney, March 2007. http://www.climateinstitute.org.au/images/reports/climatenation.pdf (viewed 22 April 2010).
Coates, Peter (Chairman, Minerals Council of Australia Standing Committee on Climate Change), 'Opening Statement to Senate Select Committee on Fuel and Energy', 8 December 2008. http://www.minerals.org.au/__data/assets/pdf_file/0008/32948/SSC_FuelEnergy.pdf (viewed 18 July 2009).
Colebatch, Tim, 'Differences slight on climate change', *Age*, 19 November 2007. http://www.theage.com.au/news/federal-election-2007-news/differences-slight-on-climate-change/2007/11/18/1195321607386html (viewed 14 July 2010).
Conley, Tom, 'The Domestic Politics of Globalisation', *Australian Journal of Political Science*, vol. 36, no. 2, July 2001, pp. 223–46.
Connection Research, *Who Are the Green Collar Workers?*, Environment Institute Australia and New Zealand, Sydney, 2009.
Connell, RW, 'Moloch mutates: global capitalism and the evolution of the Australian ruling class 1977–2002', in N Hollier (ed.), *The Power, Privilege and Politics of the New Ruling Class*, Australian Scholarly Publishing, Melbourne, 2004, pp. 1–23.
Connor, Linda, Sonia Freeman, and Nick Higgenbotham, Not just a coalmine: shifting grounds of community opposition to coal mining in southeastern Australia, unpublished paper, n.d.
Coorey, Phillip, 'Rudd pledges to create 50,000 green jobs', *Illawarra Mercury*, 30 July 2009. http://www.illawarramercury.com.au/news/national/national/general/kevin-rudd-pledges-to-create-50000-green-jobs/1582568.aspx# (viewed 6 May 2010).
Coorey, Phillip, and Stephanie Peatlin, 'Howard urges cool heads over Stern Report', *Age*, 1 November 2006. http://newsstore.theage.com.au/apps/viewDocument.ac?page=1&sy=age&kw=Stern+and+Report&pb=all_pubs&dt=selectRange&dr=5years&so=relevance&sf=text&sf=headline&rc=10&rm=200&sp=nrm&clsPage=1&docID=SMH061101GJ37E6JQ355 (viewed 14 July 2010).
Cosier, Peter, 'The planet is capable of managing carbon, but are we? Copenhagen, climate change, and the economics of nature in the 21st century', Speech to Wentworth Group of Concerned Scientists, Sydney Theatre Company, 8 February 2010. http://www.wentworthgroup.org/uploads/STC%20Speech%20Final.pdf (viewed 6 May 2010).
Courtice, Ben, 'Climate change: the case for immediate action', *GLW*, 23 August 2008. http://www.greenleft.org.au/node/40172 (viewed 2 September 2008).

Courtice, Ben, 'Don't just "say yes" to Labor's carbon price', *GLW*, 15 June 2011, p. 11.
Courtice, Ben, 'Hot scoop: socialists oppose climate change', *GLW*, 17 February 2010, p. 5.
Courtice, Ben, 'Labor Party: all hail Old King Coal!', *GLW*, 22 September 2010, p. 7.
Courtice, Ben, 'Who is really protecting power industry jobs?' *GLW*, 23 September 2009, p. 7.
Crowley, Kate, 'Environmental employment opportunities: how green are Australia's green job credentials?' *Environmental Politics*, vol. 5, no. 4, 1996, pp. 607–631.
Crowley, Kate, Climate Policy failure: was Australia's ETS more politics than policy?, Paper presented to Australian Political Studies Association Conference, University of Melbourne, 27–29 September 2010, pp. 1–14.
Cryderman, Kelly, 'Massive protest held at Copenhagen climate conference', *Calgary Herald*, 12 December 2009. http://www.vancouversun.com/technology/Massive+protest+held+Copenhagen/2334540/story.html (viewed 2 September 2010).
CSIRO and BOM, *State of the Climate*, 2010. http://www.bom.gov.au/climate (viewed 29 July 2010).
CSIRO, *Climate Change in Australia: Technical Report*, 2007. http://climatechangeinaustralia.com.au/technical_report.php (viewed 25 October 2009).
CSIRO, *The Science of Tackling Climate Change*, 30 September 2009. http://www.csiro.au/en/Outcomes/Climate/Understanding/Tackling-climate-change.aspx (viewed 21 March 2010).
Cubby, Ben, 'Coal group coy about port exposure to rising seas', *SMH*, 15 June 2009. http://www.smh.com.au (viewed 1 November 2009).
Cubby, Ben, 'Most want action now on emissions', *Age*, 29 July 2010. http://www.theage.com.au/environment/climate-change/most-want-action-now-on-emissions-20100729-10wdr.html (viewed 29 July 2010).
Cuneo, Clementine, 'Green no coal protest shuts Newcastle mine rail link, 37 arrested', *Daily Telegraph*, 14 July 2008, p. 1.
Cunich, Simon, 'Newcastle University pushes clean coal', *GLW*, 14 May 2008, p. 8.
Curson, Peter, 'Human health, climate and climate change: an Australian perspective', in TW Giambelluca and A Henderson-Sellers (eds), *Climate Change: Developing Southern Hemisphere Perspectives*, Wiley, Chichester NY, 1996.
Cutcliffe, Tony, 'Things aren't always what they seem: jumping hurdles to a post-carbon lifestyle', in J O'Brien (ed.), *Opportunities Beyond Carbon: Looking Forward to a Sustainable World*, Melbourne University Press, Melbourne, pp. 245–255.
Dall, Erima, 'Carbon price committee is a trap for The Greens', *Solidarity*, no. 28, October 2010, p. 6.
David Suzuki Foundation, A Copenhagen Climate Treaty. Version 1.0. A Proposal for an amended Kyoto Protocol and a new Copenhagen Protocol

by members of the NGO Community, WWF International, Gland, Switzerland, June 2009. http://www.greenpeace.org/international/Global/international/planet-2/report/2009/7/ngo-copenhagen-treaty.pdf (viewed 2 September 2010).

Davidson, Kenneth, 'Bush, the global crony capitalist', *Age*, 26 July 2001, p. 15.

Davidson, Kenneth, 'Editorial: Rudd's Labor's deception—a forest of fig leaves', *Dissent*, no. 30, Spring 2009, pp. 2–7.

Davies, Anne, and Adam Morton, 'Australia power to a record', *Age*, 15 November 2007, p. 2.

Davis, Mark, 'A "pox on both their houses" attitude prevails among voters', *Age*, 1 June 2010. http://www.theage.com.au/opinion/politics/a-pox-on-both-their-houses-attitude-prevails-among-voters-20100601-wsi3.html (viewed 1 June 2010).

Davis, Mark, and Marian Wilkinson, 'Rudd's renewable 2020 vision', *SMH*, 31 October 2007. http://www.smh.com.au/articles/2007/10/30/1193618883930.html?page=fullpage(viewed 14 July 2010).

Davis, Mark, *The Land of Plenty: Australia in the 2000s*, Melbourne University Press, Melbourne, 2008.

Davison, Graeme, *Car Wars: How the Car Won Our Hearts and Conquered Our Cities*, Allen & Unwin, Sydney, 2004.

Day, Michael, 'Wealthy Italians splash out on private islands', *The Independent*, 11 July 2011, p. 26.

De Boehmler, Tara, 'Beyond the Wedge', *Workers Online*, November 2004.

De Kretser, David, 'A challenge for the human race—the need for leadership', in H Sykes (ed.), *Climate Change on for Young & Old*, Future Leaders, Sydney, 2009, pp. 19–28.

Dennis, Carina, 'Culture of fear reigns at Australian research lab', *Nature*, no. 439, 2006, pp. 896–7.

Denniss, Richard, 'Can emissions trading save the planet', in H Sykes (ed.), *Climate Change on for Young & Old*, Future Leaders, Sydney, 2009, pp. 71–83.

Denniss, Richard, and David Richardson, 'Once more with feeling. Principles for reducing greenhouse gas emissions and improving the wellbeing of most Australians', Australia Institute Policy Brief no. 19, August 2010.

Department of Climate Change and Energy Efficiency, *Climate Commission—The Critical Decade: Climate Science, Risks and Responses*, May 2011.

Department of Climate Change, 'Fact Sheet: Climate change—potential costs and impacts Tasmania', 2009. http://www.climatechange.gov.au (viewed April 2009).

Department of Climate Change, *Australia's National Greenhouse Accounts. National Greenhouse Gas Inventory, accounting for the Kyoto Target*, May 2009. http://www.climatechange.gov.au/climate-change/~/media/publications/greenhouse-report/national-greenhouse-gas-inventory-pdf.ashx (viewed 17 September 2010).

Department of Climate Change, *Australian National Greenhouse Accounts. National Inventory by Economic Sector 2007*. http://www.climatechange.gov.au/climate-change/~/media/publications/greenhouse-report/NIES.

ashx (viewed 17 September 2010).
Department of Climate Change, *National Greenhouse Gas Inventory 2009.* http://ageis.climatechange.gov.au/Chart_KP.aspx?OD_ID=21308211426&TypeID=2 (viewed 9 September 2011).
Diesendorf, Mark, 'Strategies for radical climate mitigation', *Journal of Australian Political Economy,* no. 66, 2010/2011, pp. 98–117.
Diesendorf, Mark, *Climate Action: A Campaign Manual for Greenhouse Solutions,* UNSW Press, Sydney, 2009.
Diesendorf, Mark, *Greenhouse Solutions with Sustainable Energy,* UNSW Press, Sydney, 2007.
Dilworth, Craig, *Too Smart for Our Own Good: The Ecological Predicament of Humankind,* Cambridge University Press, Cambridge, 2010
Dominick, Brian, 'From here to Parecon: thoughts on strategy for economic revolution', in C Spannos (ed.), *Real Utopia: Participatory Society for the 21st Century,* AK Press, Oakland CA, 2008, pp. 380–395.
Doyle, Timothy, *Green Power: The Environment Movement in Australia,* UNSW Press, Sydney, 2000.
Dunlop, Ian, 'We can keep jobs and save the planet', *Age,* 6 March 2009. http://www.theage.com.au (viewed 7 March 2009).
Earthworker Cooperative, '100,000 Australians Campaign', leaflet, 2pp. http://www.earthworkercooperative.com (viewed 20 October 2011).
Earthworker Cooperative, 'Media Release. National Launch. 100,000 Australians Campaign. People's Power—a Union and Community Based Campaign' http://www.earthworkercooperative.com (viewed 20 October 2011).
Earthworker, 'From Fossil Fuels to Renewables: A solar, wind & water industry plan', 15 November 1999, 16pp.
Eckersley, Robyn, 'Greening the Modern State', in P James (ed.), *The State in Question,* Allen & Unwin, Sydney, 1996, pp. 74–108.
Eckersley, Robyn, 'Just carbon trading?' in J Moss (ed.), *Climate Change and Social Justice,* Melbourne University Press, Melbourne, 2009, pp. 99–127.
Eckersley, Robyn, and Peter Christoff, 'A soft start on carbon', *Age,* 11 July 2011, p. 15.
Eckersley, Robyn, *The Green State,* MIT Press, Boston, 2004.
Economou, Nick, 'Back on the "Issues Attention Cycle": Labor and environment policy from Hawke to Rudd', *Australian Political Studies Conference,* University of Melbourne, 27–29 September 2010, pp. 1–10. http://apsa2010.com.au/full-papers/pdf/APSA2010_0057.pdf (viewed 9 September 2011).
Emanuel, Kamala, 'What alternatives to the CPRS?', *GLW,* 25 March 2009, p. 3.
Enting, Ian, *Twisted: The Distorted Mathematics of Greenhouse Denial,* Australian Mathematical Sciences Institute, Melbourne, 2007.
Environmental Defender's Office, *A NSW Guide to Non-Violent Action, the Environment and the Law,* Environmental Defender's Office, Sydney, 1998.
Environmentalists for Full Employment, n.d. [c.1978], 'Covering letter for sending to prospective member-sponsor groups' (signed J Mundey), 5pp., roneod.

EV, Fast-Tracking Victoria's clean energy future to replace Hazelwood Power Station, EV, Melbourne, May 2010.
EV, The People's Climate White Paper: a community action plan for our climate, Environment Victoria, Melbourne, September 2009.
Evans, Geoff, 'A rising tide: linking local and global climate justice', *Australian Journal of Political Economy*, no. 60, 2010/2011, pp. 199–221.
Evans, Ray, *Nine Lies About Global Warming*, Lavoisier Group, Sydney, February 2006.
Fagan, Bob, and Dick Bryan, 'Australia and the changing global economy: background to social inequality in the 1990s', in J O'Leary and R Sharp, (eds*), Inequality in Australia: Slicing the Cake—The Social Justice Collective*, William Heinemann Australia, Melbourne, 1991, pp. 7–31.
Falk, Jim, and Andrew Brownlow, *The Greenhouse Challenge: What's to be Done?*, Penguin Books, Melbourne, 1989.
Ferguson, Peter, 'Anti-environmentalism and the Australian culture war', *Journal of Australian Studies*, vol. 33, no. 3, September 2009, pp. 289–304.
Firth, Stewart, *Australia in International Politics: An Introduction to Australian Foreign Policy*, 2nd edn, Allen & Unwin, Sydney, 2005.
Fitzgerald, Julian, *Lobbying in Australia: You Can't Expect Anything to Change if You Don't Speak Up*, Rosenberg, Dural Delivery Centre NSW, 2006.
Flannery, Tim, 'Burnt and hostile: just what we need', *SMH*, 21 December 2006. http://www.smh.com.au/news/opinion/burnt-and-hostile-just-what-we need/2006/12/20/1166290612689.
html?page=fullpage#contentSwap1 (viewed 5 August 2010).
Flannery, Tim, 'Now or never. a sustainable future for Australia?', *Quarterly Essay*, no. 3, 2008, pp. 1–61.
Flannery, Tim, *An Explorer's Notebook—Essays on Life, History & Climate*, Melbourne, Text Publishing, 2007.
Flannery, Tim, *Here on Earth: An Argument for Hope*, Text Publishing, Melbourne, 2010.
Flannery, Tim, *The Future Eaters: An Ecological History of the Australasian Lands and Peoples*, Reed Books, Melbourne, 1994.
Flannery, Tim, *The Weather Makers: The History and Future Impact of Climate Change*, Text Publishing, Melbourne, 2005.
FOE, *Open submission to Garnaut on the Emissions Trading Scheme*, 17 April 2008. http://www.foe.org.au/climate-justice/policy-position/garnaut-review-2008 (viewed 1 August 2008).
Foster, John Bellamy, *The Ecological Revolution: Making Peace with the Planet*, Monthly Review Press, New York, 2009.
Foster, John, *The Sustainability Mirage: Illusion and Reality in the Coming War on Climate Change*, Earthscan, London, 2008.
Frank, Christine, 'The bankruptcy of capitalist solutions to the climate crisis', *Capitalism, Nature, Socialism*, vol. 20, no. 2, 2009, pp. 32–43.
Frankel, Boris, *When the Boat Comes In: Transforming Australia in the Age of Globalisation*, Pluto Press Australia, Sydney, 2001.
Franklin, Matthew, 'Abbott slams "veiled socialism" ', *Australian*, 11 July 2011, p. 8.

Frew, Wendy, and Stephanie Peatling, 'Time to act on global warming', *SMH*, 31 October 2006. http://www.smh.com.au/news/environment/stern-warning-on-climate/2006/10/30/1162056926610.html (viewed 14 July 2010).
Fyfe, Melissa, 'Gore weighs in for green group', *Age*, 7 June 2009, p. 1.
Garnaut, Ross, 'Australia counts itself out', *Age*, 20 December 2008. http://www.theage.com.au (viewed 21 December 2008).
Garnaut, Ross, Garnaut Climate Change Review: Draft Report, June 2008.
Garnaut, Ross, *Garnaut Climate Change Review: Final Report*, Cambridge University Press, Melbourne, 2008.
Garnaut, Ross, *Garnaut Final Report. Synopsis.* http://garnautreview.org.au/synopsis.htm (viewed 1 June 2010).
Garnaut, Ross, *The Garnaut Review 2011: Australia in the Global Response to Climate Change*, Cambridge University Press, Melbourne, 2011.
Glantz, David, 'Working inside Labor: How Garrett lost his way', *Solidarity*, issue 17, August 2009, p. 16.
Glickson, Andrew (ed.), *Imagining the Real: Life on a Greenhouse Earth: Program and Abstracts*, Papers in honour of Barry Jones, ANU, Canberra, 11–12 January 2008.
Glickson, Andrew, 'Noah's ark revisited: an alliance of pro-carbon sceptics and fundamentalists in retarding 11th hour attempts at mitigation of dangerous climate change', 7 July 2009. http://webdiary.com.au (viewed 27 October 2011).
Global Humanitarian Forum, *The Anatomy of a Silent Crisis*, Global Humanitarian Forum, Geneva, 2009.
Glover, Julian, 'The lucky country?', *Guardian*, 23 November 2007. http://www/guardian.co.uk/environment/2007/nov/23/climatechange.australia (viewed 1 June 2010).
Goodin, Robert E, 'Selling Environmental Indulgences', *Kyklos*, vol. 47, 1994, pp. 573–96.
Goodman, James, 'Climate movement: a new politics of moral protest?', *Chain Reaction*, no. 107, November 2009, pp. 20–21.
Goodman, James, 'Responding to climate crisis: modernisation, limits, socialism', *Journal of Australian Political Economy*, no. 66, 2010/2011, pp. 144–165.
Goodman, James, 'Ruddism at large', *Solidarity*, issue 17, August 2009, pp. 22–24.
Goodman, James, and David Worth, 'The minerals boom and Australia's "resource curse" ', *Journal of Australian Political Economy*, no. 61, June 2008, pp. 201–219.
Goodman, James, and Stuart Rosewarne, 'Climate Policy: from carbon tax to direct action?', *Chain Reaction*, no. 110, November 2010, pp. 23–24.
Goot, Murray, 'Is the news on the internet different? Leaders, frontbenchers and other candidates in the 2007 Australian Election', *Australian Journal of Political Science*, vol. 43, no.1, 2008, pp. 99–110.
Gordon, Josh, 'Carbon offset plans not for us: Wong', *Age*, 13 September 2009. http://www.theage.com.au/environment/carbon-offset-plans-not-for-us-wong-20090912-fljq.html (viewed 15 September 2009).

Gordon, Josh, and Tom Arup, 'Victorian power firms to get 97% of carbon compo', *Age*, 19 July 2011, p. 1.
Gordon. Josh, 'ALP energy targets a "gimmick" ', *Age*, 7 April 2011. http://www.theage.com.au/victoria/alp-energy-targets-a-gimmick-20110406-1d4je.html (viewed 4 May 2011).
Gore, Al, *An Inconvenient Truth: The Planetary Emergency of Global Warming and What We Can Do About It*, Rodale, London, 2006.
Gorz, Andre, *Socialism and Revolution*, Anchor, Garden City NY, 1973.
Grattan, Michelle, 'PM's climate credibility undermined', *Age*, 28 April 2010. http://www.theage.com.au/opinion/politics/pms-climate-credibility-undermined-20100428-tq96.html (viewed 28 April 2010).
Grattan, Michelle, and David Wroe, 'Abbott's blood oath to repeal carbon tax', *Age*, 13 October 2011, p. 1.
Grattan, Michelle, and Tom Arup, 'Carbon pricing impact "minimal" ', *Age*, 7 June 2011, pp. 1–2.
Grattan, Michelle, and Tom Arup, 'Rudd in ETS backflip', *Age*, 28 April 2010, p. 1.
Green Jobs Unit. http://www.nor.com.au/environment/greenwork/gjind.htm (viewed 14 November 2000).
Green, Michael, 'Power from the ground up', *Age*, 13 December 2008, Insight, p. 5.
Greenpeace Australia, *Energy [R]evolution: A Sustainable Australia*, Greenpeace Australia, Sydney, June 2008.
Greenpeace International, *Copenhagen Climate Summit. Greenpeace Demands*, Greenpeace International, Amsterdam, March 2009. http://www.greenpeace.org/international/Global/international/planet-2/report/2009/3/climate-demands.pdf (viewed 2 September 2010).
Greenpeace, 'Fact Sheet. About Greenpeace', Sydney, June 2008, leaflet, 2pp.
Greenpeace, 'From a coal-fired economy to a renewable "Silicon Valley": ensuring a Just Transition for workers in the Hunter region', Briefing, 1 July 2008, 8pp.
Greenpeace, 'What the Oz election outcome means', 28 November 2007. http://www.greenpeace.org.au/blog/energy/?p=133 (viewed 14 July 2010).
Hall, Ashley, 'Business Council denies seeking "free ride" on climate change', ABC News, 28 August 2008. http://www.abc.net.au (viewed 29 August 2008).
Hamilton, Clive, 'Climate Change and its Implications for Australia', Annual University Lecture at the Australian Defence Force Academy, 10 September 2008, pp. 1–14. http://www.clivehamilton.net.au/cms/media/documents/articles/adfa_lecture.pdf (viewed 22 August 2011).
Hamilton, Clive, 'Climate change', in R Manne (ed.), *Dear Mr Rudd: Ideas for a Better Australia*, Black Agenda Inc., Melbourne, 2008, pp. 144–156.
Hamilton, Clive, 'Climate emergency or a crisis of democracy', 4 February 2009. http://www.crikey.com.au (viewed 5 February 2009).
Hamilton, Clive, 'Rudd at Bali and beyond', 25 November 2007. http://www.crikey.com.au (viewed 20 August 2008).
Hamilton, Clive, 'The social psychology of climate change', in T Sherratt, T Griffiths and L Robin (eds), *A Change in the Weather: Climate and Culture in Australia*, National Library of Australia Press, Canberra, 2005, pp. 187–195.

Hamilton, Clive, 'Viscount Monckton of Benchley's over-egged CV', 12 January 2010. http://www.crikey.com.au (viewed 12 January 2010).
Hamilton, Clive, and Andrew Macintosh, Taming the panda: the relationship between the WWF Australia and the Howard government, Discussion Paper no. 68, Australia Institute, July 2004.
Hamilton, Clive, and Christian Downie, University capture: Australian universities and the fossil fuel industries, Discussion Paper no. 95, Australia Institute, June 2007.
Hamilton, Clive, and Elizabeth Mail, Downshifting in Australia: a sea-change in the pursuit of happiness. Discussion paper no. 50, Australia Institute, 2003.
Hamilton, Clive, and Hal Turton, Subsidies to the aluminium industry and climate change, Background Paper no. 21, Australia Institute, November 1999.
Hamilton, Clive, and Richard Denniss, *Affluenza: When Too Much Is Never Enough*, Allen & Unwin, Sydney, 2005.
Hamilton, Clive, *Growth Fetish*, Allen & Unwin, Sydney, 2003.
Hamilton, Clive, *Requiem for a Species: Why We Resist the Truth About Climate Change*, Allen & Unwin, Sydney, 2010.
Hamilton, Clive, *Running from the Storm: The Development of Climate Change Policy in Australia*, UNSW Press, Sydney, 2001.
Hamilton, Clive, *Scorcher: The Dirty Politics of Climate Change*, Black Inc. Agenda, Melbourne, 2007.
Hamilton, Clive, The time for radical activism: speech at the Climate Action Summit, ANU, 31 January 2009.
Hansen, James, *Storms of my Grandchildren: The Truth About the Coming Climate Catastrophe and our Last Chance to Save Humanity*, Bloomsbury Publishing, New York, 2009.
Hanson, Fergus, *Australia and the World. Public Opinion and Foreign Policy*, Lowy Institute for International Policy, Sydney, 2008.
Hanson, Fergus, *Australia and the World. Public Opinion and Foreign Policy*, Lowy Institute for International Policy, Sydney, 2009.
Hardt, Michael, and Antonio Negri, *Empire*, Harvard University Press, Cambridge MA, 2000.
Harrison, Patrick, 'Students fight climate emergency', *GLW*, 26 August 2009, p. 9.
Hartcher, Peter, 'Brown leads Greens to the promised land', *SMH*, 22 May 2010. http://www.nationaltimes.com.au/opinion/politics/brown-leads-greens-to-the-promised-land-20100521-w1pd.html (viewed 1 June 2010).
Hartcher, Peter, 'How Rudd lost his nerve on an ETS election', *Age*, 15 May 2010, p. 1.
Hatfield-Dodds, Steve, Graham Turner, and Heinz Schandl, *Growing the Green Collar Economy: Skills and labour challenges in reducing our greenhouse emissions and national environmental footprint*, CSIRO, Canberra, June 2008.
Hawthorne, Mark, and Daniel Ziffer, 'Libs inconvenienced by Gore events', *Age*, 19 September 2007, p. 7.

Henderson, Troy, 'Tweet from the Hardhat Revolution', *Arena Magazine*, no. 101, August – September 2009, pp. 45–7.
Hennessey, K, B Fitzharris, BC Bates, N Harvey, SM Howden, L Hughes, J Salinger, and R Warrick, 'Australia and New Zealand', in ML Parry, OF Canziani, JP Palutikof, PJ Van der Linden and CE Hanson (eds), *Climate Change 2007: Impacts, Adaptation and Vulnerability. Contribution of Working Group II to the Fourth Assessment Report of the Intergovernmental Panel on Climate Change*, Cambridge University Press, Cambridge UK, 2007, pp. 507–540.
Henry, Don, 'Risking a new era of pollution protectionism', *Australian Options*, no. 56, Autumn 2009, p. 9.
Hepburn, John, 'Stopping Rudd's carbon plan', *Solidarity*, issue 13, 2009, p.14.
Hickson, Ken, *The ABC of Carbon: Issues and Opportunities in the Global Climate Change Environment*, ABC Carbon, Toowong Qld, 2009.
Hoegh-Guldberg O, D Karoly, I Lowe, T McMichael, C Mitchell, G Pearman, P Scaife, and A Reynolds, *Climate Change: Solutions for Australia*, WWF-Australia, Sydney, 2004.
Hogan, Naomi, 'The World People's Conference on Climate Change and the Rights of Mother Earth', *Search News*, vol. 17, no. 4, August 2010, pp. 6–7.
Hooke, Mitchell H, 'Media release—Minerals Council of Australia: White Paper on climate change: too high a price for global leadership', 15 December 2008.
Hooke, Mitchell H, 'Minerals Council of Australia—A synopsis of the Australian minerals industry—election policy issues', September 2004.
Houghton, John, *Global Warming. The Complete Briefing*, 4th edition, Cambridge University Press, Cambridge, 2009.
House, Brigitte, 'The Transition Decade—a brief history'. http://www.t10.net.au/ (viewed 11 February 2011).
Howes, Michael, 'A Climate of Change: Ecological modernisation and the politics of carbon trading in Australia', *Australian Political Studies Association Annual Conference*, Macquarie University, 27–30 September 2009.
Hutchinson, Zoe, and Holly Creenaune, 'A stifling climate: targeting social movements and policing protests', *Dissent*, no. 32, 2010, pp. 33–35.
Hutton, Drew, and Libby Connors, *A History of the Australian Environmental Movement*, Cambridge University Press, Melbourne, 1999.
India Climate Forum, 'Delhi Climate Justice Declaration', 2002. http://www.indiaresource.org (viewed 27 October 2009).
Indymedia. http://indymedia.org.au/2009/07/21/climate-change-activists-in-solidarity-with-striking-coal-workers (viewed 28 July 2009).
International Climate Conference, '4 Degrees and Beyond. Implications of a global climate change of 4+ degrees for people, ecosystems and the earth-system', 28–30 September 2009, Oxford UK. http://www.eci.ox.ac.uk/4degrees/ (viewed 22 September 2010).
International Indian Treaty Council, 'The Bonn Declaration—the Third International Forum on Indigenous People's and Local Communities on Climate Change', 14–15 July 2001. http://www/treatycouncil.org (viewed

16 October 2009).
IPCC, *Fourth Assessment Report: Climate Change 2007. Synthesis Report Summary for Policymakers*, 2007. http://www.ipcc.ch/publications_and_data/ar4/syr/en/spm.html (viewed 22 September 2010).
Iqbal, Farida, 'EPA delays decision on Margaret River coalmine', *GLW*, 26 January 2011, p. 7.
Iqbal, Farida, 'Margaret River community defeats coalmine threat', *GLW*, 15 February 2012, p. 4.
Iqbal, Farida, 'Residents fight to stop coalmine', *GLW*, 2 February 2010, p. 10.
Isla, Ana, 'Who Pays for the Kyoto Protocol? Selling Oxygen and Selling Sex in Costa Rica', in A Salleh (ed.), *Eco-Sufficiency & Global Justice. Women Write Political Ecology*, Pluto and Spinifex, Sydney and Melbourne, 2009, pp. 199–217.
Jackman, Christine, *Inside Kevin 07: The People, the Plan, the Prize*, Melbourne University Press, Melbourne, 2008.
Jackson, Stewart, 'From Social Movement to Electoral Professional Party?', *Australian Political Studies Association Conference*, Macquarie University, 27–30 September 2009. http://www.pol.mq.edu.au/apsa/papers/Refereed papers/Jackson From Social Movement to Electoral Professional Party.pdf (viewed 9 September 2011).
Johnson, Carol, 'The CPRS—a failure of the left not just the right', *On Line Opinion*, 16 March 2010, pp. 1–2. On Line Opinion - 16-3-2010_done.mht (viewed 15 October 2010).
Johnson, James, 'Non violent action and the law', in *Protest! Environmental Activism in NSW 1968–1998*, Public Houses Trust of New South Wales, Sydney, 1998, pp. 43–51.
Jones, Peter, 'Saving the planet or selling off the atmosphere? Emissions trading, capital accumulation and the carbon rent', *Marxist Interventions*, no. 1, 2009, pp. 9–22.
Jowit, Juliette, and Patrick Wintour, 'Cost of tackling global climate change has doubled, warns Stern', *Guardian*, 26 June 2008, p. 1.
Kapferer, Bruce, and Barry Morris, 'The Australian society of the state: egalitarian ideologies and new directions in exclusionary practice', *Social Analysis*, vol. 47, no. 3, 2003, pp. 80–107.
Kapferer, Bruce, *Being Equal: Identity, Difference and Australian Cultural Practice*, Berg, Oxford, 1996.
Karoly, David, 'Wrong to dismiss the dirt on Hazelwood', *Age*, 6 November 2010. http://www.theage.com.au/opinion/politics/wrong-to-dismiss-the-dirt-on-hazelwood-20101105-17hei.html (viewed 9 November 2010).
Katz, Claudio, 'Socialist struggles in Latin America', *Monthly Review*, vol. 59, no. 4, September 2007, pp. 25–41.
Keane, Bernard, 'Aluminium smelting: the best bang for your fossil-fuel subsidy buck', 10 March 2011. http://www.crikey.com.au (viewed 11 March 2011).
Kelso, James, 'Social wellbeing & happiness', *Green. Magazine of the Australian Greens*, issue 31, Winter 2010, pp. 11–14.
Ker, Peter, 'Hazelwood protest has support of key greens', *Age*, 12 September

2009. http://www.theage.com.au/environment/hazelwood-protest-has-support-of-key-greens-20090911-fkv4.html (viewed 7 May 2010).
Ker, Peter, 'Police tactics under fire at Hazelwood', *Age*, 14 September 2009, p. 1.
Ker, Peter, and Stephen Moynihan, 'The great Australian walkabout', *Age*, 12 November 2007. http://www.theage.com.au/news/federal-election-2007-news/the-great-australian-walkabout/2007/11/11/1194766508066.html (viewed 14 July 2010).
Kerin, Dave, 'Challenging the myth of Jobs vs Environment', *Chain Reaction*, no. 80, 1999, pp.19–20, 42.
Kerin, Dave, Workers creating green jobs, Lecture, Melbourne Trades Hall, 7 July 2010. Notes/audio tape in possession of V Burgmann.
Kevin, Tony, *Crunch Time: Using and Abusing Keynes to Fight the Twin Crises of Our Era*, Scribe, Melbourne, 2009.
Kininmonth, William, *Climate Change: A Natural Hazard*, Multi-Science Publishing, Brentwood UK, 2004.
Kirsch, Stuart, 'Sustainable mining', *Dialectical Anthropology*, no. 34, 2010, pp. 87–95.
Klein, Naomi, *The Shock Doctrine. The Rise of Disaster Capitalism*, Allen Lane/Penguin, New York, 2007.
Kleinman, Rachel, and Jewel Topsfield, 'Britain pressures Australia to join climate change plan', *Age*, 31 October 2006. http://newsstore.theage.com.au/apps/viewDocument.ac?page=1&sy=age&kw=Stern+and+Report&pb=all_pubs&dt=selectRange&dr=5years&so=relevance&sf=text&sf=headline&rc=10&rm=200&sp=nrm&clsPage=1&docID=AGE061031MN3H97LQ07N (viewed 14 July 2010).
Koh, Agnes, The 'Third Way' in Australia's Climate Change Mitigation, Honours thesis, School of Social and Political Sciences, University of Melbourne, June 2010.
Kovel, Joel, *The Enemy of Nature: The End of Capitalism or the End of Nature*, Zed Books, London, 2007.
Ladd, Brian, *Autophobia: Love and Hate in the Automotive Age*, University of Chicago Press, Chicago, 2008.
Langford, Vivien, 'They leave the devastation behind', *GLW*, 16 February 2011. http://www.greenleft.org.au/taxonomy/term/3459 (viewed 28 February 2011).
Lavelle, Ashley, 'The boom, the left and capitalism', *Journal of Australian Political Economy*, no. 61, 2008, pp. 296–320.
Lawrence, Geoffrey, 'Agribusiness and inequality: changes in the bush', in J O'Leary and R Sharp (eds), *Inequality in Australia: Slicing the Cake—The Social Justice Collective*, Port Melbourne, William Heinemann Australia, 1991, pp. 80–104.
Lawson, Damien, '10 lessons for the climate movement redux' in *Talk Climate*, compiled by Victorian Climate Action Centre, Melbourne, March 2010, pp. 3–7.
Lawson, Damien, 'Looking back, moving forward: ten lessons for the climate movement' in *Climate Reader*, compiled by Melbourne Climate Action

Centre and CarbonEquity, Melbourne, January 2009, pp. 3–5.
Lawton, Christopher, 'From Seattle to Copenhagen: Climate Change Summit becomes a target for protest', *SpiegelOnline*, 25 November 2009. http://www.spiegel.de/international/europe/0,1518,663142,00.html (viewed 2 September 2010).
Leviston, Zoe, Anne Leitch, Murni Greenhill, Rosemary Leonard & Iain Walker, *Australians' Views of Climate Change*, CSIRO Report, Canberra, 2011. http://www.garnautreview.org.au/update-2011/commissioned-work/australians-view-of-climate-change.pdf (viewed 9 October 2011).
Li, Minqi, 'Climate change, limits to growth and the imperative of socialism', *Monthly Review*, vol. 60, no. 3, July – August 2008, pp. 51–67.
Lindenmayer, David, *On Borrowed Time: Australia's Environmental Crisis and What We Must Do About It*, CSIRO Publishing, Melbourne, 2007.
Lohmann, Larry, Carbon Trading: A Critical Conversation on Climate Change, Privatisation and Power: Development Dialogue, no. 48, Dag Hammarskjoeld Centre, Uppsala, September 2006.
Lohrey, Amanda, 'The Rise of the Greens', *Quarterly Essay*, no. 8, 2002, pp. 1–70.
Lohrey, Amanda, 'Voting for Jesus: Christianity and politics in Australia', *Quarterly Essay*, no 22, June 2006, pp. 1–79.
Long, Stephanie, 'Market "Solutions" to Climate Change: a far cry from Climate Justice', *Chain Reaction*, no. 98, Spring 2006, pp. 20–21.
Long, Stephanie, and Cam Walker, 'The Pacific nations and the impacts of global warming', *Chain Reaction*, no. 91, September 2004, pp. 25–26.
Louviere, Jordan, Richard Carson, Edward Wei, Cam Rungie, Len Coote, Sandy Burke, Paul Wang, Bart Frischknecht, John Rose, Ricardo Scarpa, *Trading Plans that Australians Want their Government to Implement*, Centre for the Study of Choice, University of Technology, Sydney, n.d. [2010]. http://www.censoc.uts.edu.au/news/CenSoC_People_Choose_A_Carbon_Plan.pdf (viewed 29 July 2010).
Lowe, Ian, 'Reaction time: climate change and the nuclear option', *Quarterly Essay*, no. 27, 2007.
Lowe, Ian, 'Sunshine, social factors, solar energy and sustainability', in T Sherratt, T Griffiths and L Robin (eds), *A Change in the Weather: Climate and Culture in Australia*, National Museum of Australia Press, Canberra, 2005, pp. 116–125.
Lowe, Ian, 'The greenhouse effect and the politics of long-term issues', in S Bell and B Head (eds), *State, Economy and Public Policy in Australia*, Oxford University Press, Melbourne, 1994, pp. 315–331.
Lowe, Ian, 'The research community', in C Hamilton and S Maddison (eds), *Silencing Dissent: How the Australian Government is Controlling Public Opinion and Stifling Debate*, Allen & Unwin, Sydney, 2007, pp. 60–77.
Lowe, Ian, *A Big Fix: Radical Solutions for Australia's Environmental Crisis*, expanded and revised edn, Black Inc., Melbourne, 2009.
Lowe, Ian, *Living in the Hothouse: How Global Warming Affects Australia*, Scribe, Melbourne, 2005.
Macintosh, Andrew, 'Garnaut's Targets and Trajectories: a critique', ANU

Centre for Climate Law and Policy, September 2008. http://law.anu.edu.au/cclp/WP4_2008_Garnaut_critique.pdf (viewed 23 September 2011).
Macintosh, Andrew, 'Keeping warming within the 2° C limit after Copenhagen', CCLP Working Paper Series 2009/1, ANU Centre for Climate Law and Policy. http://law.anu.edu.au/cclp/WP_6_2009_1_%20Keeping_warming_within%202C_Limit_After_Copenhagen.pdf (viewed 23 September 2011).
Macintosh, Andrew, and Lailey Wallace, International aviation emissions to 2025: can emissions be stabilised without restricting demand?, CCLP Working Paper Series 2008/1, ANU Centre for Climate Law and Policy, 2008.
Macintosh, Andrew, and Oliver Woldring, Climate-carbon cycle feedbacks: the implication for Australian climate policy, CCLP Working Paper Series 2008/3, ANU Centre for Climate Law and Policy, 2008.
Mackey, Brendan G, Heather Keith, Sandra L Berry, and David B Lindenmayer, *Green Carbon: The Role of Natural Forests in Carbon Storage*, Part 1: A Green Carbon Account of Australia's South-Eastern Eucalypt Forests, and Policy Implications, ANU E-Press, Canberra, 2008, pages unnumbered.
Maddison, Sarah, and Clive Hamilton, 'Non-government organisations', in C Hamilton and S Maddison (eds), *Silencing Dissent: How the Australian Government is Controlling Public Opinion and Stifling Debate*, Allen & Unwin, Sydney, 2007, pp. 78–100.
Magdoff, Fred, and John Bellamy Foster, 'What every environmentalist needs to know about capitalism', *Monthly Review*, vol. 61, no. 19, March 2010, pp. 1–30.
Magdoff, Harry, and Fred Magdoff, 'Approaching socialism', *Monthly Review*, vol. 57, no. 3, July–August, 2005, pp. 35–47.
Maher, Sid, 'Coal and steel win promise of extra help', *Age*, 11 July 2011, p. 11.
Maher, Sid, 'Gillard's clean energy crusade', *Australian*, 11 July 2011, p. 1.
Maher, Sid, 'Green jobs dopey—union leader', *Australian*, 14 September 2009. http://www.news.com.au/green-jobs-dopey-union-leader/story-e6frfkp9-1225772486326 (viewed 7 May 2010).
Maher, Tony, 'Crib Rooms and Climate Change. Empowering Mine Workers. Presentation to NSW Minerals Council Environment Conference', 16 August 2007. http://www.cfmeu.com.au/storage/documents/NSWMC_enviro2007.pdf (viewed 21 July 2010).
Manning, Paddy, 'One climate policy, and it only comes in green', *SMH*, 1 May 2010. http://www.smh.com.au/business/one-climate-policy-and-it-only-comes-in-green-20100430-tzjz.html (viewed 20 July 2010).
Manning, Paddy, 'Protests threat to power station', *SMH*, 10 August 2009. http://business.smh.com.au/business/protests-threat-to-power-station-20090809-ee7z.html (viewed 16 August 2009).
Marr, David, 'The political journey of Kevin Rudd', *Quarterly Essay*, no. 38, 2010, pp. 1–91.
Matthews, Graham, 'How do we fix public transport?', *GLW*, 23 September 2009, p. 3.
Matthews, Graham, 'Too many cars: expand public transport', *GLW*, 1 July 2009, p. 12.

Matthews, Graham, 'What is "sustainable" transport?', *GLW*, 11 August 2010, p. 11.
McAlpine, CA, A Etter, PM Fearnside, L Seabrook, and WF Laurence, 'Increasing world consumption of beef as a driver of regional and global change: a call for policy action based on evidence from Queensland (Australia), Colombia and Brazil', *Global Environmental Change*, no.19, 2009, pp. 21–33.
McAuley, Ian, 'Taxing the miners' uncommonly large profit', *Dissent*, issue 33, Spring 2010, pp. 20–25.
McKibben, Bill, *EAArth: Making a Life on a Tough New Planet*, Black Inc., Melbourne, 2010.
McKibben, Bill, *Fight Global Warming Now: The Handbook for Taking Action in Your Community*, Holt, New York, 2007.
McMahon, Peter, 'Global crisis and Australian politics: time for the Greens to make their move', *Australian Quarterly*, vol. 78, no. 3, May – June 2006, pp. 22–27.
McMullen, Sean, 'The Precedent', *Fantasy & Science Fiction*, July/August 2010, pp. 230–256.
McNaughton, Colin, From Little Things Big Things Grow: The Greening of the Labour Movement in Australia?, BA Hons thesis, Politics Department, Monash University, 2000.
McNeil, Ben, 'The costs of introducing nuclear power to Australia', *Journal of Australian Political Economy*, no. 59, 2007, pp. 5–29.
McNeil, Ben, *The Clean Industrial Revolution. Growing Australian Prosperity in a Greenhouse Age*, Allen & Unwin, Sydney, 2009.
Meagher, Gabrielle, and Shaun Wilson, 'Richer, but more unequal: perceptions of inequality in Australia, 1987–2005', *Journal of Australian Political Economy*, no. 61, 2008, pp. 221–240.
Melucci, Alberto, *Nomads of the Present: Social Movements and Individual Needs in Contemporary Society*, Hutchinson Radius, London, 1989.
Metz, Bert, *Controlling Climate Change*, Cambridge University Press, Cambridge, 2010.
Miliband, Ralph, *Socialism for a Sceptical Age*, Courage Press/Verso, London, 1994.
Milne, Christine, 'Balance of power comes to Canberra', *Green. Magazine of the Australian Greens*, issue 32, Summer 2010, pp. 6–8.
Milne, Christine, 'Christine Milne's Speech to the Sydney Institute—the Greens, balance of power and climate politics', 28 October 2008, p. 2. http://greensmps.org.au (viewed 1 December 2008).
Milne, Christine, 'Even the Eskimos agree', *Arena Magazine*, April – May 2008, pp. 5–7.
Milner, Andrew, 'On the beach: apocalyptic hedonism and the origins of the postmodern', in Ian Craven (ed.), *Australian Popular Culture*, Cambridge University Press, Cambridge, 1994, pp.190–204.
Minchin, Liz, 'Gore's pre-election visit may add heat to politics', *Age*, 14 June 2007, p. 7.
Miragliotta, Narelle, 'One party, Two traditions: Radicalism and pragmatism

in the Australian Greens', *Australian Journal of Political Science*, vol. 41, no. 4, 2006, pp. 585–596.
Miragliotta, Narelle, 'Similar but different: a comparison of the organizational structures of Australia's Green parties', *Australian Political Studies Association Conference*, University of Melbourne, 27–29 September 2010. http://apsa2010.com.au/full-papers/pdf/APSA2010_0003.pdf (viewed 9 September 2011).
Moore, Jess, 'Carbon price: the gas agenda', *GLW*, 29 June 2011, p. 7.
Morgan, Gary, 'Election race closes—ALP (52.5%, down 2%) cf. L-NP (47.5%, up 2%)', 4 June 2010. http://www.roymorgan.com/news/polls/2010/4506 (viewed 4 June 2010).
Morrow, Helen, 'Australia's carbon price mechanism: predictability and flexibility?', 2 September 2011. http://www.grattan.edu.au/publications/106_morrow_oped_cmanz_energy.pdf (viewed 6 October 2011).
Morton, Adam, and Tom Arup, 'Solar panel companies are feeling the heat', *SMH*, 31 July 2009. http://www.smh.com.au/small-business/managing/solar-panel-companies-are-feeling-the-heat-20090731-e3ks.html (viewed 19 October 2011).
Morton, Adam, 'A glimpse of the future for Victoria's hilltops', *Age*, 1 August 2009, p. 2.
Morton, Adam, 'Climate change biggest world-health threat', *Age*, 17 September 2009. http://www.theage.com.au/environment/climate-change-biggest-worldhealth-threat-20090916-from.html (viewed 21 September 2009).
Morton, Adam, 'Climate change most dangerous event ever, says Gore', *Age*, 13 July 2009, p. 1.
Morton, Adam, 'Coming clean', *Age*, 4 November 2009, p. 13.
Morton, Adam, 'Fielding's climate mission', *Age*, 6 June 2009. http://www.theage.com.au/environment/fieldings-climate-mission-20090605-bylt.html (viewed 10 June 2009).
Morton, Adam, 'The sceptic's shadow of doubt', *Age*, 2 May 2009. http://www.theage.com.au/national/the-sceptics-shadow-of-doubt-20090501-aqa1.html (viewed 3 May 2009).
Morton, Adam, 'Warming's icy debate', *Age*, 13 February 2010, p. 5.
Morton, Adam, and Ben Schneiders, 'Abbott: the future is bright, brown', *Age*, 21 July 2011. http://www.theage.com.au/environment/climate-change/abbott-the-future-is-bright-brown-20110720-1hoxc.html (viewed 24 July 2011).
Morton, Adam, and Tom Arup, 'Gone with the Wind', *Age*, 28 April 2010. http://www.theage.com.au/environment/climate-change/gone-with-the-wind-20100427-tq52.html (viewed 1 September 2010).
Morton, David, 'Coming down to earth: encounter with David Karoly', *Age*, Insight section, 16 August 2008, p. 4.
Moss, Jeremy (ed.), *Climate Change and Social Justice*, Melbourne University Press, Melbourne, 2009.
Moss, Jeremy, 'Introduction', in J Moss (ed.), *Climate Change and Social*

Justice, Melbourne University Press, Melbourne, 2009, pp.11–21.
Mundey, Jack, 'Compatibility of the 3 "E's" vital to survival', *Environmentalists for Full Employment, Newsletter*, no. 1, December 1978.
Mundey, Jack, 'From grey to green', *Australian Left Review*, no 108, 1988/89, p.19.
Murphy, Katherine, 'Carbon trading by 2012', *Age*, 1 June 2007, p.1.
Neale, Jonathan, *Stop Global Warming: Change the World*, Bookmarks Publications, London, 2008.
Neighbour, Sally, 'Divided We Fall', *The Monthly*, February 2012, pp. 20–28.
Newman, Peter, 'Transport opportunities: towards a resilient city', in John O'Brien (ed.), *Opportunities Beyond Carbon: Looking for a Sustainable World*, Melbourne University Press, Melbourne, 2009, pp. 98–115.
Nichols, Dick, 'Five questions about a carbon tax', *GLW*, 10 March 2010, p. 11.
Norman, James, *Bob Brown: Gentle Revolutionary*, Allen & Unwin, Sydney, 2004.
NTEU, 'Climate Champions', 2010. http://www.sustainablecampus.org.au/about/?phpMyAdmin=ETKbRYZdMK8rG9AioC6djdltUu0 (viewed 22 July 2010).
O'Connor, James, 'Capitalism, Nature, Socialism: a theoretical introduction', *Capitalism, Nature, Socialism*, vol. 1, no. 1, 1988, pp. 11–38.
O'Connor, James, *Natural Causes: Essays in Ecological Marxism*, Guildford Press, New York, 1998.
O'Connor, Mark, and William J Lines, *Overloading Australia: How Governments Dither and Deny on Population*, Envirobook, Canterbury NSW, 2008.
O'Riordan, Timothy, and Andrew Jordan, 'Institutions, climate change and culture theory: towards a common analytical framework', *Global Environmental Change*, no. 9, 1999, pp. 81–93.
Oliphant, Monica, 'Australia's emissions contribution: does it matter?', in J O'Brien (ed.), *Opportunities Beyond Carbon: Looking for a Sustainable World*, Melbourne University Press, Melbourne, 2009, pp. 317–329.
Pakulski, Jan, and Stephen Crook, 'The end of the Green Cultural Revolution? Environmental activism, issues and publics in Australia', in J Pakulski and S Crook, *Ebbing of the Green Tide? Environmentalism, Public Opinion and the Media in Australia*, School of Sociology and Social Work, University of Tasmania, Hobart, 1998, pp. 1–20.
Panitch, Leo, and Colin Leys (eds), *Coming to Terms with Nature—Socialist Register 2007*, Merlin Press, London, 2006.
Papadakis, Elim, 'The Environment' in B Galligan et al (eds), *New Developments in Australian Politics*, Macmillan, Melbourne, 1997, pp.196–210.
Papadakis, Elim, and Anya Moore, 'Environment, economy and state', in S Bell and B Head (eds), *State, Economy and Public Policy in Australia*, Oxford University Press, Melbourne, 1994, pp. 334–351.
Parker, Jean, 'Unionists join climate action at Newcastle camp', *Solidarity*, issue 6, August 2008, p. 7.
Pearce, Fred, *Confessions of an Eco-Sinner: Travels to Find Where My Stuff*

Comes From, Eden Project Books, UK, 2008.
Pearse, Guy, 'Comment', *The Monthly*, July 2008.
Pearse, Guy, 'Quarry Vision: Coal, climate change and the end of the resources boom', *Quarterly Essay*, no. 33, Black Inc., Melbourne, 2009, pp.1–122.
Pearse, Guy, 'The Climate Movement: Australia's Patrons of Climate Change Activism', *The Monthly*, September 2011. http://www.themonthly.com.au/australia-s-patrons-climate-change-activism-climate-movement-guy-pearse-3786 (viewed 18 October 2011).
Pearse, Guy, *High & Dry: John Howard, Climate Change and the Selling of Australia's Future*, Viking, Melbourne, 2007.
Peterson, Ben, and Shayne Geilman, 'Climate Camp draws in coal community', *GLW*, 16 January 2010. http://www.greenleft.org.au/node/43013 (viewed 9 September 2011).
Pietsch, Juliet, and Ian McAllister, ' "A diabolical challenge": public opinion and climate change policy in Australia', *Environmental Politics*, vol. 19, no. 2, March 2010, pp. 217–236.
Pietsch, Sam, 'To have and to hold on to: wealth, power and the capitalist class', in R Kuhn (ed.), *Class and Struggle in Australia*, Pearson Longman, Sydney, 2005, pp. 21–38.
Pittock, A Barrie, *Climate Change: Turning Up the Heat*, CSIRO Publishing, Melbourne, 2005.
Pittock, Barrie (ed.), *Climate Change: An Australian Guide to the Science and Potential Impacts*, Australian Greenhouse Office, Canberra, 2003, p. 7.
Plimer, Ian, *Heaven + Earth – Global Warming: The Missing Science*, Connor Court Publishing, Ballan Vic., 2009.
Polanyi, Karl, *The Great Transformation*, Farrar & Rinehart, New York, 1944.
Ponting, Clive, *A Green History of the World: The Environment and the Collapse of Great Civilizations*, St Martin's, New York, 1992.
Popular News, 'Australia to sign Kyoto if Labor wins', 21 November 2007. http://www.climateactionprogramme.org/news/australia_to_sign_kyoto_if_labor_wins/ (viewed 14 July 2010).
Preston, BL and RN Jones, 'Climate change impacts on Australia and the benefits of early action to reduce global greenhouse gas emissions: a consultancy report for the Australian Business Roundtable on Climate Change', February 2006.
Pusey, Michael, *Economic Rationalism in Canberra: a Nation-Building State Changes its Mind*, Cambridge University Press, Melbourne, 1991.
Pusey, Michael, *The Experience of Middle Australia: The Dark Side of Economic Reform*, Cambridge University Press, Melbourne, 2003.
QPSU, 'About Climate Heroes', 2008. http://www.qpsu.org.au/climate/about.htm (viewed 22 July 2010).
QPSU, 'QPSU and ACF unite for Queensland's Climate Heroes', 2008. http://www.qpsu.org.au/climate/news/080318_launch.htm (viewed 22 July 2010).
Ratcliffe, Ruth, 'Climate Emergency Action Network begins organising', *GLW*, 25 October 2008. http://www.greenleft.org.au/node/40540 (viewed 7 November 2011).
Ratcliffe, Ruth, 'SA climate camp condemns dirty energy', *GLW*, 14 October

2009, p. 9.
Ratner, Carl, 'Cooperatism: a social, economic, and political alternative to capitalism', *Capitalism, Nature, Socialism*, vol. 20, no. 2, 200, pp. 44–73.
Reale, Elizabeth, 'ANF joins climate change campaign', *Australian Nursing Journal*, vol.17, no. 4, 2009, p. 9.
Register, Richard, *Ecocities: Building Cities in Balance of Nature*, Berkeley Hills Books, Berkeley CA, 2001.
Resistance, 'The climate movement needs freedom of speech', *GLW*, 29 July, p. 11.
Rickards, Lauren, and Karlie Tucker, 'Challenges for Australian agriculture', in H Sykes (ed.), *Climate Change on for Young and Old*, Future Leaders, Sydney, 2009, pp. 84–101.
Rintoul, Ian, '2020—the summit vision in hindsight', *Solidarity*, issue 3, May 2008, p. 8.
Rintoul, Ian, 'Greens balance of power not enough to bring change', *Solidarity*, issue 26, August 2010, p. 7.
Rootes, Chris, 'The first climate change election? The Australian general election of 24 November 2007', *Environmental Politics*, vol. 17, no. 3, 2008, pp. 473–480.
Rose, Anna, 'Young people want a voice in their future', *Newcastle Herald*, 26 September 2009, p. 19.
Rose, Fred, *Coalitions across the Class Divide*, Cornell University Press, Ithaca/London, 2000.
Rosewarne, Stuart, 'Meeting the challenge of climate change: the poverty of the dominant economic narrative and market solutions as subterfuge', *Australian Journal of Political Economy*, no. 66, Summer 2010/11, pp. 17–50.
Rucht, Dieter, 'The transnationalization of social movements: trends, causes, problems', in D Della Porta, H-P Kriesi and D Rucht (eds), *Social Movements in a Globalizing World*, St Martins Press, New York, 1999, pp. 206–22.
Rudd, Kevin, 'Budget Reply Speech 2007'. http://australianpolitics.com/2007/05/10/2007-budget-reply-speech-kevin-rudd.html (viewed 14 July 2010).
Ryan, Shobhain, '$10bn for green power program', *Age*, 11 July 2011, p. 9.
Salleh, Ariel, 'Climate strategy: making the choice between ecological modernisation or living well', *Australian Journal of Political Economy*, no. 66, 2010/2011, pp. 118–143.
Salleh, Ariel, 'Green New Deal—or Globalisation Lite?', *Arena Magazine*, no. 105, 2010, pp.15–19.
Sample, Ian, 'The father of climate change', *Guardian*, 30 June 2005. http://www.guardian.co.uk/environment/2005/jun/30/climatechange.climatechangeenvironment2 (viewed 15 June 2010).
Saunders, Ewan, 'Climate Action Summit 2010: a step forward', *GLW*, 24 March 2010, p. 11.
SCA, 'Safe climate restoration: *How fast?*', 16 August 2010, 4pp.
SCCC, *Building the Foundations for a Low Pollution, Clean Energy Economy. A Review of the Clean Energy Future Package*, August 2011. http://www.

climateinstitute.org.au/images/reports/sccc_analysiscefp_august2011.pdf (viewed 5 October 2011).
SCCC, *Stronger, Fairer, Healthier: Four Foundations for a Low Pollution, Clean Energy Economy*, November 2010. http://www.climateinstitute.org.au/images/reports/sccc_foundations.pdf (viewed 5 October 2011).
Schnaiberg, Alan, *The Environment, From Surplus to Scarcity*, Oxford University Press, New York, 1980.
Schor, Juliet, *The Overworked American: Unexpected Decline of Leisure*, Basic Books, New York, 1991.
Second Indigenous International Forum on Climate Change, 'Declaration of Indigenous Peoples on Climate Change', The Hague, 11–12 November 2000. http://www.ihrna.info (viewed 11 October 2009).
Settle, Domenica, 'That petrol emotion', *Arena Magazine*, no. 54, August/September 2001, pp. 25–27.
Sherman, David, and JW Smith, *The Climate Change Challenge and the Failure of Democracy*, Praeger, Westport CT, 2007.
Shute, Nevil, *On the Beach*, Heinemann, Melbourne, 1957.
Singer, Peter, *One World: The Ethics of Globalisation*, Text Publishing, Melbourne, 2002.
Socialist Alliance, 'Climate Charter', Impress Printing, Sydney, n.d. [c. 2010], 4pp.
Solidarity, *Climate Change is a Class Question*, Solidarity, Sydney, August 2010, 30pp.
Sparrow, Jeff, 'The worker's flag is deepest green: class struggles and the environment', in R Kuhn (ed.), *Class and Struggle in Australia*, Pearson Longman, Sydney, 2005, pp.196–209.
Spies-Butcher, Ben, 'The cost of a market solution: examining the Garnaut proposal for emissions trading', *Australian Journal of Political Economy*, no. 66, Summer 2010/11, pp. 52–70.
Spies-Butcher, Ben, and Frank Stilwell, 'Climate Change Policy and Economic Recession', *Journal of Australian Political Economy*, no. 63, Winter 2009, pp. 108–125.
Spratt, David, 'A carbon price, and then ...?', *Dissent*, issue 34, 2010/2011, pp.17–23.
Spratt, David, 'Kevin Rudd embraces climate suicide', *GLW*, 13 May 2009, p. 12.
Spratt, David, 'Suicidal sweethearts', *New Matilda*, 5 May 2009. http://newmatilda.com/2009/05/05/sweethearts (viewed 30 June 2010).
Spratt, David, 'Time for a Plan B on climate?', *Dissent*, issue 30, 2009, pp. 34–39.
Spratt, David, and Phillip Sutton, *Climate Code Red: The Case for a Sustainability Emergency*, Friends of the Earth Australia, Melbourne, 2008.
Spratt, David, and Phillip Sutton, *Climate Code Red: The Case for Emergency Action*, Scribe, Melbourne, 2008.
St Clair, Jeffrey, 'Seattle Diary: It's a Gas, Gas, Gas', *New Left Review*, no. 238, November/December 1999, pp. 81–96.
Staples, Joan, 'Australian government action in the 1980s', in H Sykes (ed.),

Climate Change on for Young and Old, Future Leaders, Sydney, 2009, pp. 112–18.
Steffin, Will, et al, *Australia's Biodiversity and Climate Change*, CSIRO Publishing, Melbourne, 2009.
Steketee, Mike, 'Defections came from both sides', *Australian*, 12 November 2001, p. 7.
Stern, Nicholas, *A Blueprint for a Safer Planet: How to Manage Climate Change and Create a New Era of Progress and Prosperity*, Bodley Head, London, 2009.
Stern, Nicholas, *Stern Review: The Economics of Climate Change*, Cambridge, University Press Cambridge, 2007.
Stern, Nicholas, *Stern Review: the Economics of Climate Change. Executive Summary*, 2006. http://www.hm-treasury.gov.uk/d/Executive_Summary.pdf (viewed 27 April 2010).
Stilwell, Frank, *Changing Track: A New Political Economic Direction for Australia*, Pluto International, Sydney, 2000.
Stock, Jenny Tilby, 'The Greens, Democrats, minor parties and independents', in A Parkin, J Summers and D Woodward (eds), *Government, Politics, Power and Policy in Australia*, 8th edition, Pearson Longman, Sydney, 2006, pp. 258–279.
Stuart, Nicholas, *Kevin Rudd: An Unauthorized Political Biography*, Scribe, Melbourne, 2007.
Supple, James, 'A power shift: from the grassroots to the government', *Solidarity*, issue 16, July 2009, p. 8.
Supple, James, 'Billions for coal expansion in NSW', *Solidarity*, issue 16, July 2009, p. 8.
Supple, James, 'Obsession with market policies dooms renewables', *Solidarity*, issue 19, November 2009, p. 11.
Taggart, Stewart, 'The best crisis we ever had!', in John O'Brien (ed.), *Opportunities Beyond Carbon: Looking for a Sustainable World*, Melbourne University Press, Melbourne, 2009, pp. 256–269.
The Greens, 'Safe Climate Bill', 22 September 2009. http://greensmps.org.au/content/the-greens-safe-climate-bill (viewed 1 November 2011).
The Transition Decade, '100% renewable electricity by 2020 is possible!', postcard.
The Transition Decade, 'The Transition Decade Strategic Framework', leaflet, n.d., 2pp.
Thomas, Amy, 'Carbon price plan: shovel billions to business', *Solidarity*, issue 32, February 2011, p. 9.
Thomas, Kerrie, 'Rural communities reject coal gas', *GLW*, 20 July 2011, p. 1.
Thompson, Frances, 'Switched on', *Newcastle Herald*, 28 March 2009, H2 p. 6.
Tietze, Tad, 'The Greens, the Crisis and the Left', *Overland*, no. 199, 2010, pp. 27–35.
Tiffen, Rodney, and Ross Gittens, *How Australia Compares*, 2nd edn, Cambridge University Press, Cambridge, 2009.
Tippet, Gary, and Mark Russell, 'Hotter, colder, wetter: it's a new world of extremes', *Age*, 16 January 2011, p. 9.

Tokar, Brian, *Toward Climate Justice. Perspectives on the Climate Crisis and Social Change*, Communalism Press, Porsgrunn, Norway, 2010.
Topsfield, Jewel, 'Gore's cinematic statement has Canberra's pontificators chirping', *Age*, 16 September 2006, p. 9.
Townsend, Terry, 'Individual versus social solutions to global warming', *GLW*, 30 April 2008, pp. 12–13.
Toyne, Phillip, 'Remember the Greens', *Age*, 11 September 1998, p. 17.
Trainer, Ted, 'A green perspective on inequality', in J O'Leary and R Sharp (eds), *Inequality in Australia: Slicing the Cake—The Social Justice Collective*, William Heinemann Australia, Port Melbourne, 1991, pp. 105–124.
Trainer, Ted, *Developed to Death: Rethinking Third World Development*, Green Print, London, 1989.
Trainer, Ted, *Renewable Energy Cannot Sustain a Consumer Society*, Springer, New York, 2007.
Trainer, Ted, *Saving the Environment: What It Will Take*, UNSW Press, Sydney, 1998.
Trainer, Ted, *The Conserver Society*, Zed Books, London, 1995.
Trainer, Ted, *The Transition: Getting to a Sustainable and Just World*, Envirobook, 2010, 330pp. http://ssis.arts.unsw2.edu.au/tsw/ (viewed 23 October 2011).
Trainer, Ted, The window of opportunity will be short, 14 June 2011, 3pp., in possession of the authors.
Trainer, Ted, *Towards a Sustainable Economy: The Need for Fundamental Change*, UNSW Press, Sydney, 1996.
Trainer, Ted, We Must Move to the Simpler Way: An Outline of the Global Situation, the Sustainable Alternative Society, and the Transition to It, 2002. http://ssis.arts.unsw.edu.au/tsw/02-The-Simpler-Way.html (viewed 2 December 2010). [also published by Critical Times Publishing, Barton SA, 2001].
Tranter, Bruce, and Jan Pakulski, 'Changing Social Bases of Environmentalism in Australia', in J Pakulski and S Crook, *Ebbing of the Green Tide? Environmentalism, Public Opinion and the Media in Australia*, School of Sociology and Social Work, University of Tasmania, Hobart, 1998, pp. 54–69.
Turnbull, Malcolm, 'Inaugural Virginia Chadwick Memorial Foundation Lecture—Sydney July 21 2011'. http://www.malcolmturnbull.com.au/homepage-speeches-articles/inaugural-virginia-chadwick-memorial-foundation-lecture-sydney-july-21-2011 (viewed 12 August 2011).
Turner, George, *The Sea and Summer*, Faber and Faber, London, 1987.
Turton, Hal, The aluminium smelting industry: structure, market power, subsidies and greenhouse gas emissions, Discussion Paper no. 44, Australia Institute, January 2002.
UCAN, leaflet, n.d. [c.November 2010), 2pp.
UNEP, 'Challenges of a Changing Earth: Global Change Open Science Conference', Amsterdam, 6 September 2001. http://www.grida.no/news/press/2187.aspx (viewed 22 September 2010).
Van Onselen, Peter, 'If Newspoll numbers hold, ALP will win', *Australian*, 2 June 2010. http://www.theaustralian.com.au/news/opinion/

if-newspoll-numbers-hold-alp-will-win/story-e6frg6zo-1225874220951 (viewed 2 June 2010).
Veness, Peter, 'Greens tout "world-leading" deal set to shake up global talks', *Age*, 11 July 2011, p. 7.
Victorian Government, *Victorian Climate Change Green Paper*, Victorian Government, Melbourne, 2009.
Vromen, Ariadne, Katherine Gelber, and Anika Gauja, *Powerscape: Contemporary Australian Politics*, Allen & Unwin, Sydney, 2009.
Wackernagel, M, and W Rees, *Our Ecological Footprint: Reducing Human Impact on the Earth*, New Society Publishers, Gabriola Island BC, 1996.
Wagar, W Warren, *A Short History of the Future*, University of Chicago Press, Chicago, 1992.
Wahlquist, Asa, *Thirsty Country: Options for Australia*, Allen & Unwin, Sydney, 2008
Walker, Cam, 'Climate refugees in Australia?', *Chain Reaction*, no. 108, March 2010, pp. 12–13.
Walker, Cam, 'Time to choose sides: the green movement and corporate power in the twenty-first century', *Arena Magazine*, no. 54, August/September 2001, pp. 22–24.
Wallerstein, Immanuel, 'Climate disasters: three obstacles to doing anything', Commentary no. 205, 15 March 2007. http://www.binghampton.edu/fbc (viewed 30 March 2007).
Walsh, Kevin, '*Climate Change: A Natural Hazard*: by William Kininmonth', *Australian Meteorological Magazine*, vol. 53, no. 4, pp. 324–5.
Wang, Tao, and Jim Watson, 'Trade, climate change, and sustainability', in L Starke (ed.), *State of the World: Confronting Climate Change*, Earthscan, London, 2009, pp. 88–95.
Watson, Ian, and Peter Browne, 'The 2007 Federal Election: Exit Poll analysis', Australian Policy Online, September 2008. http://www.apo.org.au (viewed 14 July 2010).
Wettenhall, Roger, 'Public ownership in Australia', *Political Quarterly*, no. 36, 1965, pp. 426–440.
Whelan, James, with Pru Gell and Holly Hammond, *Building the Climate Movement Online: Action Research Project*, The Change Agency and Friends of the Earth Australia, 2008. http://www.thechangeagency.org/03_enews/printable.asp?ID=105 (viewed 21 July 2009).
White, David, 'Rudd gambles with the planet', *GLW*, 13 May 2009, p. 7.
White, David, and Dick Nichols, 'Socialist Alliance and the movement against global warming', *Alliance Voices: Socialist Alliance Discussion Bulletin and National Newsletter*, April 2009. http://alliancevoices.blogspot.com/2009/03/campaaignssocialist-alliance-and.html (viewed 30 April 2009).
Wilkinson, Marian, 'Maybe the PM will now get the message', *SMH*, 13 October 2007, p. 21.
Wilkinson, Marian, 'Scientists hit back at climate scepticism', *SMH*, 1 August 2009. http://www.smh.com.au/environment/climate-change/scientists-hit-back-at-climate-scepticism-20090731-e4fd.html (viewed 31 October 2011).

Williams, Chris, 'Australia's oldest coalmine targeted', *GLW*, 14 October 2009, p. 9.
Williams, Chris, 'Thousands spell out stand against coal seam gas', *GLW*, 8 June 2011, p. 3.
Williams, Kerryn, 'Behind the shock jocks' assault on the truth', *GLW*, 4 May 2011. http://www.greenleft.org.au/node/47451 (viewed 17 May 2011).
Williams, Linda, 'Reshaping the Human Self-Image: contemporary art and climate change', in S Davies (ed.), *HEAT: Art & Climate Change*, RMIT University Gallery Publications, Melbourne, 2008.
Wiseman, John, *Global Nation? Australia and the Politics of Globalisation*, Cambridge University Press, Melbourne, 1998.
Woodward, Dennis, 'How Green was my Policy? The Rudd government and climate change', *Australian Political Studies Association Conference*, University of Melbourne, 27–29 September 2010, pp. 1–11. http://apsa2010.com.au/full-papers/pdf/APSA2010_0069.pdf (viewed 9 September 2011).
World Meteorological Association, Press Release no. 906, Geneva, 20 January 2011. http://www.nasanews.noaa.gov/stories2011/2011012_globalstats.html (viewed 20 August 2011).
Wright, Erik Olin, *Envisioning Real Utopias*, Verso, London, 2010.
WWF-Australia, *Annual & Sustainability Report 2010*. http://assets.wwfau.panda.org/downloads/mc038_annual_report_2010_27oct10.pdf (viewed 5 October 2011).
Yeates, Clancy, '$800 million wind farm in doubt', *Age*, 23 Dec 2009. http://www.theage.com.au/environment/800-million-wind-farm-in-doubt-20091222-lbre.html (viewed 27 December 2009).
Yencken, David, and Debra Wilkinson, *Resetting the Compass. Australia's Journey Towards Sustainability*, CSIRO Publishing, Melbourne, 2000.
Yu, Kimberly, Ben Sexton, and Margarita Windisch, 'Protest targets power station', *GLW*, 23 September 2009, p. 10

Index

FSC
www.fsc.org
MIX
Papier | Fördert gute Waldnutzung
FSC® C083411

Kolibri 360 GmbH
Ferdinand-Jühlke-Straße 7
99095 Erfurt, Deutschland
produktsicherheit@kolibri360.de